Technique Reference List

Technique	Section (Page)	Illustrating Experiments
Assembling apparatus	1.3B (14–17)	*
Boiling point determination (*M*)	3.3 (83)	24,*
Bumping (preventing)	1.3G (23)	*
Calibrated pipet	1.3E (18)	*
Chasers	3.2 (81)	*
Clamping equipment	1.2, 1.3C (7–8, 15)	*
Cleaning glassware	1.3K (33–34)	*
Codistillation	3.6 (112, 114)	*
Collection of crystals	2.3C–E (50–59)	*
Column chromatography	4.2A (147)	10A
Connecting pieces of apparatus	1.2, 1.3D (5, 16)	*
Craig tube, use of	2.3E (54)	3B,*
Density	3.7 (119)	^
Derivative preparation	17.3 (503–519)	
Digestion	2.3F (59)	
Disposal of chemicals	1.4 (34)	
Distillation assemblies (thermometer)	3.2 (80)	
Distillation, simple	3.1A (60–69)	
Distillation, fractional	3.1B (71)	
Distillation, reduced pressure	3.5 (103–111)	
Distillation, summary	3.4 (85)	
Drying agents	4.4A–F (175–179)	
Drying liquids	4.4G (179)	*
Drying pipet	1.3E (21)	*
Drying solids	4.4H (180)	*
Evaporation of solvent	1.3I (25)	*
Extraction	4.1C (128)	8, 9, 19–22,*
Filtering pipet	1.3E (18–21)	*
Filtration, hot	2.3E (53)	3A, C,*
Filtration, suction	2.3E (51)	3A, C,*
Flow chart	1.3A, App. B (13, 535)	*
Gas liquid chromatography (GLC)	4.2D (160)	12, 13A, 16, 24,*
Heating liquids	1.3F (21)	*
Infrared spectroscopy	5.1 (181)	8B, 14A, 16, 21, 24,*
Inlet adapters	1.3J (31–32)	*
Labels for products	1.3A, App. B (13, 532)	*
Lubricating standard Taper Joints	1.2 (5)	*
Measuring liquids	1.3E (17)	*
Melting range determination	2.1, 2.2 (37–45)	1, 2,*
Notebook	1.3A (18)	*
Nuclear magnetic resonance	5.2 (199)	14B, possible*
Paper chromatography	4.2B (151)	10B
Percentage yield	1.3A, App. B (13, 532)	*
Photochemical reactions	—	13B, 42B, (45)
Pipets and pipeting	1.3E (17–21)	*
Polarimetry	—	69 (4A, 18A)
Recrystallization	2.3, 2.3G (48, 59)	3A, B, C,*
Refluxing	1.3H (25)	*
Refractive index	3.8 (120)	*
Reports	1.3A, App. B (13, 531–541)	*
Saponification (and equivalent)	16.1 (142)	9B, 70
Scale (SM = semi-micro; M = micro)	Chap. 1 (5–12)	*
Separating liquids, microscale	4.1C (128)	8B,*
Separatory funnel	4.1C (128)	8A,*
Solvent evaporation	1.3I (25)	*
Solvent selection (recrystallization)	2.3B (49)	3C,*
Steam distillation	3.6 (111)	7A, B,*
Stoppers	1.3J (31)	*
Sublimation	4.3 (174)	9A, 23
Theoretical yield	1.3A, App. B (14, 532)	*
Thermometer adapter	see Inlet Adapters	
Thermometer placement (distillation)	3.2 (80)	*
Thin layer chromatography (TLC)	4.2C (153)	9B, 11*
Titrations	Chap. 16 (474)	8A, 70, 71
Transferring liquids	1.3E (17)	*
Unknowns	2.2, 2.3, 3.4, Chaps. 5, 16, 17	3–5, 10–12, 14, 70–76, *

*Used in many experiments.

ORGANIC CHEMISTRY EXPERIMENTS

Microscale and Semi-Microscale

ORGANIC CHEMISTRY EXPERIMENTS

Microscale and Semi-Microscale

Bruce N. Campbell, Jr.

State University of New York, Potsdam

Monica McCarthy Ali

Oxford College of Emory University

Brooks/Cole Publishing Company

Pacific Grove, California

I(T)P ™ The trademark ITP is used under license.

Brooks/Cole Publishing Company
A Division of Wadsworth, Inc.

Printed in the United States of America

10 9 8 7 6 5 4 3 2 1

Library of Congress Cataloging-in-Publication Data

Campbell, Bruce N.
   Organic chemistry experiments/Bruce N. Campbell,
Jr., Monica Ali.
     p. cm.
    Includes index.
    ISBN 0-534-17611-9
    1. Chemistry, Organic—Laboratory manuals.   2. Microchemistry—
Laboratory manuals.   I. Ali, Monica McCarthy
II. Title.
QD261.C22 1994
547'.0078—dc20   93-1509
                                              CIP

Sponsoring Editor: *Lisa Moller*
Editorial Associate: *Beth Wilbur*
Production Editor: *Penelope Sky*
Production Assistant: *Tessa A. McGlasson*
Manuscript Editor: *Jan McDearmon*
Permissions Editor: *May Clark*
Interior Design: *Publishing Principals, Inc.*
Cover Design: *Leesa Berman and Laurie Albrecht*
Cover Photo: *Lee Hocker*
Art Coordinator: *Susan Haberkorn*
Interior Illustration: *Visual Graphic Systems, Ltd.*
Typesetting: *Science Typographers*
Printing and Binding: *Arcata Graphics/Fairfield*

To Dr. E. R. Kline,
who pointed toward the delights
of teaching organic chemistry
on a very small scale
and who developed many
of these experiments
fifty years ago.

Preface

Organic chemistry experiments are performed on many scales, for which we have adopted the following conventions. Industrial and academic researchers may start with 500 g (macro), 50 g (semi-macro), 1–5 g (semi-micro), or 50–500 mg (micro). The equipment used for working with 2 g of material looks the same and is used in much the same way as the equipment used with 1000 g. Although it is often appropriate to work on a large scale, chemists frequently explore reactions on a very small scale (semi-micro or micro) rather than risk all of some precious compound they have worked hard to acquire. Because some of the equipment and techniques that are used in micro experimentation may differ dramatically from that used in larger scales, the practicing chemist must be proficient with both scales.

Organic chemists have recently begun to appreciate the considerable benefits of teaching the laboratory on the micro scale. They are attracted by the great improvements in safety, economy, timing, and reduction of waste material that result from working with a minimum of material. Students enjoy the necessary attention to detail of the speedier, cleaner experiments.

By using this text, students will gain insight into the semi-micro and micro scales. Where techniques differ according to scale, both are carefully explained and illustrated by comparable experiments (with the growing use of instrumental techniques there will be many similarities). By carefully selecting experiments and equipment the cost of converting semi-micro to micro is minimized. Some experiments can be done with common, nonspecialized equipment. Because this manual supports microscale, semi-microscale, and mixed scale programs, it will be useful during a transition period from one scale to another.

We have reduced the semi-microscale experiments as much as possible to ensure savings and safety. Most employ starting materials in the 2–5 g range. To make working with the microscale experiments easier, we have pushed them toward their upper limit: most start with quantities in the 250 mg to 2 g range.

Special Features

Beyond providing a useful tool for students in the organic laboratory, we had five specific purposes in writing this text.

1. To provide a manual that is flexible in scale and organization
2. To help students connect the lecture and laboratory parts of the course
3. To ensure that students can understand clearly and learn thoroughly techniques that are important for both micro and semi-micro scale
4. To promote the independence and self-confidence that are essential for success in a laboratory course and in research
5. To promote working in a safe environment

We have used several means to achieve these ends.

Topical approach   Each chapter includes related techniques, reactions of compounds or types of compounds, different ways to carry out similar types of reactions, advanced reaction types, and related kinds of analysis. Experiments are grouped within chapters by topics. The first five chapters introduce a variety of basic techniques, demonstrated by experiments. The middle chapters illustrate different types of reactions or of compounds and their reactions. The final chapters cover additional methods of analysis.

Organization   We begin each chapter by discussing the topic, in great detail if new techniques are involved. Most chapters contain experiments that differ in scale. Instructors thus have considerable flexibility in choosing and ordering experiments, and multistep syntheses are possible. The experiments are numbered continuously, independently of chapter numbers. They are designated SM if semi-micro equipment is used primarily and M if microscale equipment is required. Experiments that are common to both scales do not require equipment specifically associated with either scale.

References   Preceding each experiment is a list of references that are keyed to chapter subsections; for example, 1.3H refers to Chapter 1, section 3, subsection H. Graphic figures and the entries in the Technique Reference List (TRS) are similarly coded. For the experiments most likely to be done early in the course, directions are given in great detail, and include some descriptions of physical properties. Increasingly fewer details are included with later experiments because students must take the responsibility of finding appropriate data themselves. They should have access to handbooks, safety manuals, spectral collections, derivative tables, the *Merck Index*, the Material Safety Data Sheets (MSDS) required by OSHA, and the *Journal of Chemical Education*. The large number of references encourages motivated students to go beyond the text and even to design their own experiments. Instructors can use these references to foster independence and an investigatory attitude. Students will appreciate not just the specific reaction or experiment at hand, but organic chemistry as a whole.

Questions We have included experiments that use unknowns that provide differing details in the directions and that use aggressive questioning. It is important for everyone working in an organic laboratory to understand thoroughly what they are doing *before* they begin. For students working with the smallest scales this is imperative. Students must learn to ask questions about the procedure as they prepare for and carry out experiments if they are to gain the understanding and self-confidence they need to devise their own experimental schemes. Many questions are therefore interspersed in the text, not just included at the end of the chapter.

Safety Laboratory safety is predicated on thorough understanding of techniques, the use of small scale, and appropriate cautioning. Smaller amounts mean less spilling on the bench and floor and, particularly, fewer chemicals in the air. Small scale work and flameless heaters have all but eliminated fires caused by ether and other substances. We include instructions for the proper disposal of leftover chemicals (the smaller the scale the lower the waste disposal costs). It is nevertheless important that students assume the personal and social responsibility of recognizing the dangers associated with the chemicals they use or produce. We stress the benefits of consulting reference books and using computer searches. We also include many hazard warnings, clearly marked "Caution" in the margins of the text.

Note to the Student

Organic chemistry is an art and a science. It is concerned with the preparation of compounds and the study of their properties and reactions. To this end a wide variety of processes, techniques, and instruments are used, whether the goal of study is preparative or theoretical. The excitement begins when you are able to show what has been formed and how it relates to theory and prediction. Modern chemists rely heavily on instrumental techniques and on traditional methods. In the undergraduate chemistry laboratory this translates into an emphasis on laboratory techniques and instrumental methods to demonstrate what has happened during a reaction. We hope you will master techniques of preparation, separation, and analysis so that you may share the enjoyment of organic chemistry as you perform and analyze experiments.

Throughout this book is much advice on how to proceed. You will want to pay particular attention to the first five chapters, on general techniques and procedures, and to the beginnings of the chapters that follow. Although references to techniques are given at the beginning of an experiment, you should consult the Technique Reference List for help in finding more specific information.

A great many experiments are designed to prepare a specific compound, but forming the compound is less than half the battle. The mixture that results during formation contains your product and much more. Separation, purification, analysis, and identification are required to isolate the desired product and prove that you have made what you set out to make. The majority of techniques in this book are dedicated to these purposes and you will want to use them effectively.

Come to the laboratory with a mental plan, understanding what is to be done. Make it a habit to write your plan in your notebook, including the information you need. This will clarify what is to be done and how to understand what is happening. You should include in your notebook lists of equipment, chemicals with their significant properties, and the techniques you will use. Use the questions that are clearly marked in each experiment and the summary at the end to be sure you are clear about what to do. The appendix includes examples of reports, calculations, and flow charts that will help in your preparations.

Neat and safe laboratory experiments take thought; they don't happen by accident.

Use spare time to prepare pipets and other routine items, to do analyses of previous products, to clean glassware, and so on.

Included with most experiments is a section on additions and alternatives. Review this section to see how the work of the experiment could be extended. The suggestions are wide ranging, from slight modifications (doing a little more with what you have) to investigation, analysis, or creating a related experiment, evolving your own directions from the current experiment and from the references we supply. When laboratory circumstances permit, you may find it rewarding to strike off a bit on your own.

The emphasis of this book is on basic techniques that are to some extent independent of the equipment you use. Think about what you want to accomplish so you can adapt your techniques to the equipment that is available to you. This will help you to understand what you're doing now and to think for yourself later.

Master and enjoy!

Acknowledgments

We thank our families for their understanding and support. We are indebted to the teachers, coworkers, and fellow chemists who have shaped our understanding of organic chemistry. A number of students spent considerable time outside of class and during summers helping with this book. Three who deserve special mention are Maryann Wells, Lynn Lamite, and Anne Del Borgo. There is no way to thank Patricia Stone enough for her patience and typing. We are pleased to acknowledge the permission given by Ace Glass, ChemGlass, and Kontes Glass to base illustrations on their catalog material.

We also thank all the people at Brooks/Cole who made this project possible. They include Harvey Pantzis, who started the ball rolling, Maureen Allaire, Lisa Moller, Beth Wilbur, patient Penelope Sky, Leesa Berman, Susan Haberkorn, Gary Head, May Clark, Jan McDearmon, and many other unsung heroes. Finally, we appreciate the contributions of the following reviewers: Joyce Brockwell, Northwestern University; Rudi Goetz, Michigan State University; Joseph LeFevre, SUNY/Oswego; Yuzhuo Li, Clarkson University; Ingo Petersen, SUNY/Brockport; Robert Silberman, SUNY/Cortland; and Darrell Woodman, University of Washington.

We would appreciate comments from instructors and students who use this text. We are anxious to increase its utility, clarity, and accuracy, and interested in developing new experiments. Please send your suggestions to Bruce N. Campbell, Jr., Brooks/Cole Publishing Company, 511 Forest Lodge Road, Pacific Grove, CA 93950-5098.

Bruce N. Campbell, Jr.
Monica McCarthy Ali

Contents

---

*Theoretical material provided to clarify practical applications.

Note: SM is the abbreviation for semi-microscale, M stands for microscale.

CHAPTER 3

Purification and Identification of Liquids 69

_____

*Theoretical material provided to clarify practical applications.

CHAPTER 4

Other Methods of Separation and Purification 125

---

*Theoretical material provided to clarify practical applications.

CHAPTER 5

Other Methods of Identification 181

CHAPTER 6

Alcohols: Substitution and Elimination 215

CHAPTER 7

Organic Oxidation and Reduction Reactions 247

CHAPTER 8

Aromatic Substitution Reactions 279

CHAPTER 11

Acids 351

11.1 Types of Reactions 351

11.2 Displacement of Equilibrium 353

CHAPTER 12

Reactions of Benzaldehyde 377

CHAPTER 13

Natural Products 397

CHAPTER 15

Kinetics 449

CHAPTER 15

Kinetics 449

CHAPTER 1

Introduction

Very soon you will begin to witness the excitement of an organic chemistry laboratory. Although most of the work that the organic chemist does is carefully planned and executed, the results and the ways in which they come about can seem exciting. After you add a few drops of a chemical, suddenly the liquid in the test tube turns brilliant red. Or, as a mixture is carefully cooled, fine needles of shimmering crystals form. Organic chemists like to see these results, but they go a few steps further. They want to know the details of the experiment. They are concerned with exactly how a reaction occurs and with how much of a particular chemical is required to complete an experiment. But most of all, they like to know exactly what they have created. They isolate and purify the products. They analyze each product, and then begin the process again.

During your investigations, you will be using many of the experimental techniques that research organic chemists use in their laboratories. You will learn how to carry out an experiment and then how to classify the products. You will learn how to purify your products so that you can analyze or identify the chemicals you have made.

Before you begin to experiment, however, there are a few things you should know about laboratory safety, equipment, and procedures. An organic chemistry laboratory is not without its hazards. You will be working with many different chemicals, some of which are toxic. Occasionally, you may work with an open flame. And you will be working with fragile glassware. Your time in the laboratory will be much safer and more enjoyable if you heed the following advice.

1.1 Safety in the Laboratory

You should strive constantly to avoid the possibility of accidents. Paying strict attention to possible dangers that are pointed out in the text and by the instructor will prevent many accidents. Keep in mind that there is some potential danger in every exercise in this and in any other collection of laboratory assignments.

The most important aspect of safety is prevention. Spilling chemicals must be avoided. The proper use of funnels and spatulas will help you avoid spills; any that do occur must be cleaned up immediately in a manner approved by the instructor. Volatile liquids evaporate into the air and into everyone's lungs, so you should use the smallest possible amounts and work under proper hoods to minimize this problem. The hoods you use should have a good draft that improves considerably when the glass front is almost closed. The effectiveness of the draft can be observed by making a fog with dry ice and warm water and placing the front of the hood at various heights. (When you operate the hood with the glass front well down, leaving only your hands and forearms under the hood, you will also succeed to some extent in protecting your face and chest.) The presence of vapor given off by a volatile liquid is one reason why contact lenses should *not* be worn in the laboratory. Also remember that no one should be in a working laboratory without safety glasses, goggles, or a face shield. Your eyes must be protected at all times. In the laboratory, you should avoid the fire hazards of loose clothing and loose long hair. Shoes are a necessity. And you might also wish to protect your clothing with a lab coat or an apron. Other precautions are described in other sections of this chapter, especially Sections 1.3B–1.3K. These include the stability of apparatus, care in handling chemicals and equipment, and the prevention of fires.

Most instructors in the organic chemistry laboratory have learned to recognize serious injuries. If, after applying first aid, your instructor advises you to go to the infirmary for further treatment, you should follow the advice without question.

1.1A   Burns

Burns are the most common injuries that occur in the laboratory. Most less serious burns are treated by immediately icing the affected area. Burns themselves are classified according to the intensity of the damage done to the burned area. First-degree burns are those in which the skin is reddened and sensitized to heat and pressure. An anaesthetic preparation will reduce the discomfort and you may bandage the burn or not at your discretion and at the discretion of your instructor. Second-degree burns are those in which a blister is raised. They are more serious than first-degree burns and therefore must be treated with greater care. Remember that you must not, under any condition, puncture the blister, which would invite infection. You may apply a sterile bandage to the area, to protect the skin from further damage. Third-degree burns are characterized by broken skin and damage to underlying tissue. They may prove serious if infected and should never be treated (or mistreated) with a greasy ointment. This ointment would have to be removed by the physician before dressing the injury and removal is a very painful process. A proper dressing is gauze wrung out with sodium bicarbonate solution and placed over the damaged region for the trip to the infirmary. (*Note:* You must not sterilize sodium bicarbonate solutions by boiling them. Heat converts them into solutions of sodium carbonate, which is much too alkaline for contact with broken skin.)

Acid burns, usually from concentrated sulfuric and/or nitric acids, will result from careless handling of these reagents. These burns are insidious in that they produce no pain until the damage is extensive. You should immediately flood the area with plenty of clean water and then cover it with a moist sodium bicarbonate bandage. Prompt treatment of an acid burn by a physician is imperative!

Treat alkali burns by washing them with running water, covering them with noncrystalline boric acid, and bandaging them lightly. If the damage approaches that of a second-degree burn, you should obtain further treatment from a physician.

Bromine burns should immediately be cleansed with glycerol; you should also wash with water and sodium thiosulfate solution alternately and *repeatedly*. Bromine acts rapidly on the skin and must be removed quickly. Its low solubility in water makes copious washing necessary.

Phenol produces deep tissue destruction. Wash burns caused by a phenol with as much water as possible. Use soap, not an organic solvent. Tincture of iodine used after washing may render the residual phenol harmless.

1.1B   Cuts

Cuts are usually more annoying than serious, and almost always result from careless handling of glassware. Infection may result from shards of glass left in the cut. The danger of infection is greatest when a cut does not bleed. It is very important to disinfect this type of cut. If the damage is confined to the skin, apply an antibacterial cream and bandage the wound to aid clot formation. If the damage is extensive, stitches may be necessary. You should bandage the wound sufficiently to absorb the blood and go, escorted, to the infirmary. If bleeding is profuse, a compression bandage is preferable to a tourniquet unless a major artery has been cut. You can make a compression bandage by rolling a clean cork in sterile gauze and binding it over the wound in a way that applies pressure directly to the cut. A tourniquet should be used only to save a life; it may cause the loss of an arm if it is not used intelligently. Please note that cuts under the fingernail almost always become infected and require immediate treatment by a person well trained in first aid or by a physician.

1.1C   Eye Injuries

Wearing proper and approved eye protection, such as goggles with indirect venting, should prevent virtually all eye injuries.

First-aid treatment applied to the eye should always consist of immediate, copious washings with water. Speed is essential. Your instructor may then suggest the application of a *pure*, 5% sodium bicarbonate solution because this solution acts as a good buffer. It neutralizes acids and decreases the activity of bases. It is also mildly detergent and assists in the removal of tiny pieces of glass that may have been driven onto the outer surface of the eyeball. Besides the treatments discussed above, it is important for you to remember that all injuries to the eye require the attention of a physician.

1.1D   Fainting

Occasionally, a student such as yourself will become pale and feel faint while first aid is being administered. Loss of consciousness may be prevented by inhaling vapors from aromatic spirits of ammonia and lying down. If it becomes necessary to send you to the infirmary in this state, you or the accompanying person should be provided with a piece of gauze moistened with aromatic spirits or with plain ammonium hydroxide whose vapors can be inhaled as needed enroute.

1.1E   Precautions

Certain precautions will reduce the likelihood of your having an accident. For instance, you should use pick-up tongs to handle hot objects. Keep them on the desktop, conveniently at hand when you are at work. You should use face shields in those experiments involving hot acids or alkalis, particularly under conditions in which gases are evolved. *Wear splash-proof safety glasses at all times* while you are in the laboratory to guard against splashes. Avoid getting chemicals on your skin, where they may be absorbed. Consider wearing gloves.

When using separatory funnels, check to make sure that all ground stoppers fit and that glass stopcocks are properly lubricated with a suitable lubricant. You can test separatory funnels by filling them a quarter full with dilute sodium bicarbonate, adding a few drops of dilute hydrochloric acid, stoppering, and shaking. You should readily discover ill-fitting stoppers and leaky stopcocks this way. Extractions should be done in a hood.

Make sure that you exercise care in breaking glass tubing and in inserting it into the holes of stoppers. The ends of this tubing should be fire polished, cooled, and lubricated before insertion. Insert the glass tubing, which may be lubricated, by grasping it close to the stopper with a towel between your hand and the tubing itself and turning and pushing the tubing gently as you urge it in. Never point the end of a glass rod, glass tubing, or a thermometer into the palm of your hand. The tubing should not fit the holes too snugly.

Beakers and test tubes with chipped lips should be returned to be fire polished. If you notice a beaker or flask with a star-shaped crack, do not use it. This glassware will almost always leak and will frequently break.

You should use particular care in removing rubber tubing from glass apparatus. If the rubber adheres too strongly, make a gentle attempt to press it in further. This action may loosen it so that it can be removed. You may also manage to remove the tubing by prying it slightly away from the glass using a metal spatula. You can also try dropping a little glycerin or undiluted detergent between the glass and the rubber and allowing the tubing to "sit" for about an hour. If this fails, cut the tubing from the apparatus. Rubber is expendable; your fingers are not.

Further precautions are discussed in Sections 1.3B–1.3K.

1.2   Small-Scale Apparatus

Kits of standard taper glassware for work at the semi-micro scale have been available for over two decades. The numbers located on this type of equipment refer to the width at the top and the length of the joint. For instance, 14/20 means 14-mm wide and 20-mm long. Standard taper refers to the connecting joints, which are ground glass tapered at a standard angle. The inner joint fits snugly into the outer joint. Since the two ground surfaces fit so intimately, the joints can become "frozen" when exposed to heat for long periods of time. A thin coating of high-quality lubricant will reduce this problem. Apply a couple of little streaks of lubricant to the inner joint and insert that joint into the outer joint with gentle, firm turning until the joint "clarifies." You may need to practice to do this effectively. Too little grease may lead to joint freezing; too much grease may leave a gap that will allow the entry of solvent vapors, which may dissolve the grease and lead to seizing. If the joints fit well, some instructors prefer to use no lubricant. In any case, it is important for you to separate the joints as soon as practical after use.

Discussions and directions that pertain *specifically* to one or the other of the two scales used in this text will be labeled M for micro and SM for semi-micro. In preparing to do an experiment, you should list in your laboratory notebook the equipment as well the procedures to be used.

As you become more familiar with semi-microscale (SM) equipment, you will observe that the kits (Figure 1.1A) contain a variety of small round-bottom flasks (conical or pear-shaped flasks may be used instead)

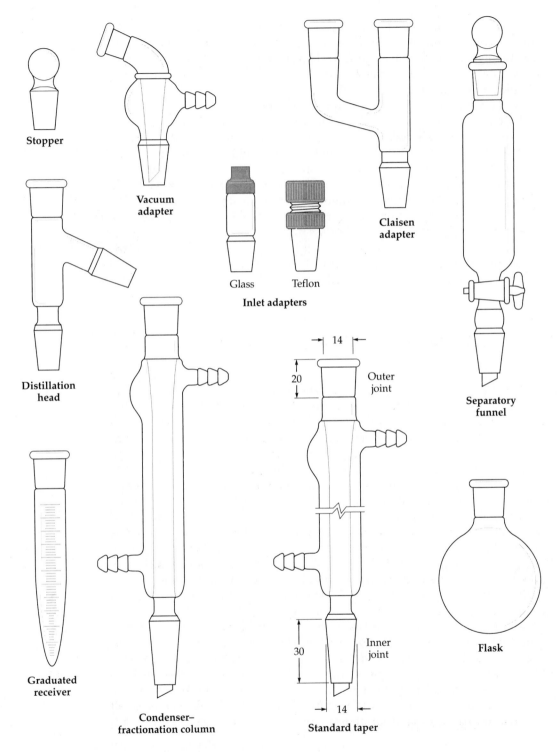

Figure 1.1A Typical semi-micro glassware with ground joints.

for carrying out reactions and boiling liquids and one or two condensers for condensing the vapors thus generated. One of these condensers can double as a fractionating column. There are inlet and outlet adapters as well as stoppers. The adapter with the downward side arm is used as a distilling head. Another type of adapter takes the condensate from distillation to the receiving vessel, and a third type can hold a thermometer, inlet tube, or homemade drying tube. A Claisen adapter or Y tube and a separatory funnel may also be included. A graduated receiver or conical tube used for collecting distilled liquids, very useful additional items, may also be found in some kits.

When you assemble this apparatus, you will find it helpful to use a number of flexible joint clamps to minimize the number of rigid clamps

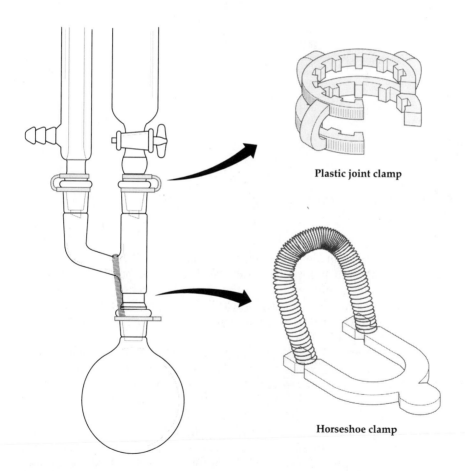

Plastic joint clamp

Horseshoe clamp

Figure 1.1B   Nonrigid or flexible clamps (FC).

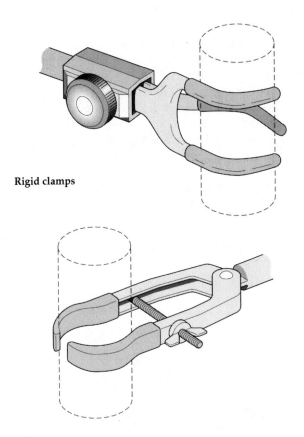

Rigid clamps

Figure 1.1C Rigid clamps (RC).

needed. One type of flexible joint clamp has a small horseshoe-shaped piece of metal with a spring joining its "legs." The horseshoe part fits under the lip of the outer joint and the spring is attached to the opposite side of the apparatus (Figure 1.1B). Rigid clamps are illustrated in Figure 1.1C. Another type of flexible joint clamp is made of plastic and can be snapped onto the joint (Figure 1.1B).

The recent renewal of interest in microscale (M) organic chemistry has encouraged glassware companies to offer a variety of kits and conversion kits to be used with microscale experiments. Most of these microkits (Figure 1.2A) use threaded conical vials with outer standard taper joints as the basic reaction vessel. Several sizes are included, with capacities of from 0.1 mL to 5 mL. Both an air condenser, B, and a water condenser are available. The Hickman still, E, is included rather than a distilling head, and a Claisen adapter, D, is offered. All of these have standard taper joints

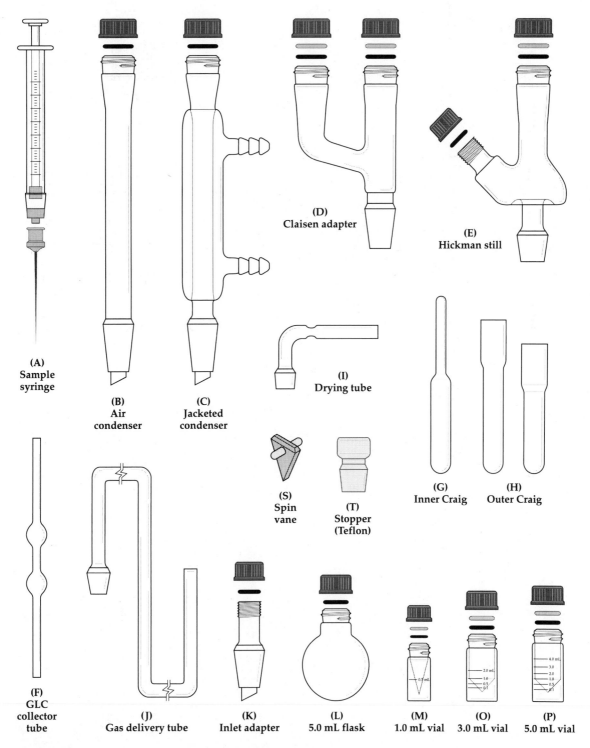

(A)
Sample
syringe

(B)
Air
condenser

(C)
Jacketed
condenser

(D)
Claisen adapter

(E)
Hickman still

(I)
Drying tube

(S)
Spin
vane

(T)
Stopper
(Teflon)

(G)
Inner Craig

(H)
Outer Craig

(F)
GLC
collector
tube

(J)
Gas delivery tube

(K)
Inlet adapter

(L)
5.0 mL flask

(M)
1.0 mL vial

(O)
3.0 mL vial

(P)
5.0 mL vial

Figure 1.2A   Typical *microscale* glassware with ground joints.

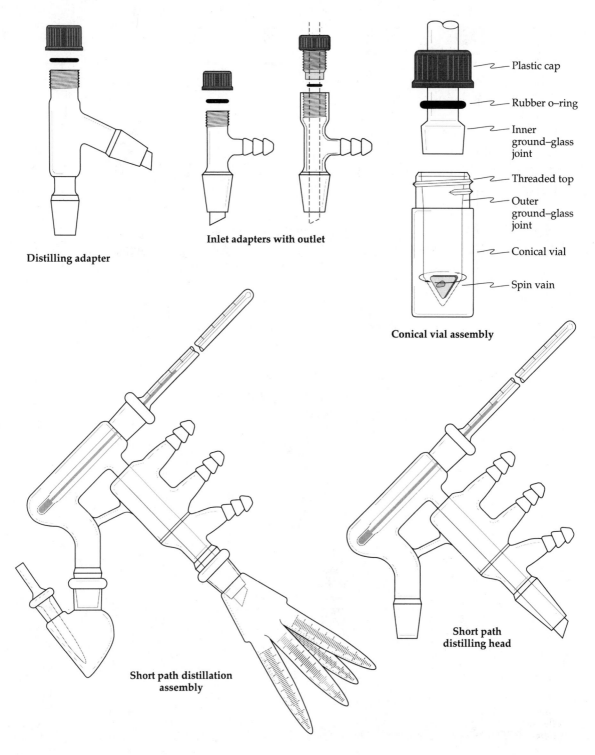

Distilling adapter

Inlet adapters with outlet

Plastic cap

Rubber o–ring

Inner ground–glass joint

Threaded top

Outer ground–glass joint

Conical vial

Spin vain

Conical vial assembly

Short path distillation assembly

Short path distilling head

Figure 1.2B More microscale equipment.

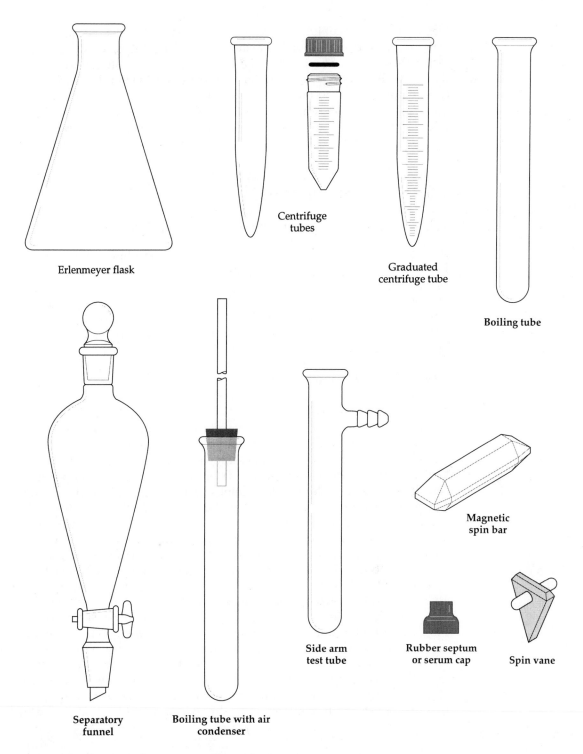

Figure 1.3 Standard or common equipment for organic experiments.

that are used without lubricant and are attached with threaded caps to the vials or to each other. Often included are Craig tubes, G/H, used for recrystallization (see Section 2.3), GLC (gas liquid chromatography) collection tubes, F, a drying tube, I, and a capillary gas delivery tube, J. With this type of apparatus, you must disconnect the joints as soon as possible after use. Additional microscale equipment and setups are shown in Figure 1.2B. In Figure 1.3 is some of the common equipment you may use.

1.3　General Procedures

1.3A　Laboratory Preparation, Records, and Reports

It is important to keep a permanent record of everything you do in the laboratory—successful or unsuccessful. Student observations have sometimes led to significant discoveries or modifications in techniques. Record all that is done and observed in the laboratory in a bound notebook and have your notebook dated and witnessed (by your instructor in an academic situation) at the conclusion of your laboratory period. Certain aspects of keeping this notebook should be clearly noted. Remember that if it should be written down, you should write it in your notebook. No loose papers are permitted. Records should be in permanent ink, and there should be no erasures. If you need to cross out an entry, do it neatly with a single line so that it can still be read. The pages should be numbered, and none should be removed. All spectra, recorder traces, etc., should be clearly identified (in detail, at the time recorded) and permanently secured in your notebook. Prepare a table of contents at the beginning or the end (work backward). Since your notebook is a potential reference work, you should make things easy to find and clearly identifiable with regard to the experiment performed. To do this, it is helpful to use wide margins with labels. Clear titles and ample sections on purpose and procedure (or approach) are also important. You must clearly state what is to be done and why as well as record your observations and reflections about what happened. The importance of a good lab notebook is reinforced by Eisenberg, *J. Chem. Educ.*, 1982, *59*, 1045.

Evidence of *pre-experimental preparation* belongs in your notebook. Before you come to the laboratory, you should read and make sure that you thoroughly understand the experiment to be performed. Planning is the key to your success in the organic chemistry laboratory. Decide what approach you will take in order to best use your time and equipment in the laboratory. List the apparatus you will use and/or draw your assembled apparatus. For your own use, determine and mark those operations that can proceed with a minimum of supervision and note that period of time. During this time, you can begin other operations, determine physical properties for the product isolated in the previous experiment, or even wash "dishes."

Many instructors require *evidence of student preparation* before they issue the chemicals needed for an experiment. Complete as much as possible of the report sheet in the back of this text (or of its equivalent) in your notebook. (See Appendix B.2.) Many instructors expect this evidence of preparation to be handed in before the experiment is begun, as a pre-lab report (which may become part of your final report). A pre-lab should include the experiment's title, its purpose, its balanced chemical equation, the procedure to be used and/or a flow chart, a table of physical properties, and calculations such as theoretical yield and those leading to the discovery of the limiting reagent. A table of physical properties will help in lab preparation by making you more familiar with the physical, chemical, and toxological properties of the materials you will be working with. You should make a list of equipment and chemicals needed, paying strict attention to each chemical's toxicity, stability, and flammability. Properties of the compounds involved can be obtained from this text, from "handbooks" (such as *The Chemical Rubber Handbook*, Lange's *Handbook of Chemistry*, *the Merck Index*, Heilbron's *Dictionary of Organic Compounds*, and Beilstein's *Handbuch der organischen Chemie*) found in the laboratory, stockroom, or library, or from computer-searchable data bases. A flow chart is highly recommended even when you are not required to do one. It is a detailed, sequential treatment of the physical and chemical processes in an experiment. It shows all of the chemical compounds used or likely to be produced, along with their ultimate disposition, and shows what happens to each of the chemical species in each step of the procedure. For examples of flow charts, see Appendixes B.1 and B.2.

The final report should include the information presented in the pre-lab as well as experimental data collected in a table, calculations and/or graphical treatment of the data, observations and changes, a discussion of the results, and answers to the questions given in the experiment and/or to those occurring to you or your instructor. The discussion should address test reactions, refractive indexes, densities, spectra, instrument readings and recordings, sources of experimental error, comparisons made with other experiments, and reasons for the magnitude or kind of data collected. You could use the report forms at the back of this text for experiments in which a product is prepared and use your notebook to report other types of experiments.

If you need to hand in a *product* with your report, it should be in a clean, dry, neatly *labeled* vial that is securely closed. The label should include your name, your laboratory period, the date, your product's name, its observed physical properties (i.e., its boiling or melting range), its actual yield, and its percent yield. (See Appendix B.1.)

Certain comments should be directed to your use of the report forms (Appendix C) provided with this text. For amounts, record those that you actually used. You can report boiling or melting points of compounds, depending upon the usual physical state of the material and on the procedure involved in the experiment. The density or specific gravity of solids will not usually be needed, but you will want to record in your notebook, solubilities of all chemical species involved in the experiment. You should calculate and record the amount of product predicted by theory. In order to find the theoretical yield, you will need to determine the limiting reagent, or the starting material that is used in the least relative molar amount required by the balanced chemical equation. Thus, if the balanced equation requires 2 moles of A to react completely with 1 mole of B and, in a particular experiment, 2 moles of B are used with 3 moles of A, A is the limiting reagent. This is true even though A is present in the greater absolute molar amount. With the limiting reagent, you can find the theoretical yield using the same methods as in general chemistry. Thus, if C is to be produced by the reaction kA + pB → mC + nD, and A is the limiting reagent, the theoretical yield of C = moles of A × m/k × the molecular weight of C (for every m moles produced, k moles are needed). The percent yield is the actual yield divided by the theoretical yield (in the same units—moles, mL, g, etc.) times 100. (See the experimental calculations in Appendixes B.1 and B.2.) These calculations should be neat and easy to follow. With these report sheets as well as with your lab notebook, neatness and clarity of organization will usually mean a better grade for you now and a more useful report for review!

1.3B   Assembling Apparatus

Although a piece of equipment should be judged by its performance rather than by its appearance, some students believe that there is little relation between a setup's appearance and its performance. This may be true to a limited extent, but there are certain precautions that you must take in order to ensure satisfactory equipment performance, which, at the same time, result in neat-appearing equipment. With greater laboratory experience, you will find that fewer accidents and mistakes occur in a neat laboratory than in a messy one. The following sections address these precautions.

Much of the rest of this chapter deals with general and specific instructions for assembling laboratory apparatus. Other chapters include detailed discussions of specific techniques as well as discussions of the experiments. Reference is made to background sections in each experiment. You may consult the index or the Technique Reference List for sections containing detailed discussions of certain types of apparatus and techniques. This reference list will be especially useful when you progress to experiments where the directions are more general. The Table of Illustrations is useful for finding out what the assembled apparatus will look like.

1.3C   Stability and Support of Equipment

Any apparatus is top heavy and needs to be supported to prevent breakage. When heating an Erlenmeyer flask, you should support it by using a clamp or ring. Filtering flasks and other apparatus on which downward pressure may be applied should also be clamped to the ringstand and rest on the base of the ringstand or on the bench. If the technique calls for you to heat, cool, or move the apparatus, you should assemble the apparatus and support it up off the base of the ringstand or up off the bench so that changes can be made without disturbing it. You may use a jack to raise or lower heaters or ice baths to or from the experimental setup. Attaching all parts of the apparatus to each other and/or to the ringstand will help ensure that no part of your setup can come loose when such changes are made.

Rigid clamps used with your ringstand should be adjusted so that they do not work against each other, straining the glassware and leading to breakage. Breakage is greatly reduced by mounting all of the items in an assembly on a single ringstand with as few rigid clamps as possible. Using this method eliminates strain and potential breakage that can arise when you shift one ringstand without a corresponding adjustment to another. You will find this method convenient, but remember that it is *imperative* to group the items so that their collective mass lies over the center of the base of the ringstand. Figure 1.4 is a view looking down upon a properly arranged apparatus setup. The square frame represents the base of the ringstand. If you have to raise and lower an assembly of this sort frequently and quickly, you may wish to arrange the clamps as shown in Figure 1.5. Make sure that you securely tighten the set screws holding the clamps to the clamp holders. The use of flexible clamps (Figure 1.1B) reduces the number of rigid clamps required and further reduces strain between parts of the apparatus. In many figures, the suggested locations for clamps are shown (RC = rigid clamp, FC = flexible clamp). If flexible clamps are not available, be sure to use rigid clamps to support the reaction or distillation vessel and any distillation receiver.

Left to themselves, things in a laboratory seldom stay secure. Thermometers tend to roll toward the nearest laboratory bench edge. Small round-bottomed flasks and vials often fall over and spill their contents. A cork ring or small beaker can provide support and stability by cradling small round-bottom flasks and by surrounding vials and small pieces of apparatus that may not stand alone.

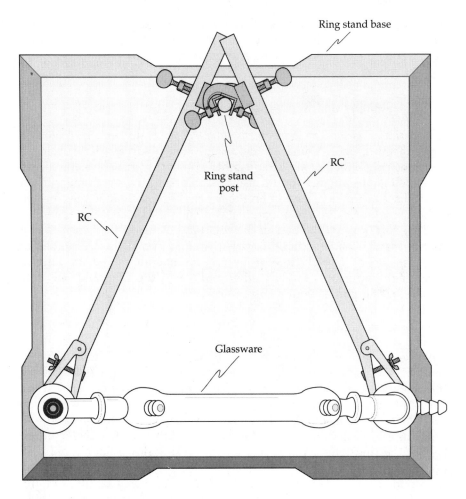

Figure 1.4 Top view of a properly balanced assembly.

1.3D Connectors

Connecting pieces of equipment together may seem like a trivial matter, but poor connecting methods can lead to product contamination, gas and/or liquid leaks, and pressure problems. For instance, when two pieces of glass tubing are joined by rubber tubing, the glass edges must be brought into contact. This contact not only improves the stability and appearance of the setup, but it also minimizes the probability of reagent attack upon the rubber, which leads to product contamination.

When you are using microscale (M) equipment, make sure that the screw caps are carefully aligned, not misthreaded, and tightened only

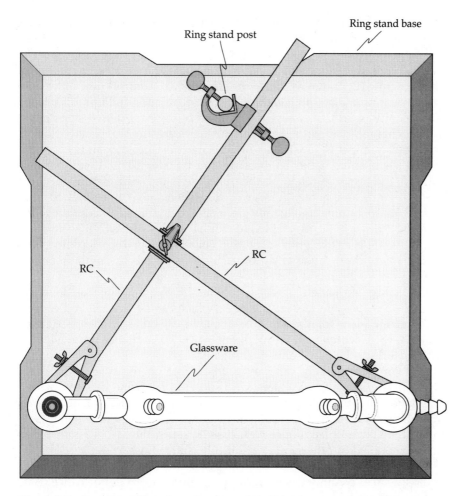

Figure 1.5 Top view of an alternative assembly. This arrangement sacrifices some stability for speed in adjustment.

enough to be snug without stripping the threads. Comments on assembling microscale and semi-microscale (SM) equipment can be found in Section 1.2.

1.3E   Measuring and Transferring Liquids

Liquids are measured and transferred using pipets and burets. A variety of pipets are shown in Figure 1.6A. Some devices to control pipets are illustrated in Figure 1.6B. Three types of pipets are employed in microscale

(M) work: Pasteur pipets, calibrated pipets, and automatic pipets (Figure 1.6A). Automatic pipets have three significant features: The volume to be delivered is or can be set, usually by a dial; the tips are disposable to prevent contamination; and the trigger or button, operated by thumb or finger while the pipet is held in your hand, controls its filling and discharge. The top button has three or four positions—full up (full release, no thumb pressure), one stop down (to fill or discharge), two stops down (full discharge), and three stops down (to release the tip). The bottom of an automatic pipet is a metal tube that is inserted into a clean tip after the old tip has been released. To measure and transfer a liquid, you partially immerse the firmly attached plastic tip in the liquid, depress the button to the first stop, and gently release it. You then move the pipet to your desired delivery area and gently depress the button to the second stop (not the third). THE AUTOMATIC PIPET MUST BE KEPT VERTICAL WHEN IT CONTAINS *ANY* LIQUID! These steps may be repeated with the same liquid or, if you wish to use a different liquid, may be repeated after you have ejected and replaced the plastic tip.

The calibrated pipets have a volume of one milliliter or less. For example, a disposable serological pipet (which can be cleaned and reused) is calibrated for a total volume of 1 mL in .01-mL increments. A 1-mL plastic syringe with a well-fitting plunger (if it has a rubber tip, use it only with water) can be used to measure many liquids.

Another pipet widely used in microorganic chemistry is the Pasteur pipet. It is essentially an enlarged, disposable medicine dropper with a long, finely drawn tip. It may be used as is for the transfer of liquids or for extraction, or, with care, rotation, and a gentle flame, you may bend its tip to reach awkward places or to condense liquids from GLC. Pasteur pipets can be modified in other ways to suit other purposes.

For extraction, preliminary filtering, or working with very volatile liquids, a small cotton (or glass wool) plug improves its performance. With this modification, a Pasteur pipet becomes a *plugged pipet*. To insert its plug, work a *small* amount (a few fibers) of the desired material into a football shape with clean, dry fingers. The circumference of this football-shaped plug should be about the same as the finely drawn tip of the Pasteur pipet. At the larger end of the pipet, insert the plug and *gently* work it into position with a piece of wire only a bit smaller than the tip's diameter (e.g., 20-gauge copper wire). Make sure that it is lodged at the outer end of the tip and that it does not extend beyond it. Also be sure that you've inserted the plug with a bit of friction as opposed to jamming it in. You may need to practice in order to learn how to make these plugged pipets.

Alternatively, you can make a plugged pipet using Rothchild's (*J. Chem. Educ.*, 1990, *67*, 425.) procedure. Attach a very short length of inert tubing containing a small cotton plug to the pipet tip before or after drawing up the liquid. Make sure that the tubing fits snugly on the tip of the pipet. If

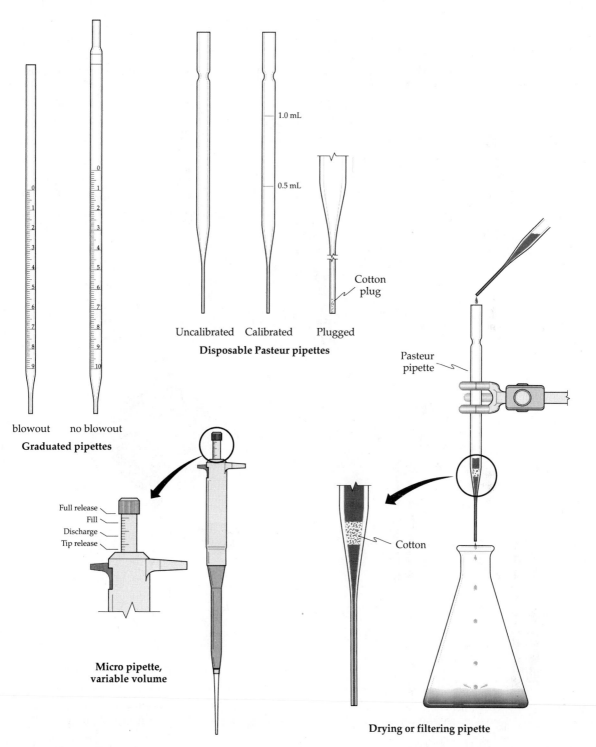

1.0 mL

0.5 mL

Uncalibrated   Calibrated   Plugged

Cotton
plug

Disposable Pasteur pipettes

blowout   no blowout

Graduated pipettes

Pasteur
pipette

Full release
Fill
Discharge
Tip release

Cotton

**Micro pipette,
variable volume**

Drying or filtering pipette

Figure 1.6A Pipets.

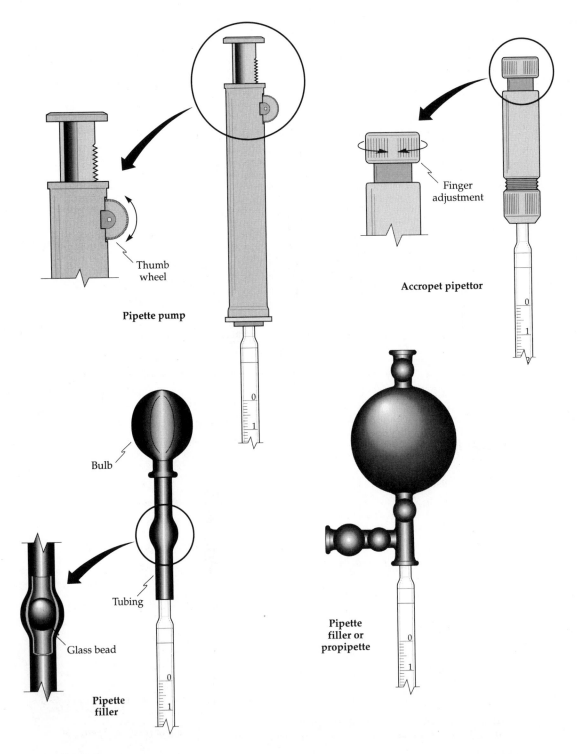

Figure 1.6B Pipet control devices.

you wish to filter solids from the liquid, place this tubing on the pipet before drawing up the liquid and remove the tubing before liquid discharge. Or place the tubing on the pipet after drawing up the liquid to remove solids during liquid discharge.

To take water out of organic liquids, you can modify a Pasteur pipet (Figure 1.6A) to make a drying pipet. Arrange a small amount of cotton in the tapering area of the pipet so that it will not move down into the tip. Layer a few crystals of drying agent (Section 4.4) on the cotton. Secure the pipet in the vertical position and add the liquid to be dried above the plug, allowing it to drip through the pipet. You may wish to push the liquid gently through the pipet using a bulb. This modification (without drying agent) can be used to filter solutions if you do not need the solid that is separated.

If your desk set does not include one or two graduated pipets, you will need to *calibrate* a few Pasteur pipets. To do this, carefully transfer 0.5 and 1 mL of water to clean, dry test tubes. First, draw up the 0.5 mL into the pipet to be calibrated and mark the height of this liquid with tape or a glass-marking pencil. Discharge the 0.5 mL and repeat the procedure with the 1-mL portion of water. If you need to calibrate for 1.5 mL, draw up both volumes of water. To check your accuracy, repeat the procedure, but make sure that the uptake and discharge of the water are quantitative (not one drop of water should be left). If you can, make a *slight* scratch at each volume point and color it in with a marking pencil. When you use these pipets, the tape or color should not be allowed to come into contact with the liquid. Alternatively, draw up a measured amount of the specific liquid to be measured and determine the number of drops per milliliter as you discharge it.

Approximate calibration of a Pasteur pipet can also be done by measuring along the length of a typical 9-inch commercial model. From the tip to a distance of 120–122 mm, there will be about 0.25 mL of a liquid, to 130 mm about 0.5 mL, and to 150 mm about 1 mL. To measure other volumes, you can use a value of 15.5 mm per 0.01 mL in the tip and 4 mm per 0.01 mL in the body of the pipet.

Devices used to fill Pasteur pipets are available (Figure 1.6B). These devices vary from mechanical pipet pumps to propipet bulbs to simple bulbs. But whatever you use to fill a Pasteur or any other type of pipet, make sure that you *NEVER* USE YOUR MOUTH TO FILL A PIPET!

To measure and transfer larger volumes of liquid, you can use graduated pipets, receivers, cylinders, or graduated centrifuge tubes (glass if you will be using organic solvents).

1.3F Heating Liquids

Heating often speeds up chemical reactions and physical processes. Different heating devices are used in the organic chemistry laboratory depending upon the situation and the nature of the material to be heated. You should not use an open flame as a heat source unless you are working with glass or taking melting and boiling points when an electrical device is not available. Flames should NEVER be used when you are working with volatile, flammable liquids, especially diethyl ether. In these cases, you should use steam baths or cones. The top of a steam bath or cone consists of a set of concentric rings that can be removed to accommodate the flask or tube being heated. This heating method avoids the use of flames but can pose a problem if the use of water must be avoided or if the temperature must be raised above that of the steam. Hot water baths have these same limitations and, unfortunately, need a source of heat. This heat source should not be a flame!

Today, there are flameless electric heaters, hot plates, heat lamps, and rigid or flexible heating mantles. The heating mantles are very useful for heating liquids contained in semi-micro standard taper apparatus. One type of heating mantle has the advantage of possessing a hard ceramic surface in which heating wires are embedded for durability. You must remember that these ceramic heaters must be at least as large as the flask used. Flexible electric heaters must fit the flask to avoid burnout.

Microscale equipment includes vessels made in a variety of shapes. These shaped vessels require a heating medium that will adapt its shape to them. A common heating medium of this sort consists of a sand bath held in a small crystallizing dish, placed on a hot plate/magnetic stirrer device. Midrange (100–250 mL) ceramic heaters also work as heated containers for sand baths, and, if they have aluminum bottoms, can be used with magnetic stirrers. They should not be heated above 50 percent capacity. The depth to which the vessel is immersed in the sand controls the rate of heat transfer (as does the control knob on the hot plate). To monitor the sand bath's temperature, position a thermometer with its bulb immersed in the sand. Heat the sand, before it is needed, to a temperature well above (ca. 50–80°C) that desired inside of the glassware.

Another type of medium used for microscale work consists of an aluminum block placed on top of a magnetic stirrer/hot plate. These blocks have a variety of holes drilled in them to accommodate a thermometer and various vial shapes. Auxiliary aluminum cuffs are also available. When you use this type of heating medium, make sure that the thermometer bulb is in the block but is not touching the hot plate. Aluminum blocks have the advantage of coming to a temperature faster than sand baths because the thermal conductivity of aluminum allows more rapid heat transfer. Start heating these blocks early too, but remember that they will not have to be as hot ($+25$–$50°C$ above the desired working temperature) as a sand bath would have to be. For more information, see *J. Chem. Educ.*, 1989, *66*, 77.

In microscale work, it is often important to estimate the temperature inside of the reaction vessel and the speed with which it can be achieved.

To get a clear idea of temperature control, you should experiment with and calibrate the heating apparatus that you will use during the course. Two things need to be determined: how fast the heating apparatus heats up and what relationship exists between the temperature of the apparatus and the temperature of the contents of the vessel being heated. To check the former, set up the heating apparatus and set the dial to a low setting. Place a thermometer in the heat transfer medium (sand, aluminum block, air) and observe and record the temperature every 3–5 minutes for 20–25 minutes or until the temperature remains stable. Record the setting, time, and temperature. Do this for several different settings. Graph your results, and keep the graph handy for future reference. To check the relationship between the temperatures inside of and outside of the vessel you wish to heat, insert another thermometer into the water-filled vessel and place this vessel in the heating medium. Pick a setting from your previous work that would predict an internal temperature of 75–100°C. Heat the medium as before. Record the temperatures inside of and outside of the vessel. You may see a difference of 10–25°C with the aluminum block and 40–80°C with the sand. The placement of the vessel and of the external thermometer will make a difference in the internal/external temperature relationship. How deep the vessel is embedded in the sand bath can influence heat transfer dramatically. With a heat lamp, exposure and distance are similarly important.

Allow the aluminum block to cool and then carefully remove the thermometer. A metal, dial thermometer may be of use here (see *J. Chem. Educ.*, 1991, *68*, A424).

1.3G   Preventing Bumping

As the temperature of a liquid increases, so does its vapor pressure. This pressure may build up as unreleased pressure at the bottom of a liquid being heated. If this buildup continues, there may be a forceful release of pressure—a bump—which can mechanically throw hot liquid all over. Bumping can and must be avoided by a gentle, gradual release of vapor pressure during heating. Using vials, tubes, and flasks with a conical shape may help, but you should *use boiling aids* as well. For most situations, you can add boiling stones (with rough points that contain trapped air) to the liquid before heating. Effective magnetic stirring is an alternative.

For open vessels (i.e., no condenser, adapter, etc.) you can use a rough glass stirring rod, a wooden splint, or stick. You might try using a small capillary tube, closed at the upper end, that will deliver a small stream of bubbles for a short period of time. These bump prevention aids should extend out of the solution being heated so they can be removed before cooling and recrystallization begin.

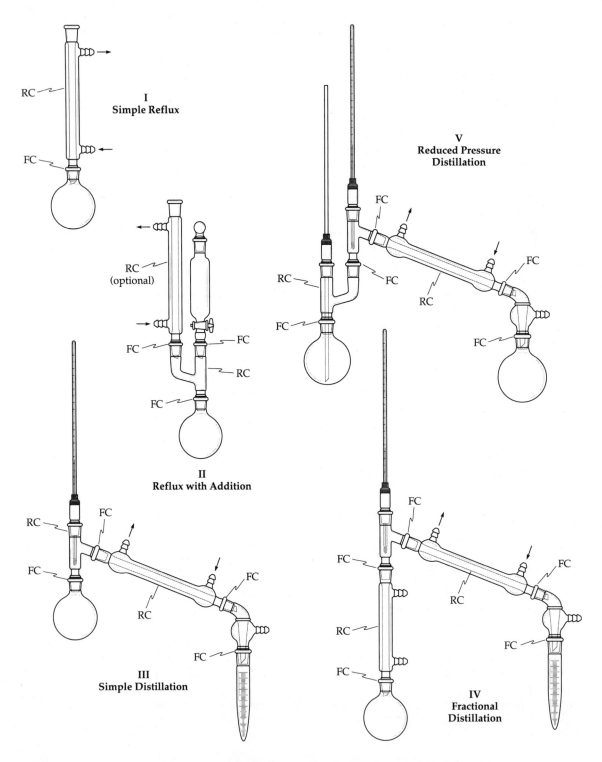

Figure 1.7 Some semi-micro glassware assemblies using standard taper ground glass joints.

When using conical vials, use boiling stones or a magnetic stirring vane (a triangle with the point down), which, as it spins, not only will mix the constituents but also will reduce the chances of their bumping. Very small magnetic spin bars are available for other flasks and tubes. Continuous stirring of the liquid will also prevent bumping. Unreactive metal sponge or glass wool has been used to break up bumping at reduced pressure.

Whenever a liquid is to be heated near its boiling point, you must take action to reduce the possibility of its bumping. The prevention aids must be added *before* you start heating.

1.3H   Refluxing

The process of refluxing is commonly used in organic chemistry to speed a reaction to completion or to equilibrium by heating it at a constant temperature. The temperature of a solution that is boiling gently and has enough heat being added to keep it doing so will not increase unless or until one of the volatile components is distilled away. The energy added as heat is used to vaporize the molecules, not to increase the temperature. The boiling also mixes the constituents. In this process the trick is to return all of the vapors to the boiling mixture as condensed liquid. To do this, you will use a condenser (Unit I, Figure 1.7 or IA, Figure 1.8A). The coolant flowing through this condenser is either cold water or room-temperature air. (Unit I, Figure 1.7 or IB, Figure 1.8A). The coolant enters the condenser at the lower inlet and leaves through the upper inlet. Make sure that you NEVER heat a closed system. A closed system has no opening for gases inside to reach the air outside the apparatus.

For semi-micro standard taper (SM) work, you can assemble a water-circulating condenser as shown in Figure 1.7. (For reactions that produce gases that do not condense, see Figure 4.5 for traps.) If you wish to use an air condenser, you can make one with an inlet adapter and a piece of glass tubing that fits it securely, or you can leave both inlets of the water condenser open.

Microscale kits (M) usually contain both a water and an air condenser (Figure 1.2A). Other microscale assemblies are shown in Figure 1.8B.

1.3I   Solvent Evaporation

The product of a synthesis or extraction often is recovered in dilute solution. When this happens, the solvent may need to be removed to isolate the product. If the solvent is high-boiling, fractional distillation or distillation at reduced pressure (Sections 3.4 and 3.5) may be used. Rotary evaporators remove large volumes of low or moderate boiling range solvents.

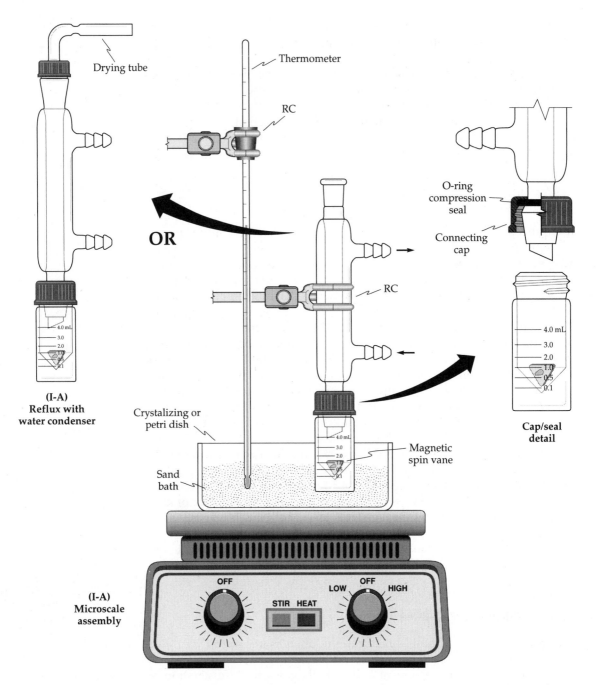

Thermometer

RC

Drying tube

OR

O-ring compression seal

Connecting cap

RC

(I-A)
Reflux with
water condenser

Crystalizing or
petri dish

Magnetic
spin vane

Sand
bath

Cap/seal
detail

(I-A)
Microscale
assembly

OFF

LOW OFF HIGH

STIR HEAT

Figure 1.8A Some microscale assemblies.

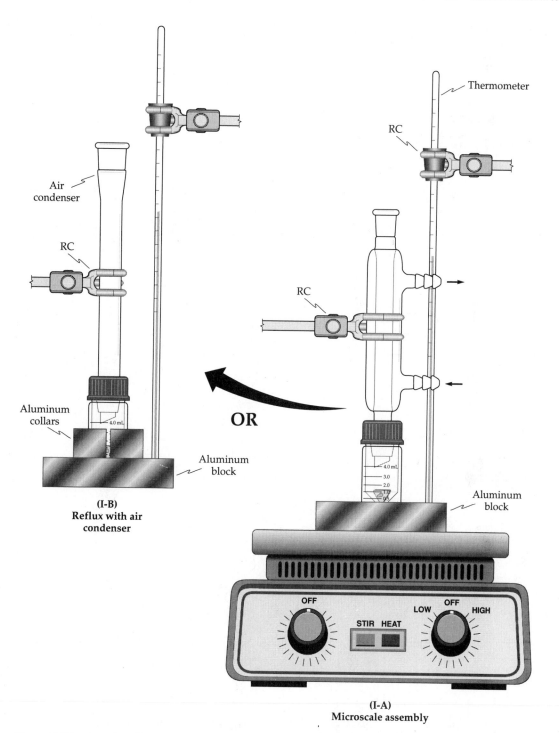

Air
condenser

RC

Aluminum
collars

4.0 mL

OR

**(I-B)
Reflux with air
condenser**

Aluminum
block

Thermometer

RC

RC

4.0 mL
3.0
2.0
1.0
0.5

Aluminum
block

OFF

LOW OFF HIGH

STIR HEAT

**(I-A)
Microscale assembly**

Figure 1.8A   *(continued)*

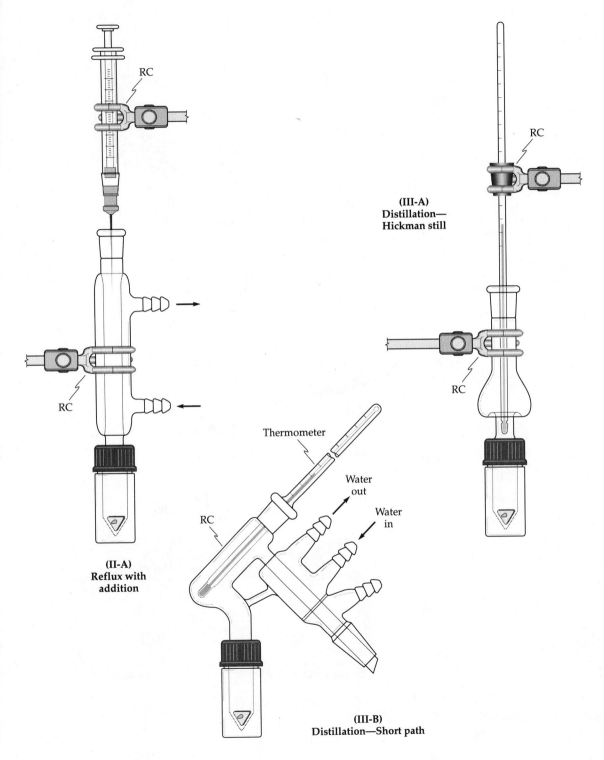

RC

**(III-A)
Distillation—
Hickman still**

RC

Thermometer

Water
out

Water
in

RC

**(II-A)
Reflux with
addition**

RC

**(III-B)
Distillation—Short path**

Figure 1.8B More microscale assemblies.

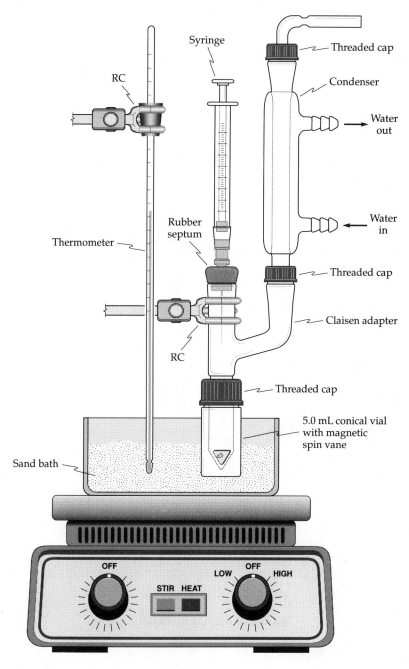

RC

Syringe

Threaded cap

Condenser

Water out

Water in

Thermometer

Rubber septum

Threaded cap

Claisen adapter

RC

Threaded cap

5.0 mL conical vial with magnetic spin vane

Sand bath

OFF

STIR HEAT

LOW OFF HIGH

(II-B)
Reflux with addition

Figure 1.8B   (*continued*)

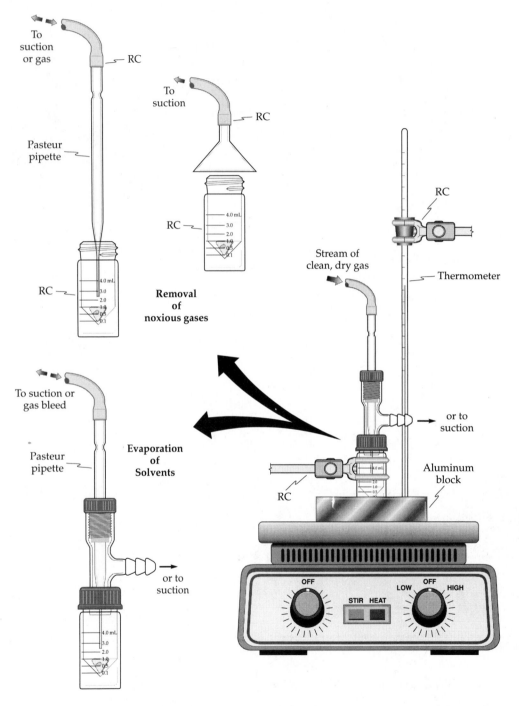

Figure 1.9 Evaporation of solvents and/or removal of noxious gases.

For microscale work (M), where volumes are modest and there is very little product, easier evaporation methods are needed. These methods are suitable for solutions consisting of low-boiling solvents (under ca. 70°C). Such solutions may be warmed using a monitored sand bath or aluminum block (or even a water bath). The purpose of the warming is *not* to boil off the solvent but to *replace* the energy required for evaporation to occur. Evaporation is promoted by stirring and by removing the vapors as they form. To assist in vapor removal, a stream of clean, dry gas, or suction, or both should be used (see Figure 1.9 and Figure 1.10).

Higher-boiling solvents can be evaporated at reduced pressure with vigorous stirring (see Figure 1.9 and Section 3.5). The recondensed solvent can be removed by inserting a syringe through the seal or septum cap on the port of the Hickman still, without an appreciable change in the pressure.

1.3J   Stoppers and Inlet Adapters

(SM) Standard taper stoppers are used to close standard taper flasks at room temperature, preventing substances from getting in or out (Figure 1.1A). Inlet adapters can be used for many purposes. The most common inlet adapter ends in a rubber sleeve that can support glass rods or tubes and thermometers. Thus this inlet adapter can be converted into a stopper, a thermometer holder, a steam or gas inlet, or a pressure bleed. Although it is easier to introduce glass tubing into an inlet adapter than into a rubber stopper, you should take the same precautions (but lubrication should usually be avoided). Additional flexibility can be gained by briskly rolling just the rubber sleeve, with pressure, between your hands or between your hand and the desktop. You may also find Teflon inlet adapters, which have the same uses, in your semi-micro kit.

(M) For microscale work, septa in screw caps serve as stoppers, but they may be attacked by volatile compounds. The TEFLON® side (dull and/or colored: red toward the reaction) should always face the contents of the container. Some kits have inlet adapters with screw caps and O rings, which can be used to support a tube, a thermometer, or a piece of glass rod used to make a stopper (Figure 1.2A, item K and Figure 1.2B). These kits may also include standard taper stoppers (Figure 1.2A, item T), which should be used rather than septums for the storage of organic compounds.

Rubber is a satisfactory stopper material for use with the vast majority of inorganic compounds. However, it is adversely affected by the halogens, by sulfuric acid and the volatile acids, and by substances that react to produce these compounds. In organic chemistry, you will seldom use rubber stoppers for two other important reasons. First, most organic solvents attack rubber; a few do so to quite an extent. Even solvents that do not attack rubber extensively may cause the stopper to swell. For

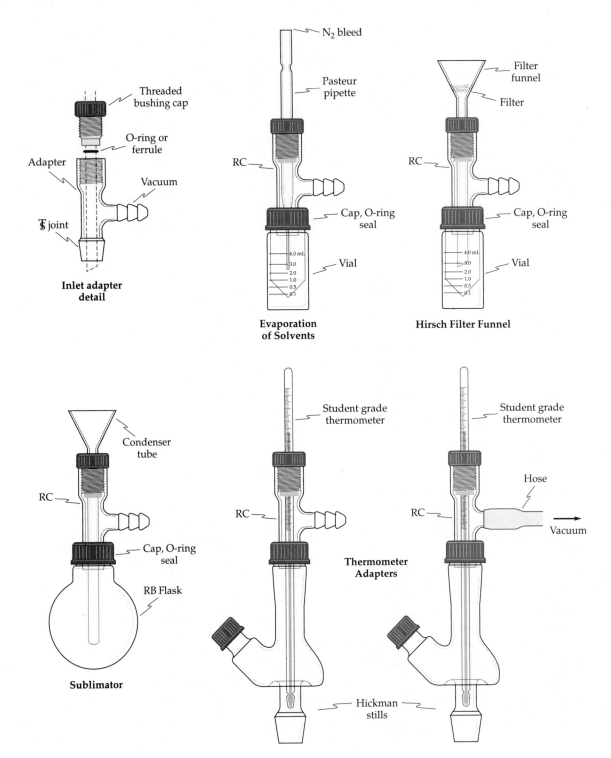

Threaded
bushing cap

O-ring or
ferrule

Adapter

Vacuum

$\bar{\underline{S}}$ joint

**Inlet adapter
detail**

N₂ bleed

Pasteur
pipette

RC

Cap, O-ring
seal

Vial

**Evaporation
of Solvents**

Filter
funnel

Filter

RC

Cap, O-ring
seal

Vial

Hirsch Filter Funnel

Condenser
tube

RC

Cap, O-ring
seal

RB Flask

Sublimator

Student grade
thermometer

RC

**Thermometer
Adapters**

Student grade
thermometer

Hose

Vacuum

RC

Hickman
stills

Figure 1.10   Inlet / Thermometer adapter uses.

example, a rubber stopper can expand when exposed to ether vapor and even cause the heavy ring neck of a flask to rupture. Other substances such as aromatic hydrocarbons, aliphatic halogen compounds, aldehydes, ketones, and esters act as active solvents for natural and synthetic rubbers and can cause a stopper to become cemented to the glass. Second, rubber deteriorates when it is heated and may become unreliable at temperatures only a little above the boiling point of water. Rubber's lack of dependability is particularly noticeable when there is a considerable pressure difference between the inside and the outside of tubing or stoppers that are also exposed to high temperatures.

Cork is, by no means, an ideal stopper material, but it is much less subject to solvent attack and heat deterioration than rubber. Also, most organic compounds exert little solvent action upon cork. Halogen-containing compounds are an exception; neither cork, rubber, nor aluminum foil can be exposed for indefinite periods to such substances. Cork is serviceable up to temperatures at which charring becomes noticeable and may be used with safety up to temperatures near 200°C.

1.3K Cleaning Glassware

In cleaning glassware, success is directly related to how quickly you begin the cleaning process. You can remove some tarry deposits, left in equipment after reactions or distillations have been completed, by washing with soap and hot water. It is good practice to disassemble glass apparatus as soon as practical after use and to soak the parts in warm/hot soapy water. This may not always help, but it costs little to try it. Do not use a test tube brush if water seems to have no effect, because its use will merely spread the deposit and soil the brush. Soaking in concentrated solutions of detergents such as Alconox® or Simple Green® will often be effective for stubborn contaminants, stopcock grease, frozen joints, or plugged syringes.

If water is ineffective, drain the glassware and pour a small amount of acetone into it. Acetone is the most effective common organic solvent. It removes most organic residues that have not become charred or polymerized. If the glass surface "comes clean" upon treatment with acetone, dispose of the resulting solution by pouring it into the proper waste bottle. Then, rinse the surface with another small portion of acetone to prevent redeposition contamination, which will occur upon evaporation of the acetone solution. You may follow this treatment by washing the surface with water or with soap solution, if drying time is available.

Solvents other than acetone may be better for specific cleaning jobs. For example, chloroform (a carcinogen), diethyl ether, or petroleum ether is a better solvent for cleaning out certain silicon stopcock and joint lubricants. Soaking in Simple Green® may also work. Just make sure that you *do not use organic solvents on any plastic* without the approval of your instructor, unless your aim is to dissolve the plastic. Also, be careful of your clothes when using these solvents for they attack certain synthetic fabrics. Use great care when working with organic cleaning solvents that are hazardous when breathed in or absorbed through the skin. Work only with supervision.

To clean standard taper glassware, avoid using strongly alkaline solutions because they may attack the ground joints. If you use these solutions, do not let them come into contact with the joints. Also avoid scratching the inner walls of flasks or tubes as this may lead to their cracking and breaking when heated.

Certain reactions leave residues that are not removable. For example, residues from pyrolysis reactions such as decarboxylations in the presence of alkalies may not be removable. In reactions forming them, the glass is overheated in the presence of the organic matter, and some of the silica in the glass is reduced to silicon or, in the case of Pyrex glass, to silicon boride. This reduction product appears as a dark deposit embedded in the glass and is not easily cleaned even by oxidation. Fortunately, the blackening itself does not seem to damage the glass, and unless strains have been introduced into the glass by heat, such equipment is still usable.

It cannot be emphasized enough that you should clean glassware immediately after its use. Delay only makes the deposits more resistant to solvents and cleaning agents. You should leave your equipment clean and support it in a position that will permit drainage when you close your storage area at the end of a laboratory period. This practice should provide clean, dry equipment for the next period so you can start without any loss of time.

1.4   Disposal of Chemicals

Last, but not least, you must properly dispose of all the materials you have used or produced in your experiment. Plan how you will dispose of all waste materials. This disposal plan should be in your notebook with the rest of your pre-planning materials (or be ready to hand it to your instructor).

Both microscale and semi-microscale experimentation are designed to minimize hazardous waste and toxic exposure. However, we must go beyond this to strain the environment as little as possible. This section contains general suggestions about disposal. Most experiments will also have general and, sometimes, specific suggestions. The general suggestions will be keyed back to this section (Table 1.1).

Waste material can be divided into five categories: nonhazardous solids, water-soluble materials that are known to be nonhazardous; waste that is clearly established as hazardous; halogenated organic materials; and non-

Table 1.1 Designation for General Disposal Categories
(See discussion for more details.)

D	Dilute and flush down the drain.
HO	Halogenated organics; dispose of in the labeled container provided.
HW	Hazardous waste; dispose of in the labeled containers provided.
I	Dispose of as directed by your instructor.
N	Neutralize and then D.
OS	Organic solvents; dispose of in individually labeled containers (e.g., acetone, ethanol).
OW	Organic waste; dispose of in the general, labeled container provided.
S	Solid waste; after evaporation, dispose of in the containers provided.

halogenated organics. The disposal methods and precautions differ by type. *Always check with your instructor* to see if specific or different directions are in force for your laboratory.

For nonhazardous solids, which are generally unreactive materials (such as paper, cork, rubber, chromatographic materials, and glass), there will be suitable containers (often a waste basket). If materials such as paper and glass are to be recycled, there will be separate containers for them. Dispose of these materials only after all organic solvents have *safely* been evaporated.

Water-soluble materials that are known to be nonhazardous should be diluted with lots of water and flushed down the drain. Acids and bases should first be neutralized (solid sodium bicarbonate or solutions of it can be used). Aqueous solutions of bleach or thiosulfate can sometimes be used to pretreat reducing or oxidizing agents.

Hazardous waste must be treated with care. Policies may vary by state, locality, and laboratory. Most often there will be one container for this waste and/or containers labeled for specific kinds of hazardous waste. *Check with your instructor for local policies and procedures.*

Commonly there will be a container in your laboratory for halogenated organic waste (—F, —Cl, —Br, —I) and another for nonhalogenated organics. If your laboratory does solvent recovery, you will find containers labeled to receive specific solvents such as acetone, methylene chloride, ethanol, and toluene. If wash acetone is to be recovered, you may be asked to use it over a *wide* funnel in the mouth of the container.

Most experiments contain a section on disposal of materials. The chemicals are listed with a disposal code that refers to Table 1.1 and thus, to the preceding paragraphs. Remember: A material is not a waste material until no further use can be found for it. Disposal is more expensive and *waste-full* than recovery and reuse!

For more information and to prepare disposal plans, consult the following sources:

Catalog/Handbook of Fine Chemicals, Aldrich Chemical Company, Milwaukee, Wis., current year.

"Chemical Management: A Method for Waste Reduction," *J. Chem. Educ.*, **1984**, *61*, A45.

Destruction of Hazardous Chemicals in the Laboratory, Lunn, G. and Sansone, E., Wiley, New York, 1990.

Hazardous Chemicals, Information and Disposal Guide, Armour, M., Browne, L., and Weir, G., 3rd Ed., Dept. of Chemistry, Univ. of Alberta, Edmonton, Alberta, Canada T6G 2G2, 1988.

Hazardous Laboratory Chemicals: Disposal Guide, Armour, A., CRC Press, Boca Raton, Fla., 1991.

NIOSH Pocket Guide to Chemical Hazards, 1990 (From U.S. Government Printing Office, Superintendent of Documents as DHHS (NIOSH) Publication No. 90-117).

Prudent Practices for the Disposal of Chemicals from the Laboratory, National Academy Press, Washington, D.C., 1983.

Safety in Academic Chemistry Laboratories, 4th Ed., American Chemical Society, Washington, D.C., 1985.

The Waste Management Manual for Laboratory Personnel, American Chemical Society, Washington, D.C., 1990.

Waste Disposal in Academic Institutions, Kaufman, J., ed., Lewis Publishers, Chelsea, Mich., 1990.

For additional resources see:

Chemical Information Sources, Wiggins, G., McGraw-Hill, New York, 1991, Ch. 12 esp. 12.4 for computer-searchable data bases such as TOXNET, TOXLINE, TOXLIT, and HSDB.

Purification and Identification of Solids

Solids are most commonly identified by determining their melting ranges and by comparing that data to what is recorded in the literature. The melting range also gives you an indication of the purity of the solid. Impure solids are usually purified by recrystallization. These two techniques will be the subject of this chapter.

2.1   Theory of Melting Ranges

You will find some theoretical background (*) useful when determining and interpreting melting ranges. For the practical aspects, see Section 2.2.

*2.1A   Relation to Vapor Pressure

In general chemistry, you learned that a pure substance has a certain tendency to escape from a phase or space in which it is placed. For solids and liquids, this property results in the phenomenon of **vapor pressure**. For gases this expansive tendency is simply pressure.

Vapor pressure results from the fact that a definite—although frequently small—proportion of the molecules of a solid or liquid possess kinetic energy (or energy of motion) sufficient to overcome the attractive forces that bind them together. An increase of average kinetic energy resulting from an increase of temperature will result in a larger proportion of available molecules near the surface acquiring the energy necessary to escape from the solid or liquid phase. This causes an increase in the vapor pressure. At any given temperature (or average kinetic energy), equilibrium results when the concentration of molecules in the vapor phase is, or becomes, high enough so that in any given period the same number of molecules of vapor are captured by the solid or liquid as escape from the surface to enter the vapor phase.

The relationship between the solid, liquid, and vapor phases of a compound is shown in Figure 2.1. The line CT represents equilibrium between the solid and its vapor; line TB represents equilibrium between the liquid and its vapor. Line TD represents equilibrium between the liquid and solid phases. These lines intersect at the "triple point" (T)

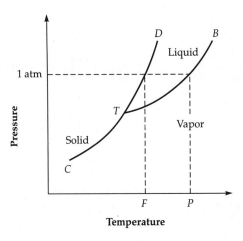

Figure 2.1 Curves showing the equilibrium between a pure solid and its pure liquid phase.

where all three phases are in equilibrium. To the left of curve *CTD*, the solid is most stable; to the right of curve *CTB*, the vapor is most stable. The liquid is most stable in the space between *DTB*. At the point where the line drawn from 1 atmosphere intersects line *TD*, the solid and liquid phases are in equilibrium at atmospheric pressure. This point, F on the temperature axis, is the normal melting point for the compound. Similarly, temperature *P*, at the intersection of the line *TB* and a line drawn from a pressure of 1 atmosphere, is the normal boiling point for the compound. If the external pressure lies below the triple point (*T*), heating would cause the solid to go directly to the gaseous phase without melting first (sublimation).

*2.1B Effect of Impurities

In order to determine the effect of impurities on the melting point of a compound, it is necessary to examine the theory of vapor formation. Only those molecules at the surface of a solid or liquid are able to leave their phases as their kinetic energies are increased by elastic collision to the point where the forces of attraction are overcome. The solid, crystalline form of a compound is not ordinarily contaminated by an impurity and so still produces its normal vapor pressure. The presence of an impurity in the liquid phase dilutes the liquid form of a compound. Consequently, fewer of its molecules are at the surface to acquire the necessary energy to leave the liquid phase, and the pressure of the vapor produced by them is reduced in proportion to this dilution. The result is illustrated in Figure 2.2. Curve *TB* again represents the vapor pressure of the pure

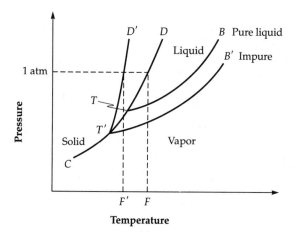

Figure 2.2 The comparison of the equilibria of a pure solid and its pure liquid phase and the same solid and its impure liquid phase.

liquid phase of the substance under consideration. Curve $T'B'$ represents the vapor pressure of the impure liquid phase of this substance. Curve $T'D'$ represents the corresponding equilibrium between solid and liquid phases. The point of intersection between a line drawn from a pressure of 1 atmosphere and curve $T'D'$, the point F' ($F' < F$), represents the new melting point of an impure compound. More simply, the melting point of a substance is lowered by the presence of an impurity.

The generalization given in the preceding paragraph does not apply to all possible mixtures of compounds. Some mixtures contain substances that can react with each other to form a third compound. In these mixtures, the resulting compound may melt at a higher temperature than either of the original substances. And, if the composition of the mixture approximates that of the resulting compound, an elevation of the melting point may be encountered.

*2.1C Eutectic Mixtures

There is a definite limit to the amount by which an impurity can lower the melting point of a compound. Each component of a mixture lowers the melting point of the other component. Two melting point–composition curves are obtained when melting points are plotted against the composition of the mixture. The intersection of these two curves represents the minimum melting point, or **eutectic temperature**, of such a mixture (see Figure 2.3). The mixture itself is called a **eutectic**.

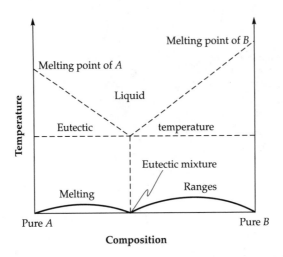

Figure 2.3 The melting point–composition diagram for a simple binary mixture.

A well-mixed eutectic will melt completely at the eutectic temperature. In this respect such a mixture resembles a pure substance. All other mixtures or impure substances melt over a range of temperatures, referred to as the **melting range**. One reason for this phenomenon is that as the solid phase undergoes progressive melting, the liquid phase, which contains the impurity, becomes less impure. Thus, during the determination, the melting point rises from a temperature at or above the eutectic temperature to a temperature which approaches that corresponding to the overall composition of the mixture (as in Figure 2.3). In an ideal case, the magnitude of the melting range is an indication of the purity of a compound. It would be greatest at a point approximately midway in composition between the pure compound and the eutectic and would disappear completely for a pure compound and for an intimately mixed eutectic mixture. Other factors that might account for a greater difference between the initial and final temperatures of melting would be imperfect heat transfer and decomposition of the compound itself.

2.2 Practical Aspects of Melting Range Determination

2.2A General Apparatus

Melting points of organic compounds are commonly determined by using thin-walled capillary tubes and an apparatus in which the thermometer accurately reflects the temperature of the sample. Electrical devices utilize a metal "block" that encloses the thermometer and the capillary in such a fashion as to illuminate and exhibit the sample clearly. The "block" is heated electrically, and the heating rate and temperature are subject to manual control. Two other electrically heated types of apparatus are of the "hot stage" variety. In one type, the crystals are placed on a hot stage between two thin, round cover slips. This has advantages for crystals that readily sublime and for crystals that do not load readily into capillary tubes. In the other type, a line of crystals is laid down upon a surface labeled with a temperature gradient. For this type of melting point apparatus, the point above which all crystals melt is the "melting point."

Air and liquid baths have also been employed to determine melting ranges. In them, a sample in a capillary tube is attached to a thermometer and positioned next to the middle of the bulb to increase the likelihood that the indicated temperature is the same as that of the sample space (Figure 2.4). Agitation and circulation are essential when using these baths and may be done mechanically (see Figure 2.5) or by convection. Thiele tubes (Figure 2.4) have long been used for liquid baths. To take advantage of mixing by convection in a Thiele tube, place the thermometer bulb and sample about midway between the two arms of the tube. The location, size, and type of flame controls the rate of heating and circulation.

2.2B   Melting Point Baths

When liquid baths are used, the liquid must be chosen with care. It should be high-boiling, of low viscosity, nontoxic, nonflammable, stable, and harmless to people and clothing. Water is an excellent choice and can be used below 90°C but, unfortunately, not above that temperature. Glycerol, some hydrogenated vegetable oils, and dibutyl phthalate have been used at temperatures below 280–300°C. Silicone fluids can also be used and some have a higher range.

2.2C   Melting Point Tubes

You can make melting point tubes from larger pieces of glass tubing, and they are also available commercially. Wall thickness should be 0.1–0.3 mm (the thickness of a sheet of writing paper). The thicker the wall, the slower the heating must be to allow for the slower heat transfer. Some tubes come sealed at one end; others are open at both ends. The latter must be sealed at one end and cooled before use. When doing this, use a small flame and constant turning to prevent sagging or clumping of the glass. A glob of glass is a very thick wall.

If capillaries are drawn from glass tubing, make them of double length with both ends sealed and store them that way until needed. When you need them, a cut in the middle of the capillary will provide two clean capillaries. If you have commercial capillaries to use, rejoice!

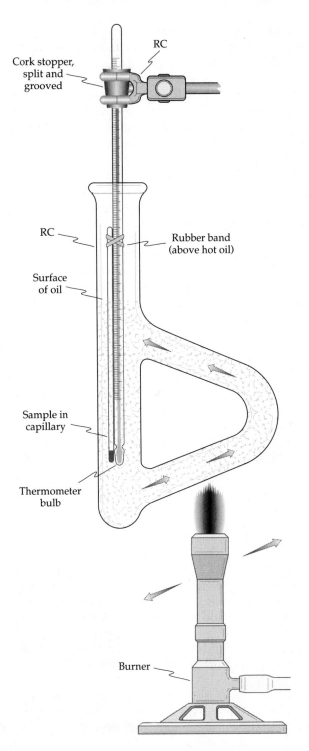

Figure 2.4 Thiele tube.

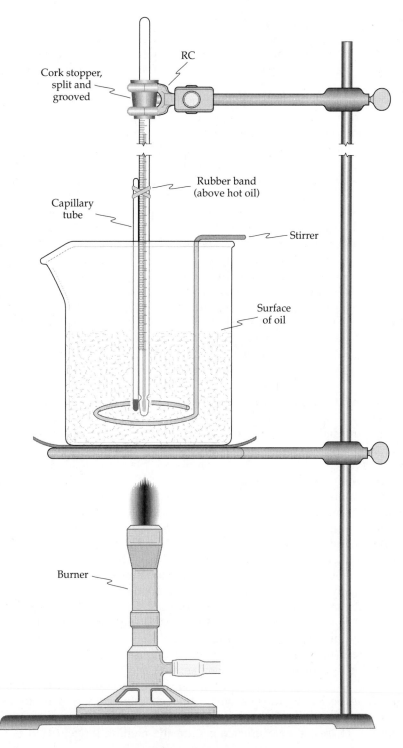

Figure 2.5 Melting point assembly with a manual stirrer.

2.2D   Methods of Determination

The solid compound should have a reasonably fine, powdery consistency before you load it into the capillary tube. Powders that are too fine, however, may trap air, causing false start and problems in heat transfer during the determination. If a piece of porous clay plate is used to finally dry the compound, rubbing and scraping the solid on the plate will usually condition the compound's consistency.

After you have gathered solid material of the desired consistency, mix a small heap of it on or in a plate, weighing paper, a watch glass, or a small beaker. Press the open end of a capillary tube into the material, downward or toward a spatula, using a circular motion. Repeat this procedure on different parts of the material until a small cylindrical load (1–3 mm) is accumulated in the open end of the melting point tube. The larger the load, the longer the melting range because of the time needed for heat transfer. With the open end up, tap the closed end of the loaded capillary tube gently on the table top. A few reverberating taps should bring the solid to the closed end. If not, draw a round file across the middle of the capillary tube, causing it to vibrate, and repeat the tapping. If the sample still refuses to move, drop the capillary down the inside of a three-foot length of glass tubing held vertical to the table top. In fact, it is a good practice to drop all samples in this manner a standard number of times (2–5) to approach consistent packing for comparable determinations. Consistent packing is important because trapped air may inhibit heat transfer as it does in insulation. If these methods fail to pack the material at the closed end, it is likely that the substance is too moist for an accurate melting point determination and that further drying or another recrystallization is necessary.

After sample packing has been completed, place the loaded capillary where it can be uniformly heated and observed. For electrical devices, this is in an illuminated interior compartment. When using a liquid for the heated medium, attach the capillary to a thermometer using a thin rubber band cut from three-tenths-inch rubber tubing. Attach the capillary so that the rubber band will be well above the level of the liquid when placed in the apparatus and so that the sample is next to the middle of the thermometer bulb. Then, position the thermometer in the middle of the bath or midway down the main stem of the Thiele tube.

Set the flame or voltage control to produce an increase in temperature of about 10°C per minute (with stirring). At 15–20° below the literature value for the melting point, cut the heating rate back so that the rate of temperature increase slows down as much as possible (but never stops or reverses!). A rate increase of 0.5° per minute as the temperature passes through the expected melting point should give a useful, narrow melting range. Note the temperature at which the solid slumps and starts to lose its crystalline nature and also at which any evidence of crystallinity just disappears. These temperatures constitute the melting range. Good technique and a pure substance should produce ranges of 1°, 0.5°, or less. Since impurities (including the solvent) cause the range to be lower and longer, the range should also reflect the purity of the sample. (Why should impurities have this effect? Consider how a melting point is lowered by a solute in a solution and the change in solution concentration as the process progresses.) A melting point is the extreme of a very short range. Some changes may be observed (e.g., movement as trapped solvent escapes) before melting really starts, and care must be taken to identify the beginning of true melting.

If you are taking the melting range of an unknown substance, do a preliminary trial, to determine the approximate melting range, without slowing the initial heating rate. You can then make second and third trials, starting each one with a fresh sample and starting slower heating at about 20°C below the melting temperature of the compound. Carefully determine the range by slowly increasing the temperature at as slow a rate as possible throughout the melting range.

2.2E   Melting Ranges of Mixtures

Two different compounds, even with the same melting point, would be mutual impurities if mixed with each other and would hence cause the melting range to be lower and longer. In fact, "mixed melting points" are widely used to determine whether two substances are the same compound or whether they are different ones. Combine equal amounts of the two compounds. Mix and, if needed, grind them together and load the intimate mixture into a capillary tube. If possible, run both the pure compound(s) and the mixture at the same time. If the melting range determined for the mixture is appreciably lower and longer than that of the pure compounds, the presumption is that they are different compounds. If the ranges appear the same, the compounds could be identical or the mixture may be eutectic. It is a good idea to do one or two more mixtures (two parts to one or three parts to one) to be certain.

E X P E R I M E N T   1

Calibration of a Thermometer or Melting Point Apparatus
References: Sections 2.1, 2.2A–D, and Experiment 2.

Because there are often small variations between thermometers, and even along the capillary of a given thermometer, it is wise to calibrate the thermometer or melting point apparatus that you will be using during the semester. To do this, determine the melting ranges for a number of pure known solids and compare the results with their expected literature

Table 2.1 Calibration Standards

Substance	Melting Point (°C)
Naphthalene	80.2
p-Dibromobenzene	87.3
m-Dinitrobenzene	89.8
Acetanilide	114
Benzoic acid	122.4
Urea	132.7
Cinnamic acid	133
Anthranilic acid	147
p-Nitrotoluene	149
Adipic acid	153
Sulfanilamide	166
p-Bromoacetanilide	167
p-Toluic acid	182
Succinic acid	189
3,5-Dinitrobenzoic acid	205

values. The melting points of these compounds should be widely separated to optimize the calibration. Select two or three compounds from those available in the laboratory (they will probably be from those listed in Table 2.1) and determine their melting ranges (see Experiment 2; HINT: From the table or the literature, you know at what point to expect the compound to melt!). The choice of knowns and the use of the apparatus should be coordinated if the apparatus must be shared. (Take turns; complete the determinations for one substance in each turn.) Start with the lowest-melting compound to be used; less time will be lost waiting for the apparatus to cool down after each melting point determination. Repeat each determination until the ranges are short and reproducible (1.0–1.5° between determinations).

Include in the report a table of ranges, a graph for corrections, and a discussion of what modifications of technique you tried in order to achieve narrower, more reproducible ranges, and the reasons for these changes. The table of ranges should include the midpoint of the range, which will be used as the observed value for graphing, and the literature value. Graph the literature values (ordinate) against the observed values (abscissas). Connect the points on this graph to be able to estimate corrections in between observations. Use a computer graphics program if one is available to plot this graph and obtain a calibration equation.

Summary

1. Determine the melting point range of the lowest-melting standard compound in duplicate, at least, until you obtain short, reproducible ranges.

2. Repeat for standards of higher range.

3. Graph your data and that of others who are using your apparatus.

Questions

1. How would the graph or calibration equation from this experiment help you in future experiments if you use the same apparatus and thermometer to determine melting points?

2. If the melting point of a substance is not known, how would a fast, rough determination of range be used?

In addition to these items, answer questions 2 and 3 from Experiment 2.

Disposal of Materials

Used melting point capillaries (nonrecycle glass)

Excess standard compounds (OW, HO, or I)

E X P E R I M E N T   2

Melting Range of an Unknown Compound
References: Sections 2.1, 2.2A–E, and Experiment 1.

Determine the approximate melting range of the unknown assigned to you. Increase the temperature at a rate of about $10°$ per minute. Record the softening and final temperatures. Repeat the determination with a fresh sample, but reduce the rate of heating, starting $10–15°$ below the observed softening point. The rate of increase as the temperature goes through the melting range should be as slow as possible without stopping. With practice this should be $0.5–1°$ per minute. Repeat the melting range determination for this pure compound until the range is short (about $0.5–1°$) and reproducible (about $1°$). If the graph or calibration equation from Experiment 1 is available, report the corrected as well as the uncorrected melting ranges. If the compound is from Table 2.1, you may be asked to identify the compound by doing a "mixed melting point" (see Section 2.2E).

Summary

1. Do a rough, rapid melting point determination.

2. Do, at least, duplicate, accurate melting range determinations.

3. If possible, do a melting point mixture with some of the pure compounds suggested by the results of the melting range determinations.

Questions

1. Why are two or more (one fast and one slow) determinations used? Why not just one slow one?

2. What effect do the following have on the melting point range:
   a. impurities in the sample?
   b. too large of a sample in the capillary tube?
   c. loose packing of the sample?
   d. a thick-walled melting point capillary tube?

3. What changes in melting point range would be expected for melting point determinations of different mixtures of compounds with similar melting points? How is this utilized in mixed melting points?
   a. How do mixed melting points demonstrate whether two samples with the same melting points are identical?
   b. Why should samples with more than one composition be tested?

Disposal of Materials

Used melting point capillaries (nonrecycle glass)

Excess unknowns (OW, HO, or I)

Additions and Alternatives

You may be asked to determine the melting ranges of several intimate mixtures with different compositions, consisting of two substances having similar melting points. If the melting ranges of enough mixtures across the entire range of composition are determined, a composite "melting point" (average of the range) diagram can be plotted (these average values are plotted against composition; see Figure 2.3). Three possible combinations are *p*-dibromobenzene with *m*-dinitrobenzene, *p*-nitrotoluene with anthranilic acid, and urea with cinnamic acid.

2.3   Recrystallization

2.3A   Conditions for Recrystallization

Organic reactions seldom yield pure compounds. In many instances, several reactions proceed simultaneously and mixtures of products result. Likewise, in most instances the reaction is not complete, and a portion of the original materials remains to be separated from the desired product. If the substances in question are solids that have high boiling points and/or that undergo decomposition at or below the boiling point, they must be separated by recrystallization.

Recrystallization of an impure substance is possible because most substances are insoluble in each other in the solid state. In other words, they do not form intermolecular compounds, nor do they crystallize in identically shaped crystals (isomorphous crystals) that can enter together into the structure of singular crystals. Thus, a solution of a mixture that is supersaturated with respect to only one compound will deposit only that one component in crystalline form. The other components remain in the solution. If any contamination remains after filtration, it is usually due to the inclusion of mother liquors and not to coprecipitation. Repetition of the process, using fresh solvent, greatly increases the purity of the desired substance.

Recrystallization, then, is the dissolution of impure crystals into a hot solution and, upon cooling, the collection of the pure component. These impure crystals are dissolved in a *minimum* volume of *hot* solvent, filtered *hot* to remove less-soluble materials, cooled to recrystallize the pure compound, and filtered cold to collect the pure compound in its crystalline form (see Summary of Recrystallization, Section 2.3G).

2.3B   Solvent Selection

An ideal solvent for use in recrystallization should possess certain properties. It will usually have different solubility properties than the crystals to be purified (like dissolves like). It should dissolve much more of the impure compound at high temperatures, usually just below the boiling point, than at room temperature. It should preferably be a much better solvent for the impurities than for the product that you desire to purify. If this is not the case, the impurities should be nearly insoluble in the solvent. The solvent should not react with the solid substance to form a new compound of any kind. Volatile solvents are convenient when crystals are to be dried quickly or when solutions are to be concentrated by solvent evaporation. Use of very flammable and/or toxic solvents should be avoided whenever possible.

Ideally, the substance to be recrystallized should not be too soluble or too insoluble in the solvent. While it is possible to work with very concentrated solutions, there is great advantage in separating a dilute filtrate from the reformed crystals. A certain amount of the solution is invariably trapped in the solid phase; this contributes the impurities that you desire to eliminate. If the solution is dilute, contamination is reduced. On the other hand, if the solubility is too low, inconveniently large quantities of solvent are required, making it necessary to use large containers and bringing up the question of solvent recovery. Quite pure products are obtained when the solvent, at its boiling point, dissolves less than one-tenth of one percent of its weight of solute, but the disadvantages of such low solubility usually outweigh the advantages. The same end is more easily achieved by using a more effective solvent and by performing a greater number of recrystallizations. Typically, moderate amounts of an appropriate solvent are used in this case. Such solvents dissolve the crystals well at elevated temperatures and only sparingly at room temperature and below.

To select a proper solvent, add 1 mL of the prospective solvent to a small amount of the solid (about 0.1 g) in a small test tube. Stir, mix, and/or grind the solid into the solvent to see if it goes into solution (observe relative quantities of solid remaining after mixing). If it goes into solution, it is probably "too good" as a solvent and they were probably too much alike. When you are satisfied that very little or none has dissolved, heat the mixture to just below the solvent's boiling point; stir, mix, and/or grind the solid into the solvent for a few minutes at that temperature. If little or none of the solid has gone into solution, add a bit more solvent and repeat the above procedure (if none goes in, try another solvent). When much or all of the solid has dissolved (if some solid still remains, decant the liquid away from it), allow the solution to cool. When cool, put the test tube into an ice bath, stir, and scratch the inside wall of the test tube with a stirring rod to initiate crystallization. In summary, if the crystals dissolve in the hot solvent and come out of the cold solvent, that's the solvent that you want. A microscale method of solvent selection is found in *J. Chem. Educ.*, 1989, 66, 88. For a description of the use of a mixed solvent, see Experiment 3C.

2.3C   Difficulty with Oils

The separation of "oils," rather than crystals, upon cooling saturated solutions complicates the purification problem. This situation often arises when the boiling point of the solvent is above and/or too close to the melting point of the material to be purified. The "oil" consists of a saturated solution of the solvent in the product, and the other layer consists of a saturated solution of the product in the solvent. Because solvents that are somewhat unlike the solute are usually chosen, the impurities present in the crude mixture may concentrate in the "oil." The result is that the product becomes less rather than more pure in the process.

Two things can be done to correct this situation. You can select a solvent that will boil at a much lower temperature, if one can be found possessing the other desired properties. As a simpler treatment, you can add more solvent to the mixture so that saturation is not reached until the temperature has fallen to a point below which only a solid phase can separate from the solution. If this involves too large a volume of solution for convenience, you can obtain the same effect by separating the "oil" from the solution and cooling the solution. Relatively pure crystals will then separate from the solution. You can collect these by filtration, and the filtrate, rather than fresh solvent, can be returned to the "oil" for reheating and a repetition of the above procedure. This process, in effect, extracts the compound from the oil.

2.3D   Failure to Crystallize

Another difficulty often encountered is the product's failure to crystallize from a supersaturated solution. This situation may arise from several causes, all of them tending to increase the viscosity of the solution. The trouble is most frequently due to a concentration of solute that is too high or to the presence of an impurity that acts as a protective colloid. Both cases yield to "seeding." In seeding, a tiny crystal of the solute is added to the cold solution, whereupon precipitation occurs. Precipitation may be very rapid or it may be slow and require refrigerator storage for several days. The difficulty with this technique lies in obtaining the seed crystal. This can sometimes be done by evaporating a drop of the solution on the end of a stirring rod that is then used to stir the solution and scratch the walls of the flask below the surface of the solution. Adding a few drops of a poorer solvent, miscible with the original solvent, to a few drops of the solution may also produce crystals.

Supersaturation may be overcome in some cases by rubbing the inside of the flask with a stirring rod. The reason for this is obscure, although it may involve alignment of molecules in a manner suitable for crystal formation. If the rod has seed crystals on it, it works even better. In a few rare cases the addition of crushed glass or sand will provide fortuitous nuclei for crystal formation. If the solution is too viscous, allowing it to warm to room temperature or adding a *little* solvent before stirring may help.

Occasionally, colloidal substances interfere with crystallization and must be removed before crystals can be obtained. This can often be accomplished by the use of adsorbents such as activated charcoal or Celite. Such adsorbents combine preferentially with larger molecular species responsible for imparting colloidal properties to solutions. Once these are removed and the charcoal filtered out, the solution is cooled and crystallization may take place. Pelletized charcoal is removed easily by filtration.

If these measures fail, the solution may require standing or evaporation before crystals will form. Even forming a solution by refluxing may help. Sometimes allowing the solution to stand, loosely covered, in a hood for a while may do the trick. More likely, the solution is not even saturated. In this case some of the solvent must be evaporated so that a supersaturated solution results upon gradual cooling.

2.3E   Recovering the Purified Crystals

(SM) On the semi-micro scale, use a small Erlenmeyer flask (50 mL, 125 mL, or 250 mL) to dissolve the solid, with heating, in the solvent selected. The shape of the flask minimizes contamination and solvent loss. A glass funnel, to be used for the hot filtration, can be placed (stem down) on the flask while it is being heated, thereby warming the funnel (without paper) while helping to condense the vapors (Figure 2.6A).

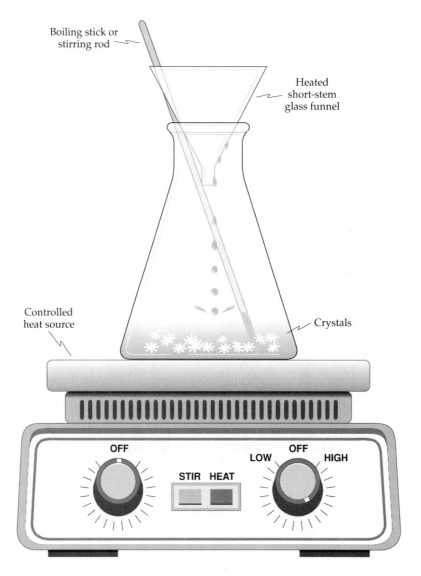

Boiling stick or
stirring rod

Heated
short-stem
glass funnel

Controlled
heat source

Crystals

OFF

STIR HEAT

LOW OFF HIGH

Figure 2.6A Heating to dissolve the solute.

When using low-boiling or unhealthful solvents, the process should be done under reflux in an apparatus such as that shown in Figure 1.7, Unit I. Start with the impure crystals, the boiling stones, and the estimated amount of solvent in the appropriate-sized flask. Reflux with water running through the condenser. Add additional solvent through the condenser, if needed. When dissolution is achieved, filter the hot solution as

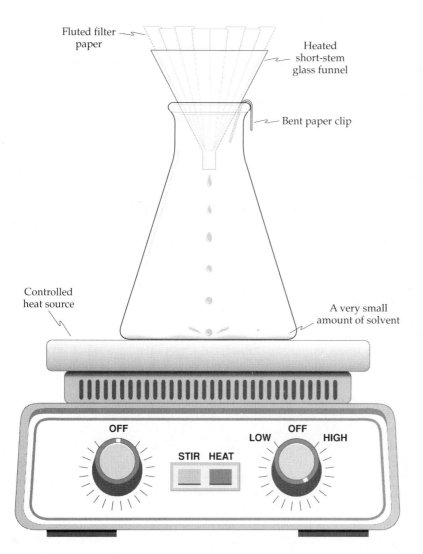

Fluted filter paper

Heated short-stem glass funnel

Bent paper clip

Controlled heat source

A very small amount of solvent

OFF

LOW OFF HIGH

STIR HEAT

Figure 2.6B Hot filtration.

described below. For the transfer, attach a rigid clamp (to be used as a handle) to the neck of the flask before removing it from the apparatus. Small amounts of solution can be transferred with a preheated pipet. If solution is poured over the ground glass, there should be no lubricant used on the joint, and as always, the joint must be cleaned as soon as practical.

The next step consists of the hot filtration, which uses an Erlenmeyer flask of the same size as above (Figure 2.6B). You must do hot filtration as rapidly as possible, keeping everything as warm as possible. The funnel should have a very short stem and should be fitted with a fluted piece of filter paper. To flute a circular piece of filter paper, fold it in half several times continuously, and then open it so as to expose a maximum amount of surface area for filtration. Hot filtration should separate the hot solution from insoluble impurities without allowing the desired product to come out of solution.

Filter the crystals that form upon cooling, using a Büchner or Hirsch funnel. Place a circle of filter paper so that it will lie flat and cover all holes on the funnel's perforated plate, and drain a bit of solvent through it by suction (Figures 2.7 and 2.8C). The filter flask and pump should be connected through a trap. Start the suction to pull the bit of solvent through and, with the filter paper securely in place, filter the cold solution with continued suction. Sometimes it helps to touch a stirring rod to the paper and pour the solution down the rod. If the crystals are to be washed, interrupt the suction by allowing air in; add a small volume of *cold* wash solvent, mix, and restart the suction. The suction can be continued after all of the solvent has gone through the filter paper to start "drying" the crystals with air.

(M) Microscale kits contain Craig tubes, which are designed to recrystallize and filter small amounts of crystals. A Craig tube consists of two parts: a bottom for recrystallization and a top or stopper to complete the apparatus as it is used for filtration (Figures 2.8A and 2.8B). If hot filtration is used, the filtrate should go directly into the bottom of the Craig tube (the volume should *not* go beyond three-fourths of the way up to the ground area). Otherwise a small amount of hot solvent can be added to a small amount of crystals in the bottom of the Craig tube. The mixture can be heated (approximately 2–5° below the boiling point of the solvent) in a sand bath or aluminum block on a hot plate. Mixing by twirling a microspatula may facilitate the dissolving process. Additional solvent should be heated at the same time and should be added a few hot drops at a time to dissolve the solid completely. When complete dissolution is achieved, you can remove the bottom of the Craig tube from the heating apparatus and allow the solution to cool. Use an ice bath to complete the reformation of the crystals. (Twirling the microspatula in the solution again may help start precipitation. When the spatula is removed, seed crystals may form as the solvent evaporates.) After thorough chilling, put the top part of the Craig tube in place with a gentle turn to seal it into the ground area. Attach a small piece of wire around or through the handle (Figure 2.8A), and cover the apparatus with a glass or unreactive plastic centrifuge tube (a bit of cotton may be placed in the bottom) so that the wire sticks out the mouth of the centrifuge tube. Invert the tube (Figure 2.8B) and, after balancing, centrifuge it. KEEP THE CENTRIFUGE COVERED WHILE IT IS SPINNING. The filtrate will be forced through the ground glass joint into the centrifuge tube, leaving the solid behind. Use the wire to

CAUTION

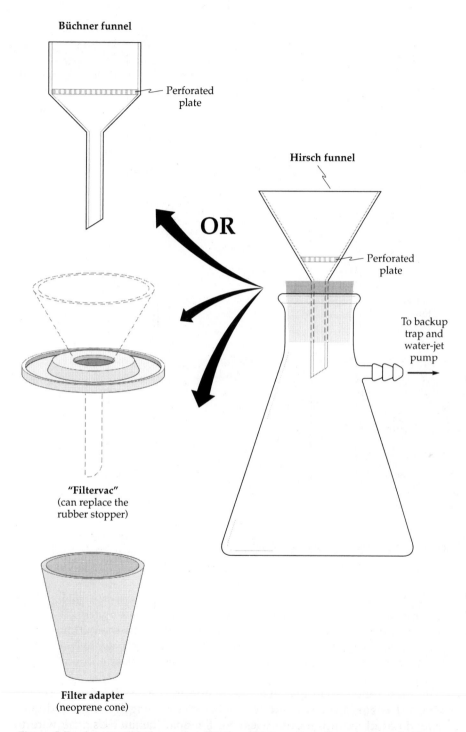

Büchner funnel

Perforated plate

OR

Hirsch funnel

Perforated plate

To backup trap and water-jet pump

"Filtervac" (can replace the rubber stopper)

Filter adapter (neoprene cone)

Figure 2.7 An assembly for suction filtration. A trap such as the one shown in Figure 3.11A, Unit D should be used when filtering with suction.

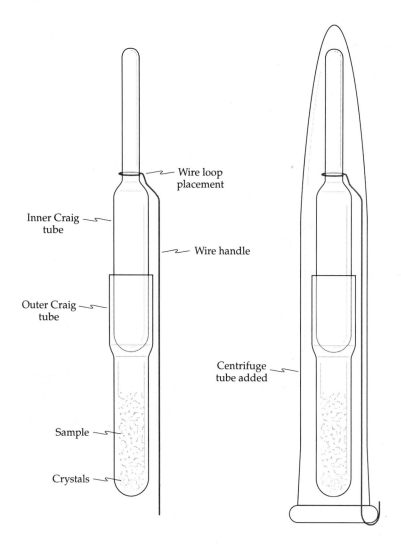

Inner Craig tube

Wire loop placement

Wire handle

Outer Craig tube

Centrifuge tube added

Sample

Crystals

Figure 2.8A Craig tube before inversion.

remove the Craig tube without disassembling it. Over a watch glass, take the Craig tube apart, carefully collecting the crystals in the bottom of the tube if you are going to wash them or recrystallize them again (*save* some for the melting point determination) or on the watch glass if they are to be dried and weighed.

If a test tube, centrifuge tube, or reaction tube is used for recrystallization, a Pasteur pipet with or without a cotton plug (no cotton should extend out of the tip) may be used for filtering. Reread Section 1.3E. The Pasteur pipet is inserted through the slurry of crystals and solution,

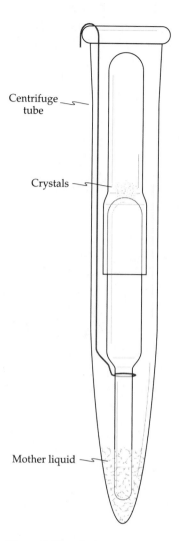

Centrifuge tube

Crystals

Mother liquid

Figure 2.8B Craig tube inverted for filtration (M).

without anything entering the pipet, to rest squarely on the bottom of the tube (Figure 2.8C). The solution is sucked gently into the pipet without allowing any crystals to enter. Centrifugation may make these separations easier. Alternatively, collect the crystals on a filter paper circle in a small Hirsch funnel by suction filtration.

Small amounts of solids may be dried (Section 4.4H) in a bottle or test tube desiccator (connected through a trap to the water aspirator). A stoppered side-arm test tube on its side provides a place to attach the

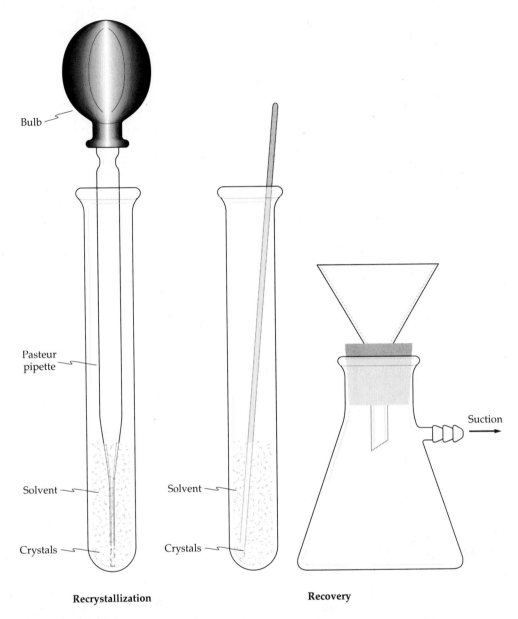

Bulb

Pasteur
pipette

Solvent

Crystals

Solvent

Crystals

Suction

Recrystallization

Recovery

Figure 2.8C Separation without a Craig tube.

vacuum tubing and easy access to the solid being dried. Also, a clay plate can be used for drying the solid as described in Section 2.2D. It is usually more convenient to let the crystals air-dry until the next period.

2.3F   Digestion

A reversed approach to the purification of impure solids is called digestion. This method can be useful when the solid is very impure. It depends on selecting a solvent in which the impurities are relatively soluble hot and the material to be purified is not. The solid mixture is treated with a small portion of the selected solvent at or near its boiling point. When saturated, the hot solvent is decanted and the process repeated. How does this work? Where is the purified material?

2.3G   Summary of Recrystallization

The need for the steps in brackets will vary with each recrystallization.

1. Choose an appropriate solvent, one in which the solute is sparingly soluble at room temperature and appreciably more soluble near the solvent's boiling point (Section 2.3B).
2. Dissolve the solute in the minimum amount of *hot* solvent. Add a slight excess of solvent (Section 2.3E).
[3. Let the solution cool slightly, and add a small amount of pelletized charcoal (if needed). Then heat it again (Section 2.3D).]
[4. Filter the hot solution by gravity through a warm funnel into a warm flask (Section 2.3E).]
[5. Evaporate the slight excess of solvent to approach saturation.]
6. Let the solution cool and allow crystals to form (Section 2.3C and D). Then cool with ice.
7. Collect the crystals by suction filtration or by centrifugation (Section 2.3E).
8. Wash the crystals (Section 2.3E).
9. Let the crystals dry.

E X P E R I M E N T   3 A

Purification of Benzoic Acid (or an Unknown) (SM)
References: Sections 2.3A–G and Experiments 3B and 3C.

The following exercise illustrates the procedure to be followed when separating a compound from impurities whose components may be more soluble or less soluble (in the solvent used) than is the compound to be purified. Benzoic acid may be separated from contaminants such as urea, charcoal, or sand. The solubility of benzoic acid in 100 mL of water is 0.27 g at 18°C and 2.5 g at 100°C. Urea is extremely soluble in water at room temperature and is infinitely soluble in water at its boiling point. Sand is insoluble in water. You may be issued a mixture whose composition is unknown; that is, either the percent of benzoic acid or both the percent and identity of the major component are not known to you. If an unknown compound is to be purified, it will be an organic compound that is moderately soluble in water.

Introduce the sample and initiate recrystallization. Record the number of the sample you receive, and weigh out a portion of it that is between 0.275 and 0.35 g. Record this weight. Place the weighed portion in a 50-mL Erlenmeyer flask, add 5–10 mL of water to it, and boil the mixture on a hot plate for at least a minute or two. During this process, preheat the funnel, which you will use for the hot filtration, by inserting it loosely into the neck of the flask so that steam may condense on it (Figure 2.6A). Should all of the solid not dissolve (and sand won't), add *small* increments of *hot* water (0.5–1.0 mL) to the system and mix. After each addition, heat the mixture to just below its boiling point for a minute or two, with mixing, and estimate the amount of solid remaining. If the amount has not decreased, proceed to the hot filtration; but if it has decreased, make another water addition until there is no noticeable decrease.

Charcoal may be employed to remove colored substances or to break up colloidal suspensions. If you use it, add a small amount of pellets only after you've added the last portion of solvent (in this case water), because it will obscure the observation of any remaining solid. DO NOT ADD CHARCOAL TO A HOT SOLUTION THAT IS BOILING OR TO ONE THAT IS CLOSE TO BOILING! Allow the solution to cool before adding charcoal and then reheat. You should completely remove the charcoal during hot filtration. When not using water as a solvent, a reflux setup is safer.

CAUTION

Filter the hot solution through a heated short-stemmed glass funnel fitted with fluted filter paper. Keep the glass funnel slightly ajar in the Erlenmeyer flask by placing a bent paper clip between the funnel and the mouth of the flask (Figure 2.6B).

Question

1. What is the purpose of the *short*-stemmed funnel (i.e., not long)? Of the paper clip?

A very small amount of boiling water (solvent) in the receiving flask may help to keep the funnel hot.

Question

2. Why is it unwise to use suction filtration (instead of gravity as directed) for hot filtration of solutions made with low-boiling solvents?

If crystals form in the filter paper or funnel during this process, the solution was probably too concentrated. Try pouring a little boiling solvent through. If that does not work, return the crystals (and the filter paper?) to the original flask, add more water (solvent), boil it to form a solution that will not be supersaturated until the hot filtration is accomplished, and hot filter with fresh filter paper.

Isolate the crystals. Allow the filtrate to cool to room temperature. Scratch and seed, as needed, to facilitate crystal formation. Complete the cooling by placing the flask in an ice bath.

3. Why should none of the cooling water be permitted to enter the flask when the flask is placed in the ice bath?

Question

Meanwhile, assemble a clean apparatus to be used for suction filtration. Connect the filtration flask to a supported backup trap (Figure 3.11A, assembly D). Use a Büchner or Hirsch funnel with a piece of filter paper that, when moistened with solvent, lies flat over the funnel's perforations. If a filtering frit is provided, filter paper is optional. The funnel should make an airtight seal with the filter flask by means of a one-holed rubber stopper, a filter adapter, or a "Filtervac" (Figure 2.7). After complete cooling, collect the crystalline product by suction filtration. Break the vacuum seal (by letting in air before the water is shut off), and wash the crystals by adding just enough cold solvent to cover them. Remove this solvent completely and dry the crystals with suction.

4. What does washing or rinsing accomplish?

Question

Prepare for recrystallization. Determine the approximate weight of the crude product. Set aside just enough of these crystals to do two melting range determinations. If the melting range is to be determined at once, you could dry a small portion of the sample by pressing it in filter paper or by rubbing it on a porous plate (or use a vacuum tube desiccator) while you continue the experiment. Determine the melting range of this sample when it is dry. If porous plates are used for drying, return them so that they may be cleaned by ignition.

Repeat the recrystallization. Recrystallize the crystals obtained above by using 2.0 mL or one-half of the estimated volume of *hot* water needed for the first recrystallization. Add more water at the boiling point in 0.4-mL increments (about 8 drops) until the product or its oil dissolves. Repeat the procedure described in the paragraphs above starting with hot filtration (only if hot filtration is needed!). After you have recollected the crystals, dry a sample of them for a melting range determination.

5. Why must a solid be dry to obtain a meaningful melting range?

Question

When the identity of the compound being recrystallized is unknown to you, it is a good idea to add the water for the first recrystallization in small portions rather than all at once.

6. Why would a too dilute hot solution cause subsequent difficulty?

Question

After each addition of fresh portions of solvent, the solution should be brought close to a boil. A little more water can be added if the last portion that has been added causes no more of the solute to go into solution at the solvent's boiling point. In some instances, the addition of a 10 percent excess of solvent beyond saturation will facilitate the hot filtration.

Interpret the results. Determine and report the melting ranges of the original crude material (remember, sand won't melt!), the product recovered from the first recrystallization, and the dry, purified product. Weigh the dry product (this weighing and a melting range determination may be done during the next lab period), and calculate the percent recovered based on the weight of the original sample.

Summary

1. Accurately weigh the sample to be purified and record the weight and identifying number; save a small sample.

2. Dissolve the weighed sample in a minimum amount of *hot* solvent (record the volume).

3. Gravity filter *hot* if necessary.

4. Cool and then chill on ice.

5. Collect the crystals by cold suction filtration and draw air through them to dry them.

6. Obtain a rough product weight for the first recrystallization and save enough solid for a melting point determination.

7. Estimate the amount of solid needed for a second recrystallization and add about one-third of your estimate.

8. Dissolve the solid in a minimum volume of *hot* solvent.

9. Cool and then chill on ice.

10. Collect the crystals by cold suction and dry them.

11. Weigh the final product; determine melting ranges for all of the samples.

Questions

QUESTIONS 1–6 IN ABOVE TEXT.

7. How would the amount of solvent needed for another recrystallization be determined from the data obtained for the first recrystallization?

8. How and where are the urea and sand (and/or charcoal) separated? Generalize the purposes of the hot and cold filtrations?

9. How does the melting range change with successive recrystallization? What would higher melting range temperatures indicate?

10. When is a hot filtration not needed as part of recrystallization?

Answer Questions 6 and 7 from Experiment 3C.

Disposal of Materials

Benzoic acid or recrystallization mixture (N, D, or I)

Aqueous solutions (D, N(NaHCO$_3$), or I)

Charcoal (S)

E X P E R I M E N T   3 B

Purification of Benzoic Acid (or an Unknown) (M)
References: Section 2.3 and Experiment 3A.

Introduce the sample and initiate recrystallization. Weigh out a 25–35 mg sample of impure benzoic acid or an unknown and record the exact weight (the sample will not contain extremely insoluble impurities, sand, or charcoal). Record the weight of a large Craig tube (bottom). Quantitatively transfer the solid to the Craig tube. Add 0.5 mL of *hot* water, washing down any clinging solid as it is added. Place the Craig tube and a small test tube containing hot water in a preheated sand bath or aluminum block (they should be warmed to a temperature above the boiling point of water so that the water in the tubes can be at about 90°C or a bit above). Allow them to come to temperature.

Add hot water (from the reserve tube) to the Craig tube in 1–3-drop portions (depending on the drop size) until all of the solid dissolves with stirring. Stir the solution in the Craig tube vigorously with a microspatula between each addition. Rinse the spatula with the next addition. Be sure that time and stirring allow complete dissolution with a minimum of *hot* solvent.

Remove the Craig tube and allow it to cool to room temperature in a small beaker or rack. If no crystals appear, stir the solution with the microspatula (there may be crystals on it). Chill it completely in an ice bath.

Isolate the crystals and interpret the results. Assemble the Craig tube (Section 2.3E, Figure 2.8B), invert, balance, and centrifuge. Remove just enough crystals from the Craig tube for a melting range determination and determine the weight of the remaining crystals by weighing the Craig tube with them in it.

Repeat the recrystallization. Use this weight and the volume of solvent used to attain it in comparison with the original weight of the crystals to estimate the quantity of solvent needed for a second recrystallization. Add one-third to one-half of the estimated volume of hot solvent to the Craig tube and recrystallize again, using the approach discussed above.

Interpret the results. Dry and weigh the crystals collected from the second recrystallization and determine melting ranges for both recrystallization products. Calculate the percent recovery for each recrystallization and overall.

Summary

1. Place a small tube containing water in the heat medium and heat it.

2. Weigh the sample to be recrystallized (record the weight and number).

3. Place the sample in a Craig tube.

4. Dissolve *as much* sample as possible by adding 0.5 mL of the *hot* water.

5. Place the Craig tube in the heat medium.

6. Add a minimum volume of hot solvent to dissolve the sample completely.

7. Cool and chill the Craig tube.

8. Centrifuge the Craig tube and collect the crystals.

9. Weigh them and save enough to make a melting range determination.

10. Recrystallize them from a minimum amount of *hot* solvent.

11. Dry and weigh them and determine melting ranges for all samples.

Questions   ANSWER QUESTIONS 3, 5–7, AND 9 FROM EXPERIMENT 3A, AND 6 AND 7 FROM EXPERIMENT 3C.

1. How could purified crystals be "washed" using this apparatus (compare with Experiment 3A)?

Disposal of Materials

Benzoic acid or recrystallization mixture (N, D, or I)

Aqueous solutions (D, N(NaHCO$_3$), or I)

Charcoal (S)

Additions and Alternatives

The mixtures you work with may contain various percentages of benzoic acid or other unknowns.

Acetylation of an Unknown (SM and M)
References: Section 2.3 (especially 2.3B) and Experiments 3A and 3B.

$$\text{Unknown}-\text{NH}_2 \atop \text{or} \atop \text{Unknown}-\text{OH} \quad + \text{CH}_3-\overset{\displaystyle O}{\overset{\|}{C}}-O-\overset{\displaystyle O}{\overset{\|}{C}}-\text{CH}_3 \longrightarrow$$

$$\text{Unknown}-\text{NHCOCH}_3 \atop \text{or} \atop \text{Unknown}-\text{OCOCH}_3 \quad + \text{CH}_3\overset{\displaystyle O}{\overset{\|}{C}}\text{OH}$$

Introduce the reagents and initiate reaction. If possible, carry out the initial reaction in the hood because some of these materials are potentially toxic. Place 1 mL of the unknown (or 1 g of crushed solid) in a small Erlenmeyer flask, a reaction tube, or a test tube (capacity of 10 mL or more) with a calibrated pipet. DISPOSE OF THE PIPET AFTER IT IS WASHED WITH 5% HCL IN SUCH A WAY THAT ALLOWS NO UNKNOWN TO TOUCH YOUR SKIN. Add 5 mL of water, then add 1.5 mL of acetic anhydride (d = 1.08 g/mL) gradually, with gentle stirring or swirling. Crystals will often form at this point or with more serious stirring and cooling. If not, heat the mixture for a few minutes with a steam or hot water bath in the hood. Do not remove the reaction mixture from the hood if there is any acetic anhydride odor remaining. Destroy the excess acetic anhydride with a few drops of dilute HCl when the crystals are formed.

CAUTION

1. What reaction occurs when you add HCl to acetic anhydride?

Question

When the crystals are formed and cooled, collect them with suction filtration using a Büchner or Hirsch funnel, and wash them with ice water.

Recrystallize. Select a solvent to use for recrystallization. The solvents available include water and one or more water-miscible solvents such as methanol, acetone, ethanol, or 2-propanol. In a series of very small test tubes, place the same amount, 0.4–0.5 mL, of each solvent (one solvent per test tube). To each, add 25–50 mg of crystals (weigh once and estimate the rest). The amount of solid and solvent should be the same in each test tube. Stir each mixture to determine the relative solubility of your crystals at room temperature. Next, where the solid has not dissolved, heat the tubes cautiously (avoid splattering the mixtures out) with a steam or hot water bath to determine your solid's solubility in the hot solvent.

Question

2. Organic solvents should not be heated with a flame. Why not?

As discussed in Section 2.3B, one of the most important factors is the difference in solubility of the solid in hot and cold solvent.

Question

3. Explain why the difference in solubility, hot versus cold, is such an important consideration.

If a satisfactory solvent is found, use it for the recrystallization as described in Section 2.3E and in Experiments 3A or 3B. If not, use a mixed solvent.

For mixed solvent recrystallization, use a "good" solvent (one in which the solid is soluble in hot) and a "poor" solvent (one in which the solid is insoluble in hot) that are miscible. Dissolve the amount of solid to be recrystallized (some could well be saved) in a minimum of the better solvent by heating it near its boiling point. Then add the poorer solvent dropwise at this elevated temperature until cloudiness is detected. BE VERY CAREFUL WITH FLAMMABLE SOLVENTS! Add more of the better solvent dropwise until the cloudiness just disappears at this temperature.

CAUTION

Question

4. What has been achieved by this adjustment of solvent amounts?

If cloudiness reappears with very little cooling, hot filtration may be difficult to achieve without adding a bit more of the better solvent and reheating.

Using the solvent or solvent mixture selected by the methods described above, recrystallize most of the remaining solid, impure, acetylated unknown as described in Experiment 3A or 3B. (A *small* amount of the solid can be saved from this recrystallization to be used as seed crystals. You can also dry this saved solid and use it to determine a preliminary melting range.) Record the amount of solvent used. If the solution is highly colored, it may be desirable to use pelletized, activated charcoal, which selectively absorbs many colored compounds. Observe the precautions mentioned previously.

See Figures 2.6A, B; 2.7; 2.8A, B, C (pages 52–53, 55, 56–58).

Question

5. What might happen if charcoal is added to a hot solution without a functioning boiling aid?

Determine the weight (rough) of the once-recrystallized material.

Recrystallize again. Do a second recrystallization using most of the product collected from the first recrystallization. From the amount of solid and solvent used above, estimate how much should be used this time.

Interpret the results. Accurately determine the weight of the final product after it is dry, and determine the melting ranges of the three dry samples. Your report should include this data and a discussion of your results. Percent recovery can be calculated with respect to the weight of the original crude product.

Summary

1. React 1 mL or 1 g of unknown (record number) with 5 mL of water and 1.5 mL of acetic anhydride.

2. Cause crystals to form, and collect them when all of the anhydride is gone.

3. Determine the solvent to be used.

4. Recrystallize twice from a minimum of *hot* solvent.

5. Dry and weigh the crystals and determine the melting ranges of both products.

ANSWER QUESTIONS 1–7 AND 9 FROM EXPERIMENT 3A AND QUESTIONS 1–5 ABOVE (THIS EXPERIMENT).

Questions

6. Suppose the solubilities in water for compound X are 6 g at 100°C and 1 g at 15°C (per 100 mL) and for compound Y they are 5 g at 100°C and 0.8 g at 15°C. These solubilities are *independent* of whether anything else is dissolved in the water. A mixture of 5 g of X and 2.5 g of Y needs to be recrystallized. How much solvent should be used for the first recrystallization? What would be the composition of the solid product? How much X is lost? How should the process be continued to obtain pure X?

7. How much X and how much Y would be lost if 100 mL of solvent was used each time for three recrystallizations? What percentage of the recovered product would be X?

Disposal of Materials

Unknown (OW, HO, HW, or I)   Product (OW, HO, or I)

Acetic anhydride (N(NaHCO$_3$))   Solvents (OW, HO, or I)

Charcoal (S)

Additions and Alternatives

You may need to employ an alternative method for solid or liquid amines that give poor results. Combine 1 g of the unknown with 15 mL of water in a small Erlenmeyer flask, and add 6N hydrochloric acid a few drops at a time, with swirling, until all of the unknown is in solution (about 1–2 mL). While dissolving the unknown as the hydrochloride salt, prepare a solution of 0.85 g of sodium acetate in 4 mL of water.

Place the acidic unknown solution on a steam or hot water bath in the hood; add 1.5 mL of acetic anhydride with swirling. Immediately add the sodium acetate solution and thoroughly mix.

Question

8. What is the function of the basic sodium acetate solution?

Follow the procedure in Experiment 3C from the point at which all of the reagents have been added (first paragraph).

For a modified approach to selecting a solvent, see Hiegel, *J. Chem. Educ.*, 1986, 63, 273.

CHAPTER 3

Purification and Identification of Liquids

Many of the liquids you will encounter in the laboratory will be impure. Most of this chapter is devoted to distillation, which is the most common method of purifying liquids. Distillation is also used to determine the approximate boiling point of the liquid as a means of identification. This chapter also discusses three physical constants that may be experimentally determined and compared to literature values for purposes of liquid identification. These constants are the boiling point, the density, and the refractive index of a liquid. Other methods of liquid identification are described in Chapter 5. Other methods of separating liquids are discussed in Chapter 4. With all of these techniques in hand, you should be able to purify and identify the liquids you encounter.

3.1   Distillation Theory

To make effective use of practical directions (Section 3.2), you need to understand the theory (*) behind distillation. We start again from the idea of vapor pressure. The boiling point of a liquid is defined as the temperature at which the vapor pressure of the liquid is equal to the total pressure of the gases above it. When it boils, some of the liquid vaporizes and can be condensed elsewhere to constitute distillation.

*3.1A   Boiling Point–Composition Curves (Simple Mixtures)

When a pure substance is partially vaporized, the composition of the resulting vapor is the same as that of the liquid that produced it. When this vapor is condensed, it yields a distillate that is identical to the distilland from which it was obtained. The situation is different, however, if a mixture of *two* volatile substances is partially vaporized. In this case the vapor is richer in the more volatile component—that which has the higher vapor pressure and the lower boiling point—than is the mixture from which it came. This is shown in Figure 3.1 and would have been experimentally determined. Curve 2 represents the composition of the vapor phases, and curve 1 represents the composition of the liquid phases

69

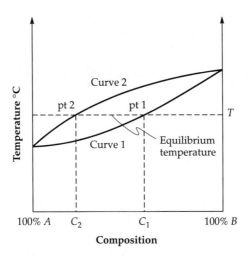

Figure 3.1 A simple liquid-vapor composition diagram.

in equilibrium with the vapor phases, at a common temperature over the temperature range bounded by the boiling points of the two substances. For each temperature T, there is a liquid of composition C_1 and vapor phase of composition C_2 that are in equilibrium with each other. These two curves (1 and 2) represent the results of analyses of binary liquid mixtures and of the binary vapor mixtures in equilibrium with them at a definite pressure, usually atmospheric, and at a definite temperature, which is the boiling point of the liquid mixture. Suppose a liquid of composition C is heated to boiling (point 1) at temperature T. Boiling liquid (C_1) will be in equilibrium with vapor (point 2) at the same temperature. The composition of this vapor and the liquid that would result if it is condensed is C_2. This is distillation in the simplest sense. Repeating this process many times with different starting compositions would be a way to construct these curves. Once this pair of curves is obtained, you can use it to study the properties and performances of various types of distilling apparatus.

Few binary systems can be used for quantitative purposes because few have been studied in sufficient detail. Certain mixtures for which analytical results are easily obtained have been studied. In the actual diagrams of these mixtures, the separation between the two curves is not as great as indicated in Figure 3.1 and in other figures used in this section. The area between the curves illustrated here is exaggerated to facilitate the insertion of details into the drawings. This fact is pointed out here to remind you that you cannot expect as good a performance from distillation equipment as the diagrams in this text might lead you to believe.

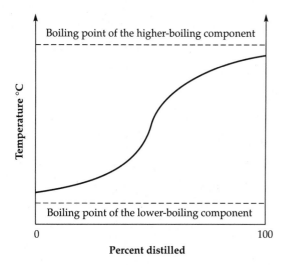

Figure 3.2 A distillation curve obtained without fractionation equipment.

Strictly speaking, the example discussed above is that of an impossible case. Figure 3.1 assumes static equilibrium, which is rarely obtained. Suppose that a mixture of two volatile substances exists with the composition C_1. This mixture is heated to its boiling point and very small amount of the mixture is allowed to vaporize. The resulting vapor will have the composition C_2, which is richer in the more volatile component, A, than is the original liquid mixture. The reason for specifying that only a very small amount of vaporization will be allowed to occur is now evident. Any extensive removal of the vapor phase would materially change the composition of the remaining liquid in the direction of higher concentrations of the component of lower volatility and higher boiling point. In an actual distillation in which vapor is removed continuously and at a constant rate, the boiling point of the mixture rises continuously, although not at a constant rate. This is illustrated in Figure 3.2, which is a typical distillation curve for a binary mixture whose vapor is obtained without the use of a fractionating column.

*3.1B Fractional Distillation

It is now possible for you to understand fractionating column operation. Figure 3.3A is a diagrammatic representation of a simple fractionating column. Vapors enter the column at a temperature T_0, substantially that of the boiling mixture. As the vapors rise in the tube, they undergo progressive cooling as a result of heat lost through the walls of the tube. By examining Figure 3.4, you can see the effects of this cooling on the vapor.

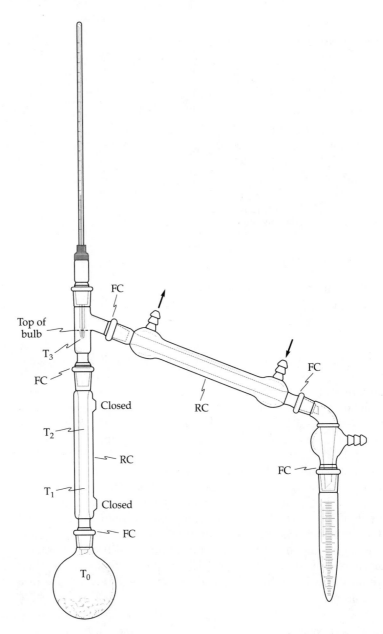

Figure 3.3A Fractionating column diagram.

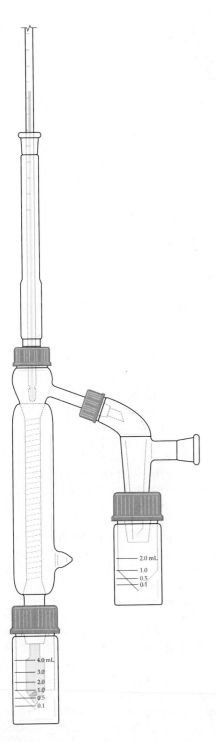

2.0 mL
1.0
0.5
0.1

4.0 mL
3.0
2.0
1.0
0.5
0.1

Figure 3.3B Microscale spinning band column.

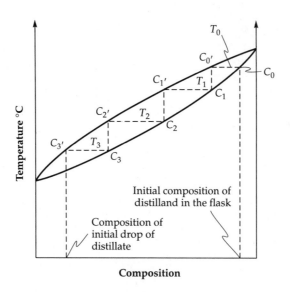

Figure 3.4 Graph showing the equilibria within a fractionating column.

It is evident from this figure that if the liquid boiling at T_0 has the composition C_0, the vapors entering the column will have the composition C_0'. If the temperature falls to T_1, T_2, T_3, etc., during the rise of the vapor and the liquid in equilibrium with it through the column, the vapor phases will have, at these temperatures, the compositions C_1', C_2', C_3', etc., and the liquid phases, as they condense upon the wall of the column and upon any "packing" in the column, will have the compositions C_1, C_2, C_3, etc. These liquid phases undergo rather complicated interactions with the ascending vapors. As the net result of these interactions, the ascending vapors lose much of their less-volatile component and increase the relative content of their more-volatile component. This is equivalent to a series of condensations and evaporations as shown in Figure 3.4. In this way, the result obtained is the same as would be obtained from several successive distillations of the mixture using only a simple distilling flask.

One way to increase separation is to increase the number of condensations and evaporations. This requires more surface area, which can be provided by packing material into the column. It can be filled with glass beads, small glass helices, or unreactive metal sponge. This increased surface increases the amount of liquid necessary to coat the surface. Thus, the increased separation is accompanied by increased loss. A good method of fractionation that increases condensations and evaporations while minimizing losses is the use of a spinning band down the center of the column. A microscale spinning band column and apparatus is shown in

Figure 3.3B. All these fractionating columns are more effective in separation than a simple distilling flask.

Another improvement can be made by attaching a condenser to the top of the column in such a way as to allow a portion of the vapors to condense and fall back upon the packing in the column. The amount of condensate allowed to fall back upon the packing divided by the amount allowed to distill into the final condenser is known as the **reflux ratio**. In other words, the reflux ratio is the volume returned to the column divided by the volume collected (RR = $V_{returned}/V_{collected}$). For ordinary purposes, a reflux ratio of ten to one is satisfactory. For more complete separations, a much greater ratio is frequently used. With a column-top condenser, or in any other setup, you must be careful not to load the column with liquid to such an extent that the spaces between the packing material are filled because, in doing this, most of the efficiency of the column is lost.

The number of complete steps required to pass from C_0, the initial composition of the liquid distilland (see Figure 3.4), to C_3', the composition of the initial distillate that issues from the still, is known as the number of **theoretical plates** of the column and still pot. You can assume that one theoretical plate is due to the initial vaporization, that is, to the still pot. Thus, the number of theoretical plates due to the column is one fewer than the total number of steps. In the present example, C_{final} is taken to be C_3', and the column has three theoretical plates. Since the height of a fractionating column is limited by the space in which it is set up, the usefulness of a column is frequently expressed as the height equivalent to a theoretical plate, **HETP** (HETP is the height of one theoretical plate). You can obtain this characteristic by dividing the height of the column by the number of theoretical plates that it exhibits in use (HETP = H_{column}/no. of theoretical plates). A smaller HETP is better!

In practice, fractional distillation should produce a distillation curve more like Figure 3.5 than Figure 3.2.

*3.1C Minimum- and Maximum-Boiling Mixtures

The foregoing discussion of boiling behavior and distillation was illustrated by a relatively simple and ideal type of binary mixture: one for which the boiling point at any composition lies between the boiling points of the two constituents. There are, however, other binary mixtures that exhibit more complex intercomponent interaction and less simple boiling behavior. For one type, there is a range in composition over which the boiling points of the mixtures fall below that of even the lower-boiling of the two pure components. At some composition in this range, the mixture will have its lowest boiling point and is called a "minimum-boiling" or **azeotropic mixture**. Systems exhibiting minimum-boiling points at some point on their composition curve are very common—more common, in fact, than those that do not. You may be familiar with the azeotrope of

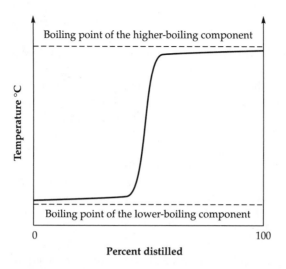

Figure 3.5 A distillation curve obtained with fractionation equipment.

ethanol and water already. Known systems that exhibit maxima in their
boiling point curves are relatively rare and are referred to as maximum-
boiling azeotropes.

A water and isopropyl alcohol system constitutes an interesting example in which a minimum-boiling mixture occurs. Pure isopropyl alcohol
boils at 82.5°C. Water boils at 100°C. A mixture of water and isopropyl
alcohol having an alcohol content of about 87.9% by weight or 90.6% by
volume boils at 80.4°C (Figure 3.6A). In minimum-boiling mixtures such as
this one, the liquid and vapor phases have the same composition at this
temperature, and it is impossible to obtain the lower-boiling component in
pure condition by distillation because the composition of the minimum-
boiling mixture intervenes between the composition of the original mixture and the coordinate of the desired pure compound. The line at M, the
composition of the minimum-boiling mixture, can be regarded as an
effective barrier. The concentrations of the components of the mixture
undergoing distillation cannot pass this point. If a mixture of the composition C_1 is subjected to distillation, vapor of the composition C_1' is produced. By the use of a high-efficiency column or by many redistillations, a
distillate with the composition C_1'' is obtained. The residue in the distilling
flask becomes water-enriched. Its composition shifts in the direction of
C_1'''. The composition shift in the liquid and vapor phases is indicated by
arrows. Redistillation of the distillate eventually produces a distillate of
composition M (a constant-boiling mixture). If the original mixture has a
composition C_2, richer in isopropyl alcohol than a mixture of composition

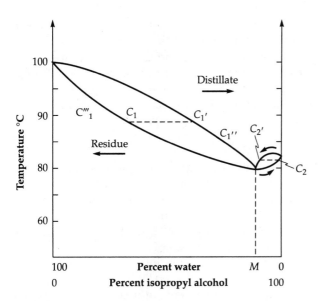

Figure 3.6A A composition diagram for a system with a minimum boiling point.

M, the vapor rising from and in equilibrium with the liquid at its boiling point will have the composition C_2'. The use of an efficient fractionating column will cause the vapors delivered to the condenser to undergo a concentration shift in the direction of M. Repeated distillations will result in nearly complete separation of the mixture into isopropyl alcohol and a mixture of composition M. The mixture of composition M will contain practically all of the water that was present in the original mixture of composition C_2.

The only mixtures of isopropyl alcohol and water that can yield pure isopropyl alcohol by simple distillation are those in which the initial composition lies on the alcohol side of composition M. In such cases, the alcohol appears not as the distillate but as the residue remaining in the distilling flask after the removal of the more volatile minimum-boiling mixture of composition M. Of course, if the distillation is continued, pure isopropyl alcohol may be collected as a final fraction. For all practical purposes, systems of the type illustrated in Figure 3.6A can be regarded as made up of two independent parts separated by the composition M. Such systems, upon distillation, behave as if the components were water and a mixture of M or alcohol and a mixture of M, rather than alcohol and water.

A similar analysis could be made of the distillation of dilute ethanol–water mixtures produced by the fermentation of grain or grapes.

These mixtures are often distilled to make solutions richer in ethanol. Whiskey is prepared, in the former case, and brandy in the latter. Since the yeast that do this job usually die before the ethanol content reaches 15%, distillation increases the value to some purchasers. It is said that some buyers of enriched spirits devised a test to avoid being cheated. The trick was to put a pile of gun powder on a stump, pour the solution to be purchased over it, and then try to light the gun powder. If it lit, that was "proof" that the solution was good stuff. A solution of about 50% ethanol/50% water would give 100% proof; thus it was called 100-proof solution. Absolute alcohol (no water) is 200 proof. The best that can be done by fractional distillation is 95% ethanol. To get absolute ethanol, benzene is added and a tertiary azeotrope employed, but that leaves traces of the benzene (which is toxic) in the alcohol. The removal or reduction of water is an example of the use of azeotropic distillation to reduce the amount of higher-boiling solvent in a mixture.

Minimum-boiling mixtures are very common in organic chemistry (see Table 3.1). Sometimes the minimum is depressed only slightly below the boiling point of the more-volatile component, and, in such cases, the composition of the ultimate distillate lies close to the pure, relatively volatile component. In other cases, the depression is relatively great and the less-volatile component is present in much larger amounts. This situation makes the separation of the components to even a moderate degree of purity extremely difficult.

Table 3.1 Examples of Minimum-Boiling Azeotropes

Component A	Component B	b.p. A	b.p. of azeotrope	%A (W/W)
Binary				
Ethanol	Water	78.1	74.8	95.5
1-propanol	Water	97.2	87.7	71.7
2-propanol	Water	82.5	80.4	87.9
2-methyl-2-butanol	Water	82.8	79.9	88.3
2-methyl-2-pentanol	Water	102.3	87.4	72.5
Benzene	Water	80.2	69.3	91.1
Toluene	Water	110.8	84.1	80.4
Anisole	Water	153.9	95.5	59.5
2-methylpropyl propanoate	Water	136.9	92.8	67.8
Benzene	Ethanol	80.2	68.2	67.6
Toluene	Ethanol	110.8	76.7	32.0
Toluene	2-propanol	110.8	81.3	21.0
Tertiary				
Benzene	Ethanol/water	80.2	64.9	(74.1/18.5/7.4)

Maximum-boiling mixtures rarely occur in organic chemistry but are much more common in inorganic chemistry. The systems hydrochloric acid–water, hydrobromic acid–water, and nitric acid–water may be familiar to you as the concentrated acids. In such systems, considerable interaction occurs between the components involved. Organic compounds that form maximum-boiling mixtures seldom have boiling points more than a few degrees apart. The four systems listed in Lange's *Handbook of Chemistry, Fourth Edition* were:

Benzyl alcohol, b.p., 205.2°
m-Cresol, b.p., 202.2°
Maximum, 207.1°

Formic acid, b.p., 100.8°
Diethyl ketone, b.p., 102.2°
Maximum, 105.4°

Formic acid, b.p., 100.8°
Methyl *n*-propyl ketone, b.p., 102.3°
Maximum, 105.3°

Formic acid, b.p., 100.8°
Water, b.p., 100°
Maximum, 107.3°

The formic acid–water system can be regarded as typical of those having a maximum boiling point. Pure formic acid boils at 100.8°C, but a mixture of formic acid and water containing 77.5% formic acid boils at 107.3°C. This situation is illustrated in Figure 3.6B. Any mixture containing a greater concentration of water than a mixture of composition M' will, upon distillation, evolve a vapor richer in water than the original mixture. At the same time the contents of the distilling flask will become progressively richer in formic acid until the composition M' is reached. At this concentration the vapor and liquid phases that are in equilibrium have the same composition, and, from this point on, no changes in composition or boiling point (i.e., a constant-boiling mixture) occur during the distillation. If, on the other hand, the original mixture had been richer in formic acid than in M', distillation would have produced a distillate richer in formic acid, and again the contents of the distilling flask would have approached the composition M'.

The results obtained when two mutually soluble components form a constant-boiling mixture during fractional distillation can be summed up in three statements:

1. Complete separation of both components cannot be done by distillation alone.

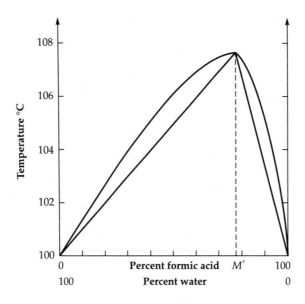

Figure 3.6B A composition diagram for a system with a maximum boiling point.

2. If the boiling point of the constant-boiling mixture is lower than that of either of the two pure components, a perfect column would deliver a distillate having initially the composition of the azeotrope. The final fraction would consist of whichever pure component was in excess.

3. If the boiling point of the constant-boiling mixture is higher than that of either of the two pure components, a perfect column would deliver a distillate that consists of the component in excess. The final distillate would have the composition of the azeotrope.

3.2 Practical Aspects

In assembling any distillation apparatus, one of many important considerations is thermometer placement. A thermometer's placement may have a major influence on the "accuracy" of the distillation temperature and on how it compares to the substance's boiling point. Therefore, whatever the apparatus, you must arrange it so that the bulb of the thermometer is fully bathed in vapor before any of the vapor condenses in an area from which it cannot return to the pot. When a distilling head is used (SM), the *top of the bulb* should be even with (not above) the bottom of the side arm. With the Hickman still (M), the top of the bulb should extend down into the throat far enough to be even with the bottom of the expansion chamber, if that is possible, *without* the bulb touching glass.

You can support thermometers by using inlet adapters or O rings in the top of screw cap apparatus. Be sure to provide an opening to the outside atmosphere. *Do not heat a closed system.* When they are supported external to the apparatus, a bored, split, or notched cork in a clamp or an inlet adapter in a clamp should work.

Thermometer position isn't the only important consideration as far as the distillation process is concerned. You must also take care to ensure that the vapor is in thermal equilibrium with its environment before the vapor passes to the condenser. Also remember to use boiling chips or stones, but no more than one or two at a time. Do NOT add boiling aids at or near the boiling temperature of the liquid. (What might this sudden release of pressure do?) Magnetic stirring is an alternative boiling aid.

The distilling head, the fractionating column, and the Hickman still must be completely vertical (90° from horizontal). Anything else will undermine separation. It is good practice to clamp the fractionating column in position first (true vertical) and make the rest of the apparatus conform to it. Be sure the flask or vial is securely attached with a flexible clamp or screw cap. Gravity affects all items in the chemistry laboratory. If you don't use a fractionating column, you should position the distilling head or Hickman still first. (See Unit I, Figure 1.7 and IA, Figure 1.8A.)

You should also observe two more precautions. Never boil off all of the liquid so that a dry flask or vial is being heated. This may cause the glass to break, possibly violently; any peroxides present to explode; and residues to become intractable tars. Also, allow water to circulate in the condenser for liquids boiling up to 150–160°C. For liquids boiling above 150°C, use an air or a water condenser with both the inlet and outlet open (i.e., without any water).

One of the principle difficulties that you may encounter in small-scale distillations is the loss of material that clings to surfaces or that is needed to fill the vapor space in a column. Much of this loss can be avoided by the use of "chasers." The chasers used should be relatively inert liquids boiling at 50–100° above the substances being distilled. To use a chaser, add 10–20 percent of the chaser (by volume) to the boiling flask or vial before the distillation begins. When the temperature reading shows clear signs of rising above the desired distillation temperature, or if it falls despite continuing or slightly increasing the heating rate (insufficient vapor around the thermometer may be caused by exhaustion of the lower-boiling component), you should discontinue collection of the pure distillate.

You may wonder when to change receivers during fractionation or how much loss you may have if you do not use a chaser. Imagine that the data in Table 3.2 came from the distillation of a binary mixture (compare to

Table 3.2 Sample Results

Temperature (°C)	Volume (mL)	Total volume (mL)
25	0	0
48	0	0
74	0	0 (1st drop)
75	.5	.5
76	.5	1.0
76	.5	1.5
76.2	.5	2.0
76.3	.5	2.5
76.9	.5	3.0
79	.2	3.2
85	.2	3.4
93	.2	3.6
94.5	.2	3.8
96	.2	4.0
96	.5	4.5
96.2	.5	5.0
96.3	.5	5.5
96.4	.5	6.0

(Distillation was discontinued, and the apparatus was cooled and drained.)

Total collected	6.2

The volume left in the distilling flask was 1.8 mL.

Experiment 4C). First, plot a graph of this data with volume on the x-axis and temperature on the y-axis. From this graph and from the data, determine where relatively pure compounds were being collected (hint: a pure compound should have a consistent distillation temperature). What are the distillation temperatures (boiling points) of the two compounds being distilled? The distillation began with 9 mL of a mixture of two pure compounds. What percent of the original mixture was lost (unrecovered)? What might be the composition of the distillate collected between the two plateau regions? Of the liquid left in the distilling flask? Estimate the original composition of the mixture. Where should receivers be changed to catch the "pure" samples?

Most distillations should be completed with some analysis to demonstrate identity and purity. The infrared spectrum, NMR spectrum, gas liquid chromatograph (GLC), refractive index, and density are the most common methods of analysis used. The use of the GLC is discussed in Chapter 4, and the use of the others is described in this chapter and in Chapter 5. When a refractometer is accessible to you, also try to determine the refractive index of each substance that you purify by distillation.

3.3 Micro Boiling Point Determination

If you must identify a very small amount of relatively pure liquid, a micro boiling point determination is recommended. To take a micro boiling point, place a drop or two of the sample in a capillary tube with a bore large enough to contain a melting point tube. Then, insert an inverted

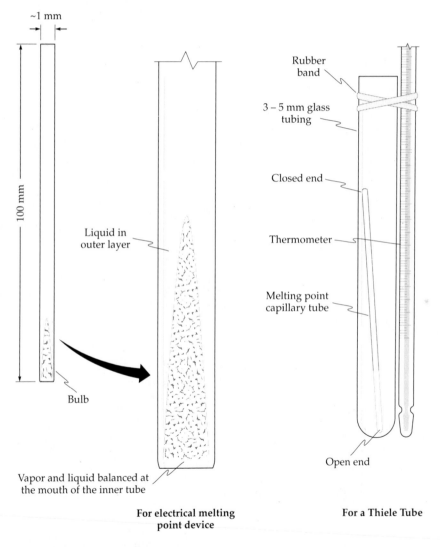

For electrical melting point device

For a Thiele Tube

Figure 3.7A Micro boiling point assembly.

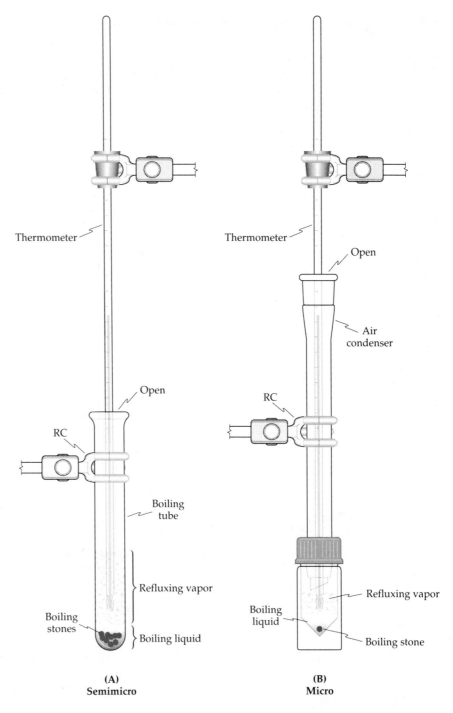

Thermometer

Thermometer

Open

Air
condenser

Open

RC

RC

Boiling
tube

Refluxing vapor

Refluxing vapor

Boiling
stones

Boiling liquid

Boiling
liquid

Boiling stone

(A)
Semimicro

(B)
Micro

Figure 3.7B Rapid boiling point determination (A) semi-micro (B) micro.

melting point tube so that the open end is below the surface of the liquid (Figure 3.7A). Attach this assembly to a thermometer as in a melting point determination and place it in a melting point bath (see Section 2.2B). Heat the bath gradually until a stream of bubbles begins to issue from the inverted melting point tube. Then, by careful, slow cooling and heating, determine the temperature at which the pressure of the vapor inside the melting point tube is just equal to the atmospheric pressure. This is done by noting when the liquid level is just at the entrance of the melting point tube and/or by noting the cessation and initiation of bubbling. Repeat cooling and heating if possible. You might also wish to consider the reason for the contrast between the temperature for this determination and that determined during a distillation.

If an electrical melting point device (Chapter 2) is to be used, a melting point capillary tube will serve as the outer tube. The inner tube is made from glass tubing of still smaller bore and is closed carefully at what will be the upper end. This can be done by heating and fusing two capillaries and then drawing them to a narrower diameter away from the seal. This smaller capillary, including the seal, must move freely within the other one. Use a syringe with a needle to load the liquid to be tested into the outer capillary. If no syringe is available, warm the closed end of the capillary and then place the open end in the liquid to draw up a sample as the capillary cools. A procedure similar to that outlined above is then followed. See Mayo, et al., *J. Chem. Educ.*, 1985, 62, 1114 for a discussion of micro boiling point determinations.

A quicker, if potentially less accurate, method uses a boiling tube, reaction tube, or small test tube, depending upon the amount of liquid available. This method requires a larger sample of liquid. The liquid should not occupy more than about one-tenth of the tube; the thermometer should be suspended so as not to touch the sides and should be four-tenths of the way up, above the liquid. The top of the bulb should not be much more than half way up the tube. The idea is to make sure that hot liquid can reach the bulb but that enough space remains, when the bulb is bathed in vapor, to assure that all of the liquid will condense and not escape. The liquid (with a boiling stone) is then gently heated, and the ring of condensate is caused to rise slowly (with a gradual increase in heat as needed) until the bulb of the thermometer is bathed in vapor and is in equilibrium with the vapor (Figure 3.7B). If the vapor is removed with a Pasteur pipet, that would constitute distillation.

3.4 Outline of Steps for Distillation

1. Set up the proper type of apparatus, correctly aligned and securely clamped.
2. Position the thermometer properly.

3. Put a boiling stone or a magnetic spin vane or bar in the distilling vessel, add the liquid with a funnel or pipet, and attach the vessel securely to the apparatus.

4. If you're using water in the condenser, start it flowing. (Start the stirrer if any.)

5. Heat to initiate gentle boiling.

6. Make the condensing vapors move up the apparatus gradually by controlling the heating rate.

7. Make sure that the thermometer bulb is bathed in vapor before any distillate is collected.

8. Make frequent temperature readings as the distillation progresses.

9. Collect fractions that have a fairly constant distillation temperature range.

10. Change collection vessels when the temperature rises relatively rapidly or levels off.

11. Discontinue heating when the last fraction has been collected or before all the liquid has gone from the flask, vial, or "pot."

12. Disassemble the apparatus as soon as possible to prevent frozen joints.

13. Clean the "pot" before any tar solidifies.

E X P E R I M E N T   4 A

Simple Distillation of a Water and 2-Propanol Mixture (SM)
References: Sections 1.3; 3.1, 2, 7, 8; and 4.2D.

Assemble the apparatus. Set up the apparatus for a simple distillation (Figure 3.8) using a 25-mL flask and a graduated receiver (or a graduated centrifuge tube or graduated cylinder). Do not lubricate the adapter–receiver joint. Have another graduated receiver ready. Make sure the bulb of the thermometer is placed in the distilling head adapter so that it will be bathed in vapor before any vapor reaches the vapor-condensing surface.

Introduce the reagents. Place 8 mL of 2-propanol, 8 mL of water (or 16 mL of another mixture or unknown) and a boiling stone or two into the 25-mL flask. Reassemble the apparatus.

Initiate the distillation. Make sure that you have cold water flowing gently through the condenser (bottom to top) before heating is begun. Heat the flask gently. As the heating progresses, hot vapors will condense on the colder parts of the glass and stream back to their starting point. In a

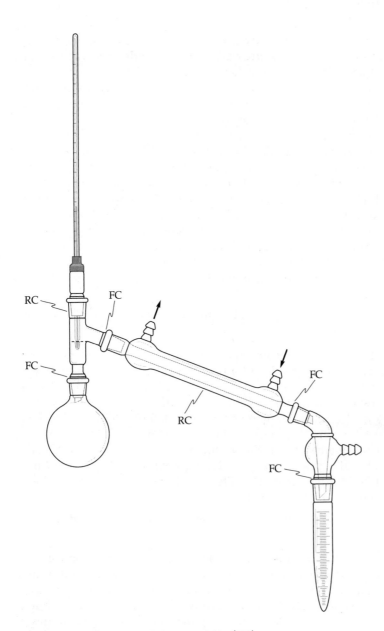

Figure 3.8 Simple distillation assembly (SM).

tube-shaped area, such as the neck of the flask or the bottom of the distilling head/adapter, the leading edge of these condensing vapors will form a ring. Carefully control the heating rate so that this condensate moves gradually up to and around the thermometer bulb. The bulb of the thermometer and the vapor should be at the same temperature (thermal equilibrium) before the vapor reaches the condenser.

Record the temperature when the first drop condenses, and, from then on, record it for every 0.5 mL or 10 drops collected. Record the temperature more often when it is rising rapidly. With care, the temperature should remain steady near the boiling point of the more volatile compound (see Table 3.1) for the collection of 6–8 mL. Increase the heating rate very gradually as the amount of 2-propanol in the pot becomes low. This will supply enough energy to keep the vapors up to and in thermal equilibrium with the thermometer. When the temperature starts rising rapidly (above 83–87°C), change receivers. To do this, release the flexible clamp while holding both receivers; switch them so not a drop is lost, and replace the clamp.

Question

1. Why is it more effective to complete the distillation without interruption?

Collect this fraction and record volumes and temperatures until the temperature remains essentially steady just below or near the boiling point of the second component (water). Change receivers and collect a milliliter or two. Stop at this point; remove the heat, and allow the apparatus to cool and drain completely. Record the volume in the first receiver as the fraction 1F (*F*irst), that in the second as the intermediate fraction 1M (Inter*m*ediate), and that in the third receiver plus the distillation flask (pot), after draining and excluding boiling stones, as the third fraction 1R (*R*esidue). Report the difference between the sum of these three and the original volume as the loss.

Question

2. Suggest the compositions of 1F, 1M, and 1R based on the distillation temperatures.

The process up to this point represents a one-plate distillation for two liquids boiling 20 degrees apart. If time permits, clean the apparatus and redistill fraction 1F and/or fraction 1M, and collect the same three fractions (2F, 2M, 2R) and report loss 2. Fraction 1R is typically pure enough so that redistillation is not justified. The redistillation of 1F constitutes a second theoretical plate.

Interpret the results. Determine the density, refractive index, and/or infrared spectrum of 1F, 1M, and 1R or their composition using the GLC (Chapter 4) as directed by your instructor. (The density of water is 1 g/mL and that of 2-propanol is 0.785 g/mL. The refractive indexes of 2-propanol and water are 1.3776 and 1.3330, respectively, at 20°C.)

Plot the temperature recorded against the total volume collected to analyze the effectiveness of this operation. See Table 3.2 and the discussion of it. To convert from drops to volume, find the conversion factors by measuring the volume in each fraction and dividing by the total drops in that fraction. Water is about 1 mL for 20 drops but other liquids may vary. You may use a computer graphics program to make the plot. Answer all questions posed in this text to improve your understanding of the material. They may also be part of your report.

Summary

1. Assemble the apparatus, including the sample and a boiling stone (start the water flowing).

2. Heat the system to move the vapor to the thermometer.

3. Record temperature and volume from the first drop.

4. Change the receiver when the fraction temperature changes.

5. Stop heating when the second component is clearly distilling.

6. Cool and drain the setup; collect the rest of the third fraction.

7. Evaluate the composition of the fractions and the mixture.

QUESTIONS 1 AND 2 IN ABOVE TEXT.

Questions

3. From your plot, where *should* the receiver change be made to collect just the first fraction (1F)? Where should a *second* receiver change have been made to collect the pure, distilled second component?

4. Account for the loss of material as a result of distillation. How would redistillation affect the final yield?

5. If the original mixture was of unknown composition, determine its composition two ways, based on all material recovered and on the initial volume. Make explicit assumptions about the nature of all volumes involved in this calculation.

6. Compare your redistillation results with the initial distillation results.

7. What factors affect the efficiency of separation in a distillation? What improvements could be made here?

8. Are the first and third fractions pure compounds? Explain. Could they ever be?

9. Why is the distillation temperature of a liquid often different from the reported boiling point? What causes this?

Disposal of Materials

Boiling stones (S)   First fractions (1F, 2F, etc.) (I or D)

Intermediate fractions (D)   Residues (1R, 2R, etc.) (D)

2-Propanol (OS)   Other organic compounds used (OW)

Additions and Alternatives

You may also use ethanol and methanol with water. If all three alcohols are used, the comparison of results will be quite instructive. Other mixtures could be distilled in the same fashion, e.g., methanol, *t*-butyl alcohol, or ethyl acetate with toluene. Also, the composition and/or the components of the mixture issued to you in this experiment may be unknown. You might wish to make use of gas liquid chromatography in order to analyze the composition of a fraction resulting from the distillation of this unknown mixture.

Determine the density and refractive index for several mixtures of the two components you actually use. Where a plot of these properties versus composition is linear, they may be used to determine the composition of distillation fractions.

E X P E R I M E N T   4 B

Simple Distillation of a Water and 2-Propanol Mixture (M)
References: Sections 1.3; 3.1, 2, 7, 8; 4.2D; and Experiment 4A.

Assemble the apparatus. Place 0.8 mL of water and 0.8 mL of 2-propanol into a 3-mL vial with a boiling stone or spin vane, and attach it firmly to a Hickman still. Clamp the Hickman still to a ringstand in a completely vertical position. Place the vial in a sand bath (or in an aluminum block) so that the liquid in the vial is below the surface of the sand (or the aluminum) (Figure 3.9A). Position the hotplate/stirrer directly beneath the heating bath or block. Clamp the thermometer (in a cork or in an inlet adapter) so that the bulb is in the throat of the still and the apparatus is not closed to the atmosphere (Section 3.2). Leave an opening; never heat a closed system. Another thermometer, its bulb positioned completely in the sand bath (or aluminum block), can be used if you are monitoring the bath temperature. If a spin vane is used, adjust the stirrer's motor to give it a slow, steady spin.

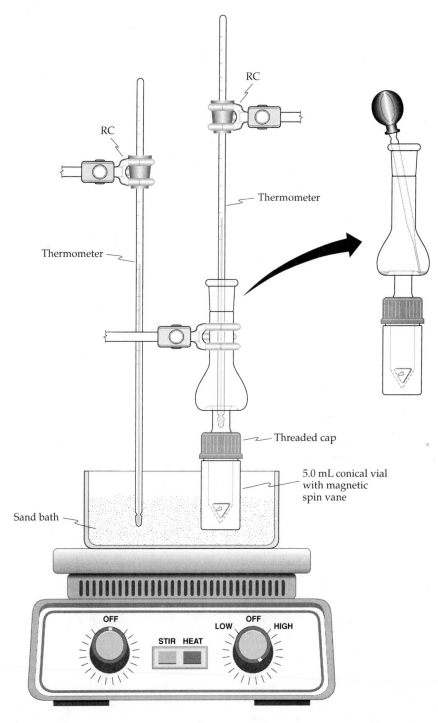

RC

RC

Thermometer

Thermometer

Threaded cap

5.0 mL conical vial
with magnetic
spin vane

Sand bath

OFF

OFF
LOW   HIGH

STIR   HEAT

Figure 3.9A   Simple distillation (M).

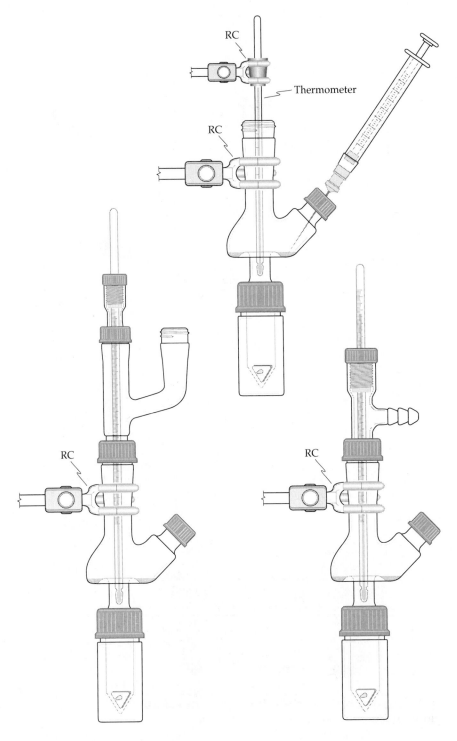

Figure 3.9A *(continued)*

The temperature readings given below are very dependent on where the thermometer bulb is positioned. You may change fractions at different temperatures, depending on how your thermometer is positioned.

Before beginning distillation, be sure that you can remove the distillate as it collects without disturbing the distillation. Practice using the Pasteur pipet. If you find it easier to use a bent-tipped pipet, use a small flame and turn and bend the tip of the Pasteur pipet 30–45° so that it fits into the collection area (the side pocket) of the Hickman still. You will need a separate pipet (or two syringes and a pipet) for each of the three fractions, as well as a labeled container for each fraction.

Initiate distillation. Initiate heating at a reasonable rate, and reduce it a bit when a ring of condensate appears in the vial, so that the condensate ring rises slowly through the throat of the still. Continue a *slow*, steady heating until the temperature of the inside thermometer rises rapidly and then reaches a more or less steady value. Then, conduct the distillation, collecting two fractions *without interruption*.

Collect the fractions. Remove the first fraction (1F) as it collects, with its own pipet, while the temperature remains essentially steady. This fraction can also be removed with a syringe through the septum of the side port. When the temperature begins rising again (above 83–86°C), use the pipet reserved for the intermediate fraction (1M) to remove the distillate that collects during this rise. Continue the distillation until the temperature levels off above 93–95°C. Stop heating, remove the heat, and allow the apparatus to cool and drain. Remove any more intermediate fraction with the pipet reserved for that fraction. Remove the third fraction (1R) from the vial with its pipet. Measure the volumes of all the fractions, record them, and calculate the loss.

Interpret the results. The fractions may be analyzed by gas liquid chromatography (GLC). From the distillation temperatures and the volumes of the fractions, suggest the composition of the fractions and of the original mixture.

Redistill the fractions. If time and volume permit, redistill (you must clean the apparatus and remove all traces of liquids before reuse) the original intermediate fraction (1M). Add the first fraction (2F) collected to the original first (1F) and the new third (2R) collected to the original third (1R). Make fraction 2M as small as possible. Redistill the total first fraction (1F and 2F) once or twice more, keeping and redistilling only the first fraction each time (record volume and composition). With your instructor's permission, two people with two kits can work together on this redistillation. Include a table of volumes and compositions in your report to focus your discussion of fractionation. Estimate the equivalent of your work in theoretical plates. (Remember that each simple distillation is the equivalent of one plate.) Continued redistillation of the first and third fractions is equivalent to fractional distillation. You could also redistill the combined middle fraction in order to increase the yield.

Summary

1. Assemble the apparatus, including the sample, and a spin vane, or a boiling stone.

2. Heat to move the vapor up to the thermometer.

3. Remove fractions as they distill.

4. Stop heating when the second component is clearly distilling.

5. Cool and drain the setup; collect the rest of the third fraction.

6. Evaluate the composition of the fractions and of the mixture.

7. Redistill fractions as appropriate.

Questions

1. The temperatures at which you change pipets and collection vessels are approximate, depending on where the thermometer is placed. How should the decision to change them be made?

2. In the redistillation of combined 1F and 2F, what would be the expected relative size of 3F, 3M, and 3R? 4F, 4M, and 4R?

3. If the original residue were redistilled, what fractions (temperature and relative size) might be observed?

4. How would you decide whether there is enough volume to redistill if you knew the loss from the first distillation?

Answer Questions 4 and 7–9 from Experiment 4A when appropriate.

Disposal of Materials

Boiling stones (S)	First fractions (1F, 2F, etc.) (I or D)
Intermediate fractions (1M, 2M, etc.) (D)	Residues (1R, 2R, etc.) (D)
2-Propanol (OS)	Other organic compounds used (OW)

Additions and Alternatives

See Additions and Alternatives for Experiment 4A. If a short path distilling head is available, assemble it as shown in Figure 3.9B, and increase the total volume to 2 mL. Follow the directions for Experiment 4A. Also, plan on changing receivers at about 0.8–1.0 mL. Methanol can be substituted for 2-propanol and the change from 1F to 1M made at about 66–69°C. For a microscale distillation that calculated HETP, see *J. Chem. Educ.*, 1992, 69, A127.

The initial distillation in this experiment constitutes a one-plate distillation. Further distillation of the fractions parallels fractional distillation. The

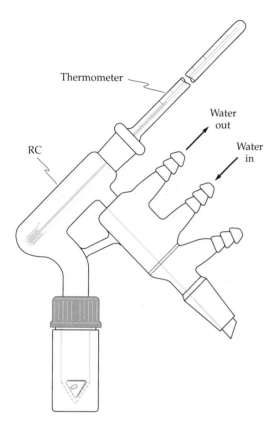

Figure 3.9B Simple distillation assembly (M).

results could be compared with separating your mixture by GLC and collecting the components.

EXPERIMENT 4C

Fractional Distillation of a Mixture (SM)
References: Sections 1.3; 3.1, 2, 7, 8; and Experiments 4A and 4B.

DISCUSSION   In fractional distillation, packing and equilibrium are important considerations. If the micro spinning band is not available, three basic setups are usable. The first setup, using the semi-micro kit and a packed column, is the basis of Experiment 4C (Figure 3.10A). The packing in the basic setup increases not only the surface area and the efficiency of separation, but also the loss through coating surfaces with liquid. Thus, it is not as practical at the micro scale as it is at the semi-micro scale.

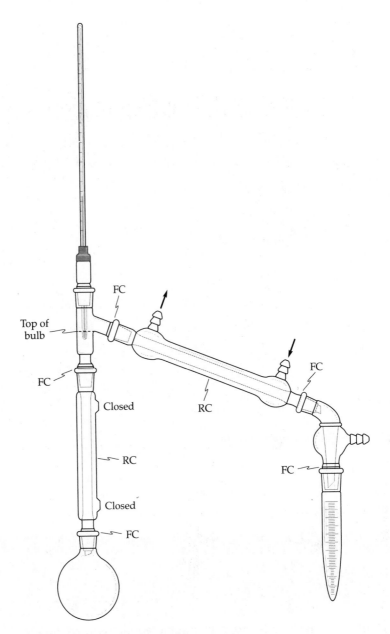

Figure 3.10A   Fractional distillation (SM).

Procedure

The Liebig condenser may have indentations toward the bottom end; if it does, you may use glass beads or helices instead of metal sponge as packing. If you are using metal sponge, it should be an unreactive material such as stainless steel. Beware of halogenated organics, which may react with almost any metal. The metal sponge should be introduced *gently* into the body of the condenser (using a hooked wire to pull and/or a glass rod to push). Make sure that the wire packing does *not* extend beyond the body of the condenser into the joints. Too thin packing reduces separation; too dense packing invites flooding, pressure changes, and larger losses. Do not overpack. The water connectors on the packed column should be closed with dropper bulbs, rubber tubing (from one to the other), or serum caps (reversed end). The outer shell of the column/condenser should be dry. (Alternately, the jacket space could be partially evacuated with a water aspirator.) This type of condenser sealing stops it from acting as an air condenser and provides a bit of thermal insulation that assists in maintaining thermal equilibrium in the column. During distillation, gradually increase heating as needed to keep the thermometer bulb surrounded by vapor, to maintain thermal equilibrium, and to provide uninterrupted operation.

Assemble the apparatus for fractional distillation (Figure 3.10A) starting with the *packed column* (above), which must be clamped completely vertical with a rigid clamp and must have enough clearance for the flask, heater, and support. Allow space for rapid removal of the heater without disturbing the rest of the apparatus should cooling the flask suddenly become necessary. Attach the distilling head with a flexible clamp (Figure 1.1B). Clamp the condenser to the head so it will conform and fit snugly without strain. (Use both rigid and flexible clamps.) The vacuum adapter and graduated receiver can also be attached with flexible clamps. When positioning the thermometer, make sure that the top of the bulb is just at the bottom of the side arm.

Introduce the reagents. Place in a 25-mL flask, a boiling stone or two or spin vane, and, using a funnel, add 7–8 mL each of water and 2-propanol (or 16 mL of an unknown mixture of them). *Record their accurate volume(s).*

Initiate the distillation. With the heater in place, begin brisk heating until boiling is evident. Then, regulate the rate of heating so that the ring of condensate moves slowly and steadily up through the packed column. Make sure that the bulb is bathed with vapor to achieve equilibrium before vapors pass to the condenser. Adjust the heat now and throughout the distillation to maintain a *slow, steady* drip rate (about 5–10 drops per minute).

Collect and analyse the fractions. Record the temperature when the first drop condenses and, from then on, record it every 0.5 mL (10 drops) and every 0.2 mL (4 drops) when the temperature is changing rapidly. When the temperature begins the rise to the intermediate fraction (85–86°C), change receivers without losing a drop. (Practice this before you start. See Experiment 4A.) When the temperature appears to level out near the boiling point of water, stop heating, remove the heater, and patiently wait for the apparatus to cool and *completely* drain. (This time may be used to analyze the first fraction by gas liquid chromatography.) Record the volume of the liquid in the first receiver as fraction 1F, that in the second as 1M, that which is in the distilling flask as 1R, and the difference between the sum of these and the original total volume as the loss. The composition of the fractions can be evaluated by boiling point and gas liquid chromatography. Density (for 2-propanol, density = 0.785 g/mL) and refractive index (for water, it is 1.3330, for 2-propanol 1.3776 at 20°C) may also be recorded.

Interpret the results. Plot the temperature recorded against the volume collected (in milliliters or drops) to analyze the effectiveness of separation. For an unknown mixture, estimate the composition of all fractions by distillation temperature and/or gas liquid chromatography. Make an explicit assumption about the composition of each fraction and the loss, in your estimation, of the composition of the original mixture. Show your work. To convert from drops to volume, find the conversion factors by measuring the volume in each fraction and dividing by the total drops in that fraction. Water is about 1 mL for 20 drops, but other liquids may vary.

Redistill the fractions. Determine if redistillation of fraction 1F and/or 1R is needed or is practical on the basis of their composition and volume. The volume of 1M should be too small for practical redistillation using this apparatus. If redistillation is to be done, the apparatus (especially the packing) must be rinsed clean, drained thoroughly, and air-dried before reuse.

Summary

1. Assemble the apparatus for fractional distillation.

2. Place 16 mL of the mixture, along with boiling stones, in the 25-mL flask.

3. Heat gradually to maintain thermal equilibrium throughout the apparatus.

4. Record volumes and temperatures of collection.

5. Change receivers based on temperature.

6. Stop heating when the second component is clearly distilling.

7. Cool and drain the setup; collect and measure the second component.

8. Evaluate the results.

9. Redistill as appropriate.

1. Why is the distillation stopped before the third fraction is completely distilled?

2. How would you decide if there is enough volume for a redistillation?

ANSWER QUESTIONS 1–5 AND 7–9 FROM EXPERIMENT 4A AS APPROPRIATE.

Disposal of Materials

Boiling stones (S)

Intermediate fractions (1M, 2M, etc.) (D)

2-Propanol (OS)

First fractions (1F, 2F, etc.) (I or D)

Residues (1R, 2R, etc.) (D)

Other organic compounds used (OW)

Additions and Alternatives

Prepare graphs of refractive index or density versus composition by measuring the properties of 5–10 known mixtures of the two components. These can then be used to analyze the fractions.

The mixture of 2-propanol (b.p. = 82.5°C), *t*-butyl alcohol (82.6°C), or cyclohexane (80.7°C) with toluene (b.p. = 110.8°C) works well and is an interesting contrast if part of your class distills different mixtures (the temperatures will be similar). Or try using a mixture of toluene and cyclohexanol (b.p. = 160.7°C). Other mixtures for unknowns are suggested in Experiment 4A. You or other members of the class might want to try different packing materials or a *small* amount of packing and compare the results.

The graphs for this experiment may be done with a computer.

E X P E R I M E N T 4 D

Fractional Distillation of a Mixture (M)
References: Sections 1.3; 3.1, 2, 7, 8; 4.2D;
             and Experiments 4A–4C.

Assemble the apparatus. A microscale apparatus can be assembled as an analogy of the equipment used in Experiment 4C (see Figure 3.10B). It consists of a 5-mL vial or a 5 or 10 mL flask connected to the air condenser (no packing) topped with a short path distilling head or adapter. Position the thermometer so that it is bathed in vapor before any vapor condenses for collection. The receivers can be vials or graduated tubes.

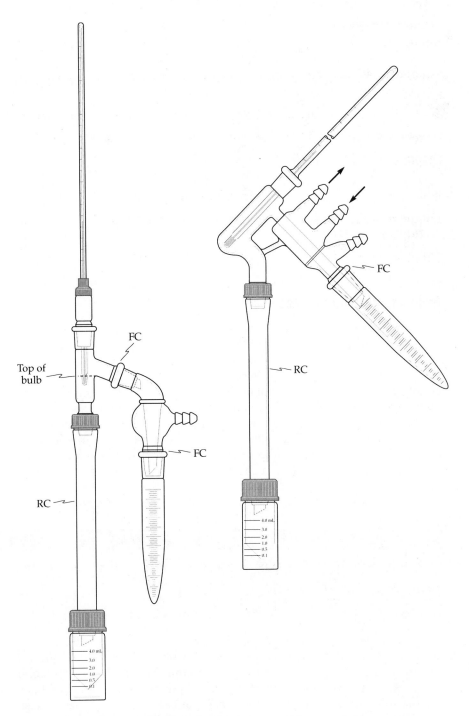

Figure 3.10B Fractional distillation (M).

Introduce the reagents. The sample that you will use consists of 3.6 mL of an unknown liquid mixture or 1.8 mL each of water and 2-propanol.

Initiate the distillation and isolate the products. Use the procedure in Experiment 4C but record the temperature for each 0.2 mL increment. If you use the water condenser as the unpacked fractionating column, close the water inlets, and do *not* have any water in the outer shell. If you use the air condenser as the fractionating column, use a "loose" coat of aluminum foil about it so that heat loss is minimized. If water is one of the components, noting when the beading of the droplets changes will reveal when "pure" water reaches the top of the column.

Interpret the results. For the unknown mixture, estimate the composition of all fractions by distillation temperature and/or gas liquid chromatography. From this data, deduce the composition of the original mixture. Show your work. Determine the percent loss and, if you did the previous experiment, compare the percent loss and effectiveness of separation for the two scales.

Summary

1. Assemble the apparatus for fractional distillation.
2. Place 3.6 mL of the mixture and a boiling stone or spin vane in a 5-mL vessel.
3. Heat gradually to maintain thermal equilibrium throughout the apparatus.
4. Record volumes and temperatures of collection.
5. Change receivers based on temperature.
6. Stop heating when the second component is clearly distilling.
7. Cool and drain the setup; collect and measure the second component.
8. Evaluate the results.
9. Redistill as appropriate.

ANSWER THE QUESTIONS FROM EXPERIMENT 4C.

Questions

Disposal of Materials

Boiling stones (S)	First fractions (1F, 2F, etc.) (I or D)
Intermediate fractions (1M, 2M, etc.) (D)	Residues (1R, 2R, etc.) (D)
2-Propanol (OS)	Other organic compounds (OW)

Additions and Alternatives

Please see those for Experiment 4C.

EXPERIMENT 5

Distillation of an Unknown Binary Mixture (SM or M)
References: Sections 1.3; 3.1, 2; and Experiments 4A–4D.

Assemble the apparatus from Experiment 4C or 4D (Figure 3.10A or 3.10B) (fractionating column first!) using a 25-mL flask (5-mL vial or 5-mL or 10-mL flask). Prepare a data table in your notebook.

Introduce the reagents. Remove the distilling vessel, add a boiling stone or stirring bar or vane, and introduce (using a funnel) a 15–16 mL sample of your unknown (3.6 mL for a 5-mL vessel and 6 mL for a 10-mL vessel). Record the exact volume and unknown number in your notebook.

Initiate the distillation. as in Experiment 4C (4D) and record the temperature for the first condensate and for each 0.5 mL (0.2 mL on the micro scale) collected thereafter, recording more frequently when the temperature is changing rapidly.

Collect the distillate. Change the receiver *only* when and if the graduated volume of the first receiver is almost filled. When the temperature stabilizes for the second plateau and some of the higher-boiling material has been collected, stop heating, remove the heat source, and allow the apparatus to cool and drain completely. When this is complete, measure and record the volume of the liquid remaining in the distilling vessel (do not discard it if it will be used later).

Interpret the results. Determine the loss (unrecoverable liquid). Plot the observed temperature against the volume collected (Section 3.2). Observe the plateau regions (horizontal portions) of the graph. Estimate the distillation temperatures of the two components. Calculate the composition of the unknown by assigning all the volumes (first plateau, intermediate, second plateau, residue, and loss) to one or both of the components and by dividing the total volume of each component by the original volume. If the plateau regions are not sufficiently flat or if the intermediate volume is appreciable, repeat the distillation with a fresh sample of your unknown. Evaluate the residue by gas liquid chromatography, density, or refractive index, if directed to do so.

Summary

See the summaries for Experiments 4C and 4D.

Questions

1. What assumption about composition (explain each) should (could) be made for the intermediate?

2. For the residue?

3. For the loss?

4. What factors affect the narrowness of the distillation ranges and the volume of the intermediate fraction?

5. Why might the distillation temperatures observed here differ from recorded boiling points? How would this affect your use of distillation temperatures to identify unknowns?

6. Why is a loss expected? How would it compare with that experienced in simple distillation?

7. What is the advantage of distillation without interruption?

8. If you were seeking the best samples with respect to amount and purity, where would you change receivers? Explain.

Disposal of Materials

Boiling stones (S)   Excess unknown, distillates, residue (I)

Additions and Alternatives

You could take small samples from the center of each plateau (heart cuts) without interruption, by careful receiver manipulation. This permits fraction analysis using infrared, nuclear magnetic resonance, gas liquid chromatography, density, or refractive index. See the sections of Chapters 3, 4, and 5 where these five techniques are discussed. Depending on the method(s) to be used, a few drops or 0.5–1.0 mL, may be taken. In the latter case, the volume must be recorded for use in the graph and in the calculations. Your instructor may furnish a list of possible unknowns and their properties so that the components can be identified. Your approach will depend on the properties given. Have your instructor check your preliminary results *before* beginning any other procedures. See Experiments 4B and 4D for other ways of doing this experiment.

3.5   Distillation at Reduced Pressure

The boiling point of a compound, and hence the temperature at which it should distill, is the temperature at which the vapor pressure of the liquid equals the atmospheric pressure above the surface of the liquid. Thus, if the external pressure above the liquid is reduced (from 760 mm), the liquid will boil, or distill, at a temperature below its normal boiling point.

1. Why? What if the atmospheric pressure is higher?   **Question**

Such distillations are used for high-boiling liquids because they minimize energy required and heat losses due to radiation and often avoid the

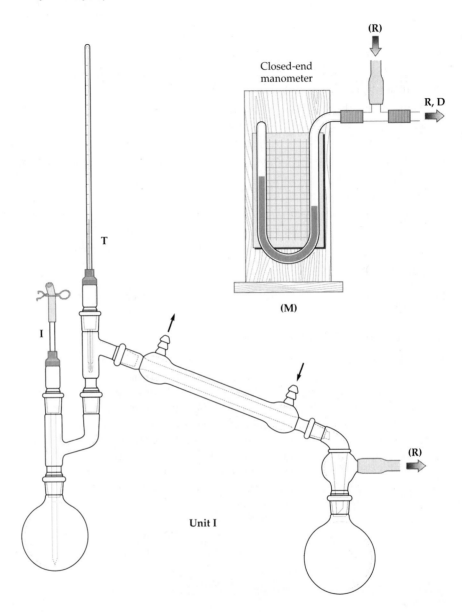

Figure 3.11A Distillation at reduced pressure (SM).

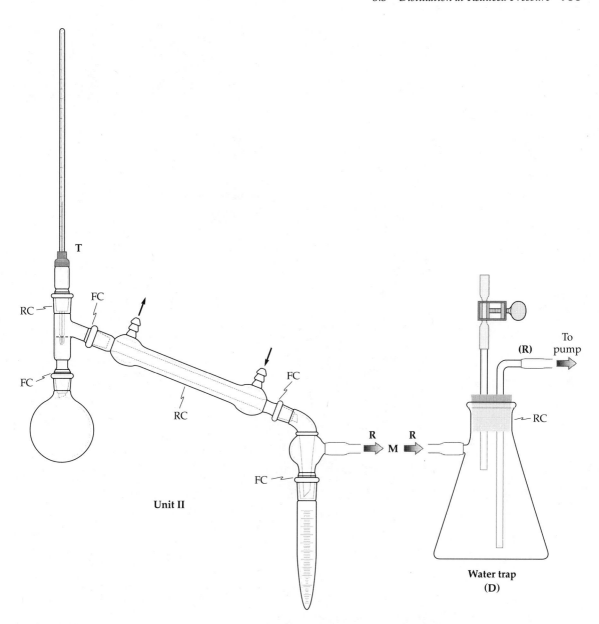

Figure 3.11A *(continued)*

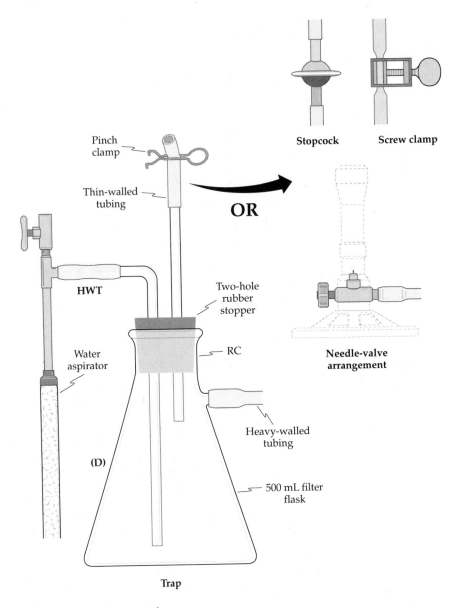

Figure 3.11A (*continued*)

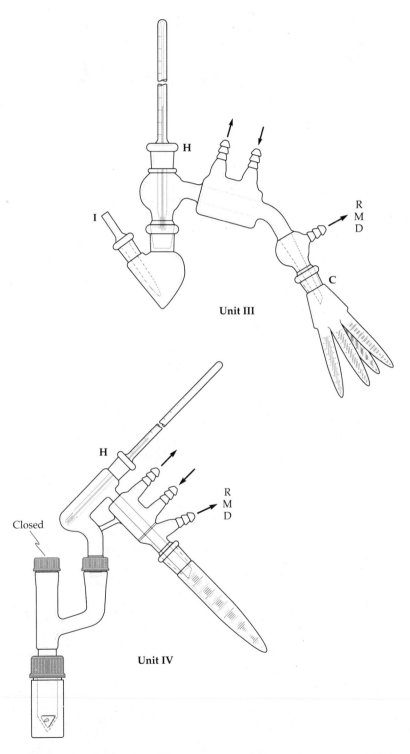

Figure 3.11B Distillation at reduced pressure (M).

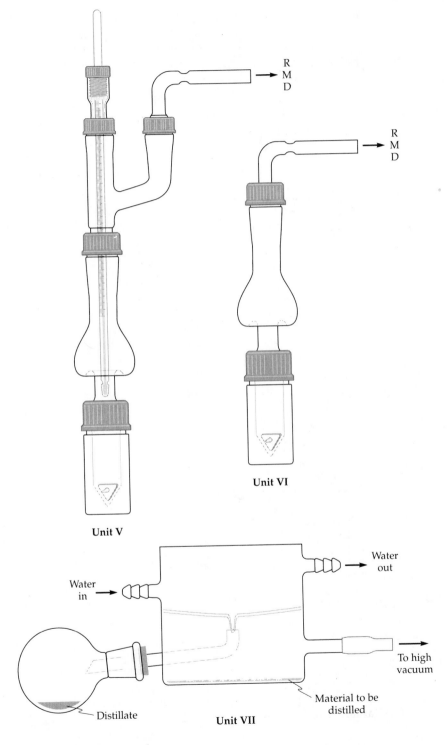

Unit V

Unit VI

Unit VII

Distillate

Water in

Water out

Material to be distilled

To high vacuum

Figure 3.11B (*continued*)

decomposition that takes place at higher temperatures. The assemblies for distillation at reduced pressure are shown in Figures 3.11A and 3.11B.

For Unit I in Figure 3.11A, draw a very fine flexible capillary tube from a length of 6-mm soda-lime glass tubing or a Pasteur pipet. Position the body of this tubing in an inlet adapter so that the tip just reaches the bottom of the flask or vial. Place a short length of three-sixteenths-inch rubber tubing over the larger end of the glass tubing. Place a screw clamp on the rubber tubing to control the passage of air through the capillary. This portion of the assembly is shown in Unit I at (I). Air or inert gas can be bled into the system during distillation to release pressure, agitate the solution, and minimize bumping.

Position a thermometer (T) with an inlet adapter in the distilling head so that the top of its bulb is just below the bottom of the delivery arm and the bulb is not in contact with the glass walls at any point. All joints must be tight and properly greased, or it will not be possible to obtain a satisfactory reduction in pressure.

It is very desirable, but not necessary, to have some kind of manometer in the system as close to the distillation as possible. A commercial closed-end manometer (M) is shown in Figure 3.11A.

Trap (D), used with a water aspirator, is a large filter flask carrying a two-holed rubber stopper. Through one hole, there is a short length of glass tubing closed with a piece of three-sixteenths-inch rubber tubing and a screw clamp used for opening the system. Through the other, there is an L-tube, which must extend close to the bottom of the flask (to draw out the water, backed up in the event of low water pressure and drawn out at

Table 3.3 High Boiling Liquids

Compound	b.p., 760mm	est. b.p., 10mm
Ethyl α-bromo-propionate (H)	159–161°C	(56–58°C/14mm)*
t-Amylbenzene	189–190°C	69°C
t-Amyltoluene	208–210°C	87°C
Butyl α-bromo-propionate (H)	192–196°C	75°C
Methyl benzoate**	199°C	80°C**
Acetophenone	202°C	86°C
Benzyl alcohol	205°C	93°C
Nitrobenzene	210°C	87°C
Hexanoic acid	205.4°C	94°C
Ethyl benzoate**	212°C	89°C
Quinoline (H)	237°C	(120°C/20mm)*
p-Iodotoluene	211.5°C	88°C

*Values in parentheses are actual.
**Prepared by the general method used for Butyl propanoate.
(H) Hazardous handle with care.

its normal pressure). This protects the system from backup flooding. The rubber tubing (R), which is used in lengths of six inches or more to connect pieces of this apparatus, must be vacuum tubing, which will not collapse under reduced pressure.

The other units in Figures 3.11A and 3.11B may be used when bumping can be minimized by other means. Magnetic stirring will avoid most bumping. Unreactive metal sponge or glass wool may break up some bumping. Special fraction cutters (C) are available for fractional distillation at reduced pressure. The assembly used for this purpose could be as shown in Figure 3.11B. Short path heads (H) are very useful.

An experiment on distillation under reduced pressure may be used alone or as an integral part of another experiment. Some suggested compounds for use in distillation experiments are given in Table 3.3. Some of these compounds are produced in experiments described in this manual or by modifications of them. Your instructor may prefer to have you work in pairs to expedite setting up the apparatus.

The normal boiling points of liquids are more readily available than those at different reduced pressures. A number of devices are available to estimate these distillation temperatures. A computer program to estimate boiling points versus reduced pressure is discussed in *J. Chem. Educ.*, 1990, 67, 505.

E X P E R I M E N T 6

Distillation at Reduced Pressure (SM or M)
References: Sections 3.2, 3, and 5.

Assemble the apparatus shown in Figure 3.11A, Unit I with a manometer, trap, and a 25-mL flask. Omit the bubbler and use the apparatus as shown in Unit I, or use Unit II if a magnetic stirrer is used. Weigh the sample to be used if the density is not known.

Introduce the reagents. Introduce 10–12 mL of sample into the 25-mL flask (using a funnel) or 12–16 mL if a low boiler is to be removed before distillation at reduced pressure. If you use units III or IV, Figure 3.11B, introduce 1–4 mL of sample into a 3-mL or 5-mL vial or flask containing a magnetic stirring vane. (Unit V should be used if you are instructed to use a bubbler.)

Initiate the distillation and isolate the products. If there is a solvent and/or a low-boiling reactant mixed with the compound to be distilled, you must first remove them by distillation at atmospheric pressure using an open system and another receiver. This preliminary distillation should not be carried to above 120°C and should be discontinued below this temperature whenever the rate of distillation falls off markedly. Remove the heat when this happens. Attach a fresh receiver to the condenser and start suction when the system has *cooled*. If the liquid is still hot, it may bump or boil too rapidly. Loosen the screw clamp (I) a very little so that a slow stream of tiny bubbles issues from the end of the capillary. This reduces "bumping." If the bubbler is not used, establish a steady mixing with the stirrer. Boiling stones and other aids do not work effectively at reduced pressure. When the bubbling and pressure have stabilized, initiate gradual heating. The temperature may rise rapidly. When it becomes approximately constant, record the temperature and pressure.

When the distillation is essentially complete, remove the heat while retaining the reduced pressure. Remove the heat if liquid in the flask is almost gone or the temperature changes considerably from the distillation temperature. When the flask has reached room temperature or close to it, gradually let air into the system until the inside and outside pressures are equal. Then shut off the pump. This approach should prevent pyrolysis of material left in the flask.

Interpret the results. If the density of the product is unknown, you may obtain the yield by weighing the receiver before and after emptying it. Turn in the product with the report, and calculate the recovery from the amount of original sample.

Questions

QUESTION 1 IN SECTION 3.5.

2. What is bumping?

3. How does frothing differ from "bumping"?

4. What effect would a stream of air be expected to have upon a frothing liquid?

5. Why does a water-jet pump "suck back" when the water pressure falls? How does the trap protect the system?

6. Suggest a mixture selected from Table 3.3 that might be separated by fractional distillation at reduced pressure.

Disposal of Materials

Boiling stones (S)   Excess unknown, distillates, residue (I)

All liquids (I)

3.6   Steam Distillation

If a substance doesn't dissolve in another substance, a solution is not formed. The vapor pressure of each substance does not change. Therefore, the vapor pressure of a liquid is unaffected by contact with a liquid with which it is not mutually soluble. Each of the liquids exerts the vapor pressure characteristic of itself at the temperature involved. By the law of partial pressures, the total pressure produced by the mixed vapor is equal to the sum of the partial pressures of the components.

$$P = P_1 + P_2 \tag{1}$$

Since these partial pressures are identical with the vapor pressures, the total pressure is the sum of the vapor pressures of the individual liquids. It becomes evident that, as the temperature is increased, this sum will reach the value of one atmosphere before either of the vapor pressures alone does so. Thus, a mixture of two immiscible liquids will reach its boiling point at a temperature below that at which the more volatile liquid alone boils.

According to Avogadro's Law, equal volumes of gases under the same conditions of pressure and temperature contain the same number of molecules. If two gases occupy the same volume and exist at the same temperature, their molecular concentrations will be proportional to their partial pressures, which, in this instance, are the vapor pressures of the liquids.

$$C_1 = kP_1 \quad \text{and} \quad C_2 = kP_2 \tag{2}$$

Mass concentrations equal molar concentrations multiplied by molecular mass.

$$W_1 = C_1 M_1 \quad \text{and} \quad W_2 = C_2 M_2 \tag{3}$$

Solving for the molecular concentrations C_1 and C_2 in order to eliminate them,

$$C_1 = W_1 / M_1 \quad \text{and} \quad C_2 = W_2 / M_2 \tag{4}$$

From these a relationship between the vapor pressures and the masses of the immiscible liquids codistilling can be obtained. Expressions (4) are substituted in expressions (2).

$$W_1 / M_1 = kP_1 \quad \text{and} \quad W_2 / M_2 = kP_2 \tag{5}$$

Dividing the first expression by the second (5),

$$\frac{W_1}{M_1} \times \frac{M_2}{W_2} = \frac{P_1}{P_2} \tag{6}$$

and multiplying both sides of the equation by M_1/M_2,

$$W_1 / W_2 = (P_1 M_1) / (P_2 M_2) \tag{7}$$

Thus, the number of grams of the first substance distilling with each gram of the second will be equal to the quotient obtained by dividing the product of the vapor pressure and the molecular weight of the second substance into the corresponding product for the first.

This calculation can be illustrated by a specific example. Benzene and water together boil at 69.3°C. The vapor pressure of benzene at this temperature is 534 mm and that of water is 226 mm. The molecular weight of benzene is 78 g and that of water is 18 g. If benzene is indicated by subscript (1) and water by (2), $W_1/W_2 = (534 \times 78)/(226 \times 18) = 10.2$ grams of benzene distilling with each gram of water. This azeotrope is listed in Lange's *Handbook of Chemistry* as containing 91.1% benzene and 8.9% water by weight—a ratio of 10.2 : 1. This is a particularly fortuitous example.

In spite of the example just given, values calculated in this way are not particularly accurate. They are considered useful approximations. Incomplete immiscibility results in hybrid systems that deviate more or less from the ideal. There are probably other factors that interfere with the accuracy of the calculations. The brief table given here illustrates this point.

Ratio of Grams of Substance to Grams of Water

Substance	From Tables	Calculated
Ethylene chloride	11.0	10.9
n-butyl alcohol	1.63	1.30
Toluene	6.4	4.2
Methyl n-butyrate	7.7	5.2

As the boiling point of the nonaqueous substance increases, its vapor pressure, compared with that of water, decreases. There is ordinarily an increase in the molecular weight (see Equation 7), which partially compensates for this reduction in vapor pressure, although this may be very slight. Even substances like nitrobenzene and aniline, which possess molecular weights little greater than benzene and toluene but which have boiling points of 211°C and 184°C, will distill with water in practical amounts. Assuming that these liquids have vapor pressures of 20 mm and 45 mm, respectively, at about 99°C, steam distillation conducted at 760 mm will produce distillates containing, respectively, 0.18 g and 0.32 g of these substances per gram of water.

Steam distillation is useful in many ways. Because the temperature cannot normally rise above the boiling point of water, the process makes it possible to remove thermally unstable, water-insoluble substances of appreciable volatility from reaction mixtures without causing their decomposition. Certain mixtures, consisting largely of solids, can be stripped of a small quantity of volatile water-insoluble materials. Volatile materials that cannot be recovered easily by direct distillation, can be separated from tarry reaction mixtures by steam distillation. Some triglycerides (fats) can be steam distilled.

The process can be reversed. Benzene (hazardous) or toluene vapors can be passed through solids from which it is necessary to remove water. In this way, oxalic acid dihydrate can be rendered anhydrous, and biological materials can be dried preliminary to further treatment. In the esterification of butyl alcohol with propanoic acid, water, which is formed as a by-product, can be removed with benzene (hazardous) or toluene to improve the yield. Steam distillation is employed in several experiments in this manual but is particularly well illustrated by Experiment 47A, which utilizes it in two ways.

Finally, it should be noted that steam distillation is a specific case of a general type, codistillation. In the above discussion, it is not necessary that one of the liquids be water, though indeed this is the case most frequently encountered in the laboratory.

EXPERIMENT 7A

Isolation and Characterization of an Organic Oil (SM)
References: Sections 1.3E, I; 3.6; 4.1; and Experiment 7B.

Assemble the apparatus for steam distillation as shown in Figure 3.12A, using a 250-mL flask. The thermometer may be replaced with a glass stopper. Almost fill the separatory funnel with water.

Introduce the reagents. Introduce two-thirds of an orange or grapefruit peel (cut up about 12 g, weigh in a small beaker, and remove the soft pithy layer that clings to the rind—chopping the peel in a blender helps) and about 50 mL of water into the flask. Start water flowing through the condenser, and attach a graduated receiver. Have four more graduated receivers available.

Initiate the distillation. Note the water level in the flask and initiate *brisk* boiling. When distillate is condensing and the water level has fallen to about one-half to two-thirds of the original level, begin to drip water into the reaction vessel at a rate to just maintain the original level. Add water to the funnel as it is needed.

Isolate the products. Collect 10 mL of distillate in *each* 12–15-mL receiver until 50 mL is collected. If you have the patience to do so, collect more. Add 2 mL of saturated salt solution to each receiver.

Separate the oily organic layer with a Pasteur pipet. To remove the top layer, tilt the tube, and place the tip of the pipet against the wall just above the interface; gently pull up liquid so that none of the bottom layer is brought with it. In this fashion, transfer the product layer carefully and discharge it into a small vial or tube containing a bit of anhydrous sodium

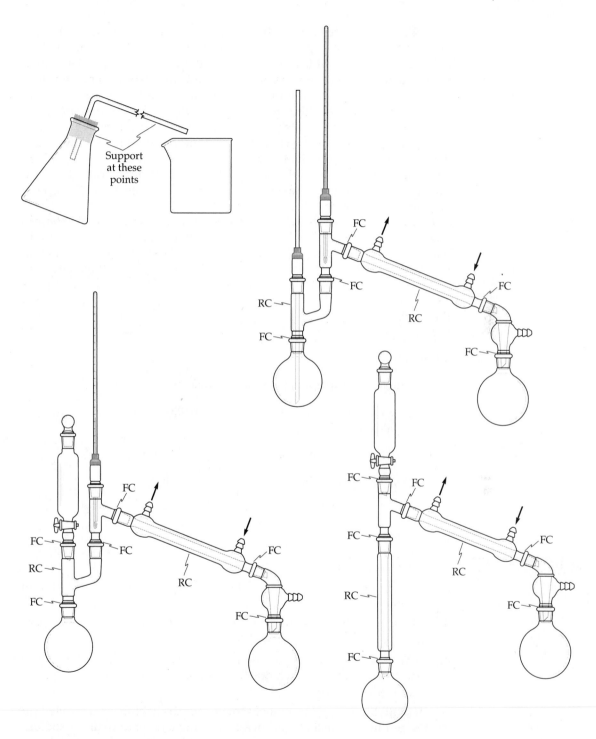

Figure 3.12A Steam distillation (internal source).

or magnesium sulfate (see Section 4.4 on drying agents). If the top layer is too thin for this, use a pipet to transfer the top 1.0–1.2 mL from each of the receivers to a small test tube. Remove the *bottom*, aqueous layer from the combined samples in the small test tube with a pipet and add a small amount of anhydrous sodium or magnesium sulfate to the top layer remaining in the test tube. In either case, drain the product from the drying agent through a filter pipet. If the layers cannot be separated or seen, you should extract (Section 4.1) the aqueous product with two equal portions (by volume) of dichloromethane (toxic—do not inhale or get it on your skin) and evaporate the solvent (Section 1.3I). Ether can be used if suitable precautions are taken in evaporation. Note the difference in densities of these two solvents.

Interpret the results. Weigh the product in a preweighed vial. You can determine the density, boiling point, refractive index, optical activity (it will probably take the samples from the whole class), infrared, and nuclear magnetic resonance. The probable major product is R(+) limonene with the properties: b.p., 175°C; d_4^{20}, 0.8403; n_D^{20}, 1.4735; and $[\alpha]_D^{20}$, +125. Turn in the properly labeled product with your report. Report the percent recovery.

Summary

1. Assemble the apparatus for steam distillation.

2. Place water in the funnel and water, a boiling stone, and the starting material in the flask.

3. Distill with vigorous boiling.

4. Supply water as needed.

5. Separate or extract the product from the aqueous product.

6. Evaluate the product.

Questions

1. Where is the steam in this "steam" distillation?

2. Why should you expect the distillate to separate into an oily and an aqueous layer?

3. Why should the distillation be stopped shortly after you can no longer detect two layers in the new distillate?

4. If the pipet separation does not work, it will be necessary for you to extract with methylene chloride. See Chapter 4 and outline the procedure from this point.

5. How do the properties you observed compare with those expected for the product? Explain.

6. Draw the stereochemically correct structure for R(+) limonene.

7. What would happen if the orange peel were heated in the flask without water?

Disposal of Materials

Orange peel pieces (S)   Methylene chloride (HO, OS)

Drying agent (S)   Ethyl ether (OS, OW)

Limonene (D)

Additions and Alternatives

You can steam distill many of the products from other experiments in this manual. This setup (7A) is convenient for solids and can be used with whole or ground cloves.

You can do this experiment with a larger flask and a steam generator/source/trap. This procedure (7A) is effective for oils boiling under 200°C (Live steam gives an extra boost for high-boiling compounds. Why?). Another alternative is to use a 500–1000-mL Erlenmeyer or Florence flask arranged as in Figure 3.12A, 100–250 mL of water, and the skin of one or more oranges.

For additional characterization methods, transformations, and information see *J. Chem. Educ.*, 1968, 45, 537; 1980, 57, 741; and 1983, 60, 79. You could also test the product with Br_2 and $KMnO_4$ (Chapter 17, Experiment 74, Exercises 1 and 2).

If you extract the organic layer with a solvent (ether/top or methylene chloride/bottom), you need to separate the organic layers, combine, dry, and gently evaporate the solvent. See Experiment 9A.

EXPERIMENT 7B

Isolation of Anethole (SM)
References: Sections 1.3l, 3.6, 4.1, and Experiment 7A.

Assemble the apparatus for steam distillation as shown in Figure 3.12A, using a 100-mL flask.

Introduce the reagents. Fill the addition funnel with water and place 1 g of *freshly ground* anise and 30 mL of water in the flask with a boiling stone.

Initiate the distillation and isolate the products. Note the water level in the flask and initiate *brisk* boiling with no water in the condenser (run warm water in it if a solid forms). Add water during distillation to maintain about two-thirds of the original volume in the flask. Too much water inhibits the extraction by requiring so much heat. Add water to the funnel as needed. Collect and isolate the product as was done in Experiment 7A.

Interpret the results. Report percent recovery, and properly label the product (anethole: m.p. = 22.5°C; b.p. = 235°C; d^{20} = .99; n_d^{20} = 1.5614). Evaluate the product by collecting some of the following types of data: infrared, nuclear magnetic resonance, melting point, boiling point, density, and refractive index.

Summary

See the summary for Experiment 7A.

Questions

FROM EXPERIMENT 7A, ANSWER QUESTIONS 1–5.

6. What is the structure of the product? What chemical tests could be used to identify the compound (see Chapter 18)?

7. What advantages and disadvantages does the external generation of steam have compared with internal generation of steam?

Disposal of Materials

Anise (S)

Methylene chloride (HO, OS)

Anisic acid (N, D)

Ethyl ether (OS, OW)

Additions and Alternatives

Consider extracting other freshly ground spices (caraway, clove, cumin, nutmeg) or oils (lemon grass) using this method or using a Soxhlet apparatus. The product (anethole) may be converted to anisic acid on one-fifth the scale using the method of Garin (*J. Chem. Educ.*, 1980, 57, 138). Anisic acid can be purified by sublimation.

This isolation (and 7A) can be done using external steam. To do this, assemble a steam trap made from a large filter flask with side arm closed (clamp it to a support) and a two-holed rubber stopper containing two bent glass tubes. The tube to be attached to the steam line or generator should reach almost to the bottom of the filter flask. The one attached to the system comes just through the stopper (Figure 3.12B). This should trap liquid water in the flask. The trap can be drained through the side arm. If steam lines are not available, assemble an external generator (the "source" in Figure 3.12B) and arrange to keep the trap warm. Assemble the apparatus as shown in Figure 3.12B, using a 50-mL flask. Place 1 g of *freshly ground* anise and 20 mL of water in the flask. Start the production of steam. When only steam and not hot water is being produced from the generator or steam line (STEAM IS MEAN!), connect the steam line to the

CAUTION

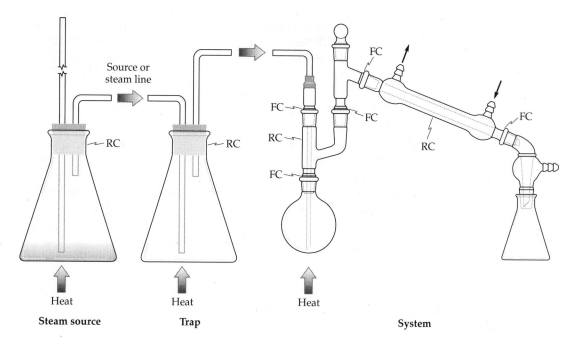

Figure 3.12B Steam distillation (external source).

trap (keep the trap and the reaction flask heated!). Begin boiling the water in the flask and then allow *steam* to enter the apparatus. Follow the general directions for collection given in Experiment 7A. Do not allow the volume of water to exceed two-thirds of the vessel's capacity. Recover and characterize the product as described above. If solid begins to form in the condenser, drain the water and use it as an air condenser (no hoses). It may even be necessary to run warm water through the condenser. When the distillation is complete, lift the steam inlet tube above the surface of the water in the distillation flask, then shut off the steam. For a more elaborate steam trap see *J. Chem. Educ.*, 1983, 60, 424.

3.7 **Determination of Density**

The density of a substance is defined as its mass per unit volume (d = mass/volume). The most common units used to express density are grams per milliliter (g/mL). Densities are commonly reported at 20° or 25°C.

To determine the density, weigh a small vessel (a 1-mL or 3-mL vial) to the nearest milligram. Pipet, quantitatively, 1 mL or 0.5 mL of the compound to be measured into the weighed vessel. Reweigh the vessel. The net weight is the difference in weights. Divide the net weight by the volume to determine the density.

3.8 Refractive Index

The refractive index, like the boiling point, melting point, or density, is a physical property used to either identify a substance or check its purity. Light travels at different velocities in different media. The refractive index, n, may be defined as the ratio of the velocity of light in a vacuum to the velocity of light in a condensed phase, for example in a liquid.

$$n = \frac{\text{Velocity of light in a vacuum}}{\text{Velocity of light in a condensed phase}}$$

For reasons of practicality, the velocity of light in air rather than in a vacuum is compared to its velocity in a condensed phase.

$$n_{obs} = \frac{V_{air}}{V_{liquid}} = \frac{\sin A}{\sin B}$$

According to Snell's Law, the ratio of these velocities is equal to sin A over sin B, where A is the angle of incidence of light and B is the angle of refraction of light as indicated in the diagram below.

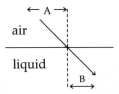

The refractive index is dependent upon two variables: the temperature at which the measurement is taken and the wavelength of the light used. Temperature affects the refractive index by altering the liquid's density, consequently changing the velocity of light in the liquid. The refractive index is usually reported at 20°C. If the refractive index is determined at another temperature, an approximate correction factor of 0.00045 units per degree Celsius is used for most substances other than water. Refractive index varies inversely with temperature; that is, as the temperature increases, the refractive index decreases, and vice versa. Suppose the refractive index of toluene is measured at 23°C and is found to be 1.4954. The literature value is 1.4969 at 20°C. As you can see, three times 0.00045 units per degree Celsius is added to the observed value in order to correlate it to the literature value. If the temperature had been below 20°C, the correction would have been subtracted.

$$1.4954 + 3°C(0.00045(°C)^{-1}) = 1.4969$$

Different wavelengths of light used in this measurement alter the refractive index by being refracted to different extents in the same medium and by giving different refractive indexes for that medium.

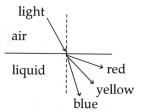

The refractive index is usually reported using a sodium discharge lamp as the light source. (Yellow light of a wavelength of 589 nm is called the "sodium D line." One nanometer = 1 nm = 10^{-9} m.)

$$n^{20}_D = 1.3874$$

where 20 refers to the temperature of the measurement and D refers to the wavelength of light used in the measurement. If another wavelength of light is used, the D is replaced by the value used in nanometers.

The refractive index is very sensitive to the presence of impurities. Unless a substance is extremely pure, it is usually not possible to have the last two decimal places in the observed value match those given in the literature value.

Most organic liquids have a refractive index between 1.3400 and 1.5600. Water has a refractive index of 1.3330 at 20°C and can be used to calibrate a refractometer.

$$n^{20}_D = 1.3330 \quad \text{for } H_2O$$

The refractive index for water decreases 0.0001 unit per degree Celsius for an increase in temperature between 20° and 30°C.

The Abbé refractometer (Figure 3.13) is commonly used to determine the refractive indexes of liquids. A few drops of liquid from an eyedropper is placed on a prism (do NOT touch the prism, because it scratches easily). A second prism is closed down upon the first to produce a thin layer of unbroken liquid. By viewing through the eyepiece, you can adjust the dividing line between the light and dark halves of the circle until the division of the two coincides with the center of the cross hairs present in the refractometer viewing field.

After the measurement is made, the prisms should be cleaned using 95% ethanol or methanol (which evaporates quickly) and a soft tissue or Kimwipe.

You may wish to consult *American Laboratory*, March 1991, 80.

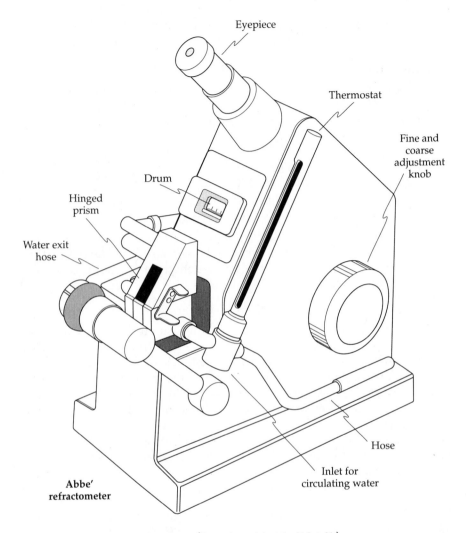

Eyepiece

Thermostat

Fine and coarse adjustment knob

Drum

Hinged prism

Water exit hose

Hose

Inlet for circulating water

Abbe'
refractometer

Figure 3.13 Abbé refractometer (Bausch and Lomb Abbé 3L).

3.8A Possible Experiments

Determine the refractive indexes of two pure liquids, of a mixture of the two liquids, and/or of each liquid after separation by simple distillation and after separation by fractional distillation. Compare the refractive indexes of the distilled samples with those of the pure samples. *Which is the more efficient means of liquid separation, simple or fractional distillation?* Justify your answer.

Alternatively, determine the refractive indexes of two pure liquids and of several mixtures of them. Plot the refractive indexes versus the compositions of the mixtures. Determine the refractive index of an unknown mixture of these two liquids. If your plot is linear, use it to determine the composition of the unknown. This experimental graph is very helpful in the analysis of distillation fractions.

You may also determine the density, refractive index, and boiling point of an unknown pure liquid. These values can be compared to a list of possible unknowns. You may wish to determine the "boiling point" by distillation and/or by taking a micro boiling point. The distilled material can then be used to determine the density and the refractive index.

CHAPTER 4

Other Methods of Separation and Purification

One of the most common tasks in the organic laboratory is the separation of a desired compound from impurities. This chapter introduces several techniques that are often used for that purpose, including extraction, four types of chromatography, sublimation, and drying. Some are useful with liquids, some with solids, and many with both. You will find them useful no matter what your scale of operation.

4.1   Extraction

In order to extract a substance dissolved in one solvent into another solvent, you can use one large volume of the new solvent or divide the volume of extracting solvent into several smaller volumes. Sections 4.1A and 4.1B give a theoretical demonstration (*) that more small portions of extracting solvent will remove the solute more completely. There are, of course, practical limitations to extractions and these include the number of times you can extract a solution and how small the volume of extracting solvent can be. The discussion of practical considerations is given in Section 4.1C.

*4.1A   Distribution Law

If two immiscible liquids are shaken together in the presence of a third substance that is soluble in each, this third substance will dissolve in the layers, now acting as solvents, in such a way that the concentrations of the third substance (dissolved in the two different layers) that result are proportional to the solubilities of the solute in each of the two solvents.

$$C' = KC'^* \tag{1}$$

and

$$C'' = KC''^* \tag{2}$$

where C' is the concentration of the same solute in solvent S'; C" is the concentration of the solute in solvent S"; C'* is the concentration of the solute in a saturated solution of the solvent S". K is the proportionality constant. C'* and C"* are also constants at any given temperature. Dividing (2) by (1) yields:

$$C''/C' = D \qquad (3)$$

where D is the so-called distribution or **partition coefficient**.

If 1 gram of some substance A is initially present in enough solvent, S', to make v' milliliters of solution, the concentration of A is given by

$$C'_1 = a_1/v' \qquad (4)$$

where a_1 represents the amount of solute in S' before the first extraction. If v" mL of an immiscible solvent S" is added, a quantity of solute A, say x grams, leaves solvent S' and enters solvent S", resulting in the attainment of equilibrium and the satisfying of relationship (3). Then the concentration of A in S" is

$$C'' = x/v'' \qquad (5)$$

and the concentration of A remaining in solvent S' is

$$C' = (a_1 - x)/v' \qquad (6)$$

The distribution coefficient now becomes

$$D = C''/C' = xv'/(a_1 - x)v'' \qquad (7)$$

This can be solved for x, the amount of A extracted in a single extraction from v' mL of solvent S' by the use of v" mL of solvent S",

$$x = a_1 Dv''/(v' + Dv'') \qquad (8)$$

*4.1B   Repeated Extractions

If several extractions are to be performed by the successive uses of solvent S" in volumes v", it is more convenient to solve for the amount of substance A remaining in S' after each extraction. If this quantity is represented by

$$r_1 = a_1 - x_1 \qquad (9)$$

where a_1 is the amount of solute in S' before the first extraction and x_1 is the amount of solute removed by this extraction, the quantity r_1 becomes

$$r_1 = a_1 - x_1 = a_1 - a_1 Dv''/(v' + Dv'') = a_1 v'/(v' + Dv'') \quad (10)$$

for the amount of A remaining in S' after the first extraction.

In the second extraction, the quantity r_1 of solute A remaining in S' is partitioned between S' and S''. In this case, r_1 replaces a_1 in (10). The remainder r_2 of A in S' after the second extraction is now

$$r_2 = r_1 v'/(v' + Dv'')$$

$$= [a_1 v'/(v' + Dv'')] \; [v'/(v' + Dv'')]$$

$$= a_1 \left[\frac{v'}{v' + Dv''} \right]^2 \quad (11)$$

It now becomes evident that each extraction leaves a remainder smaller than that of the previous extraction by the factor $v'/(v' + Dv'')$ so that after the nth extraction, r_n, the amount of A remaining in S', becomes

$$r_n = a_1 \left[\frac{v'}{v' + Dv''} \right]^n \quad (12)$$

Is it better to use a given quantity V'' of solvent S'' in one extraction or to subdivide it into n equal portions and to employ these portions in successive extractions? By substituting V''/n for v'' in (12) there results

$$r_n = a_1 \left[\frac{v'}{v' + DV''/n} \right]^n$$

$$= a_1 \left[\frac{v'n}{v'n + DV''} \right]^n \quad (13)$$

where V'' is the total volume of extractant used. Two courses of action are now open. It can be shown analytically that the function r_n has no minimum for any finite value of n. The student will find it much simpler, however, to assign arbitrary values to v', V'', and D and to plot the resulting values of r_n against their corresponding values of n. In this way, it will become evident that the greater the number of times a quantity of extractant is subdivided, the more nearly complete the removal of the solute becomes. For a different presentation of this problem, read Sharefkin and Wolfe, *J. Chem. Educ.*, 1944, 21, 449.

4.1C   Practical Limitations

In practice, extractions do not work as well as is described in Section 4.1B. Because of limited laboratory time, it is very difficult to mix the two liquid layers thoroughly enough to ensure complete equilibrium between the solute concentrations in these two immiscible solvents. It also takes some time for the two solutions to separate into two distinctive layers. Unavoidable losses of the liquids occur when droplets cling to glass walls. These losses can be a real problem when volumes are small as in micro or even in semi-micro work. Also, in Section 4.1B, it was assumed that the two solvents were completely immiscible. They rarely are. When they dissolve in each other to any extent, the partition coefficient, D, becomes smaller, and the extraction is less efficient. Thus, you must use special apparatus and techniques to reduce the impact of these problems.

It is important for you to be effective in the mixing process. You can approach thorough mixing mechanically by vigorous and repeated shaking and stirring. Separation is expected to occur on the assumption that the liquids are insoluble in each other and that they have different densities. In practice, separation is slow at best. Vigorous mixing, required to ensure complete equilibrium, often produces droplets that are slow to coalesce or emulsions that break up slowly. You should support (clamp) the mixing vessels securely as you wait for separation (Figure 4.1). Gentle swirling or gentle mixing with a stirring rod may help.

Unfortunately, the problems may be compounded if the solvent densities are similar and/or if the liquids are somewhat soluble in each other. To remedy the situation, you can change solvents so that there is a greater difference between the densities and so that the solvents are less soluble in each other. When using water, you can add sodium chloride or calcium chloride to the aqueous layer, or you can use a saturated sodium chloride solution in place of water. Also, if you mix the solvents in a centrifuge tube, centrifugation will aid in separation.

Often, you will need to determine which type of layer, polar or nonpolar (aqueous or organic), is on the bottom (as the more dense solution). Knowledge of the densities for the pure solvents will lead you in the right direction. Some organic liquids are more dense than water, but most are less dense. You can expect an aqueous solution, especially a salt solution, to be more dense than water. If you can't tell which layer is which, a simple test will help determine which layer is aqueous: Mix a few drops of water in a small test tube with a few drops (an equal amount) of the suspected aqueous layer. *What should you see if the layer is aqueous? What should you see if the layer is not aqueous?* Always testing will help you avoid problems.

(SM) To extract on the semi-micro scale (with 8–10 mL or more), use variously sized conical separatory funnels (Figures 1.3 and 4.1) supported by a padded iron ring or clamp. The funnel size that you use should be a

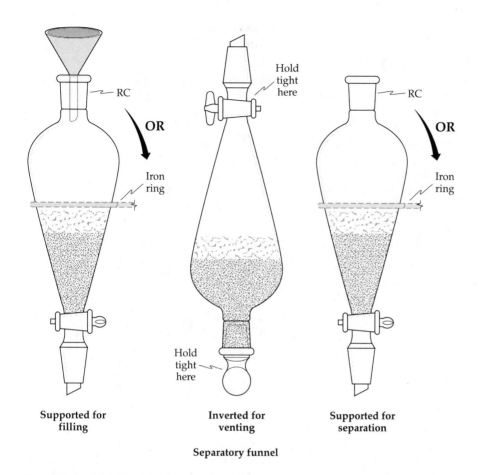

Figure 4.1 Extraction with a separatory funnel.

bit larger than the total amount of liquids that will be in it at one time. Because of the vapors from the solvents, the operations involved in extraction should be done in a hood. To load a separatory funnel, close the stopcock, and introduce liquids through a funnel. Stopper the separatory funnel securely after your addition (hold the stopper in place with one hand and the stopcock with the other), invert it, shake it *briefly*, and vent it by tilting the stopcock up toward the ceiling (away from everyone) and opening it. *Be sure to keep the stopper and stopcock firmly in place*. After a short time of shaking, invert the funnel and vent it again. When equilibrium has been established, support the funnel securely (stem down) to allow for layer separation. Hints for improving separation are discussed above. To remove the bottom layer, remove the stopper before you release that layer through the stopcock. To remove the upper layer after removing the bottom layer, pour it out through the unstoppered top to minimize further contamination.

(M) For combined volumes of less than 5 mL, you can use a conical vial or centrifuge tube and a pipet for extraction. To mix the two solvents, securely seal the vessel, shake it, and carefully vent it. They can also be mixed with a pipet (see below). Then allow the layers to separate. (Centrifugation will often assist separation!) To remove the bottom layer (Figure 4.2A), introduce a clean pipet or plugged pipet through the two layers to the bottom of the vessel. The cotton plug in the pipet is often helpful. It keeps out small particles and assists in the transfer of volatile solvents. Seat the tip of the pipet so that it contacts the bottom squarely with no gaps. Then gently draw the bottom layer up into the pipet. To separate the top layer, draw both layers into the pipet or plugged pipet, allow them to separate, release the bottom layer, and transfer the top layer. To mix small volumes of liquids, draw both layers up and release them briskly. The last time that they are drawn up, gently discharge the layers into separate vessels (Figures 4.2B and 4.2C) after they have separated. Alternatively, remove the bottom layer and transfer the top one.

The solubilities of acidic and basic substances can be modified by converting them into salts, which are usually more soluble in water than are the original materials. This possibility is of value when you desire to extract such a product from impurities in a complex reaction mixture. To do this separation, first convert the acid or base to a salt, and dissolve it in water. Remove the impurities using a solvent immiscible with water. (You could also extract the organic solution with an aqueous acid or base.) Then regenerate the acid or base by adding an unreactive strong acid or base, respectively, to the aqueous extract. Extraction with a nonaqueous solvent will remove the regenerated free acid or base from the aqueous layer.

EXPERIMENT 8A

Extraction of Propanoic Acid (SM)
References: Section 4.1.

a. *Titrate the stock solution.* Place a 4-mL portion (measured with a pipet) of a 1N solution of propanoic acid into a 125-mL Erlenmeyer flask, dilute it with 10–20 mL of water, and titrate it, as directed by your instructor, with standardized NaOH (about 0.1N). Use phenolphthalein as the indicator. Calculate the number of millimoles (or milliequivalents) and milligrams of acid in a milliliter of the acid solution.

CAUTION WHEN ETHER IS TO BE USED IN THE EXPERIMENTS THAT FOLLOW, THERE SHOULD BE NO FLAMES ON THE BENCHES WHERE ETHER EXTRACTIONS ARE IN PROGRESS! ETHER IS VERY FLAMMABLE.

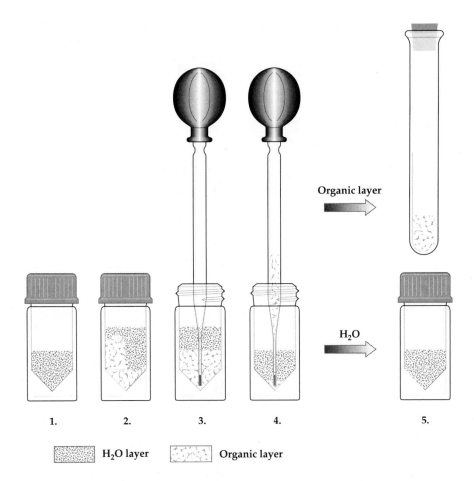

H₂O layer **Organic layer**

Figure 4.2A Extraction of a product from an aqueous solution with a solvent more dense than water.

1. The product is in an aqueous solution.
2. A denser organic solvent is used to extract it from the aqueous solution by thorough mixing.
3. The Pasteur or plugged pipet is placed at the bottom of the conical vessel.
4. The lower organic layer is removed without taking any of the aqueous layer.
5. The organic layer is transferred to a dry test tube or conical vessel. The aqueous layer remains in the original vessel.

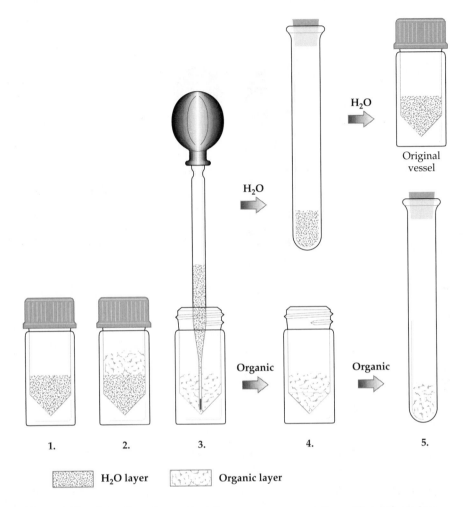

Figure 4.2B Extraction of a product from an aqueous solution with a solvent less dense than water.

1. The product is in an aqueous solution.
2. A lighter organic solvent is used to extract it from the aqueous solution.
3. The lower aqueous layer is removed from the organic layer without taking any of the organic layer in the Pasteur pipet.
4. The aqueous layer is transferred to a test tube or conical vessel. The organic layer is left in the original vessel.
5. The organic layer is transferred with a pipet to a test tube for storage. The aqueous layer is transferred back into the original vessel with the original Pasteur pipet.

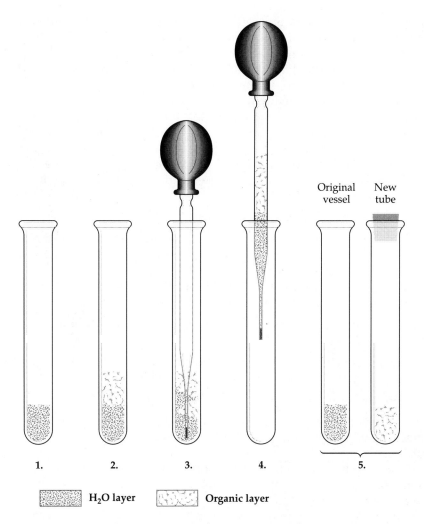

Figure 4.2C Extraction with a pipet (alternative to 4.2B).

1. The product is in an aqueous solution in a conical vessel.
2. The lighter organic solvent is added.
3. The two layers are mixed thoroughly by drawing into and discharging from a Pasteur pipet.
4. The two layers are drawn into the pipet and allowed to separate.
5. The lower aqueous layer is carefully discharged into the original vessel and the organic layer into a storage tube.

b. *Assemble the apparatus.* Support a small separatory funnel (Figure 1.1 or 4.1) (stem down) by a protected ring (rubber bumpers can be made by splitting short pieces of rubber tubing) or by a clamp.

Introduce the reagents. Place a 10–12-mL portion of propanoic acid solution, carefully measured, in the small separatory funnel. Add exactly three times that volume of ether (30–36 mL) to the funnel.

Initiate the extraction. Shake the liquids together to ensure that they are mixed completely. Use both hands to keep both the stopper and the stopcock firmly in place during this process. Release the pressure frequently during this operation (after a shake or two) by opening the stopcock while the funnel is inverted, stem up (away from anybody or toward the back of a hood).

Question

1. Why should the stem be up for venting?

Repeat this procedure with more vigor as the pressure buildup diminishes. Then return the funnel to its upright supported position to allow the layers to completely separate.

Isolate the products. Carefully and completely separate the lower aqueous layer by drawing it off into a clean 125-mL Erlenmeyer flask.

Question

2. Why must you remove the stopper before draining the lower aqueous layer?

If you are in doubt about the nature of the bottom layer, add a few drops of this layer to a few drops of water to test miscibility.

Question

3. Explain this test.

Titrate the aqueous layer as you titrated the portion in part (a). Determine the amount and the percentage of acid extracted by the ether (show your calculations).

c. Place a fresh sample of the acid solution, with the same volume as used in part (b), in a clean separatory funnel and extract it with an equal volume of ether. Separate the aqueous layer, return it to the separatory funnel after the ether is removed through the top, and extract a second time with the same volume of fresh ether. Repeat this process with fresh ether and titrate the aqueous layer after this third extraction, as previously described. Determine the amount and the percentage of acid extracted by the ether as before. It should be noted that the same volume of aqueous acid and the same total volume of ether were used in parts (b) and (c).

Question

4. Calculate the efficiency (% extracted) of single and multiple extractions.

d. Extract a 12-mL portion of a solution containing 44.5 g of propanoic acid per liter of ether (about 0.6N) with 10–12 mL of 1N NaOH or $NaHCO_3$ (carefully measured). Take particular care in releasing the pressure during the basic extraction, especially if $NaHCO_3$ is used.

Q u e s t i o n

5. Why must special care be taken when you use $NaHCO_3$?

Titrate the aqueous layer with 0.1N HCl, using methyl orange as the indicator.

Interpret the results. Calculate the amount and percentage of acid extracted by the aqueous solution from the ether. Note that the excess base is being titrated.

Summary

a. (1) Standardize the stock solution of propanoic acid.

b. (1) Extract a portion of the stock propanoic acid solution with one large portion of organic solvent.

(2) Titrate the aqueous layer as in part (a).

(3) Calculate how much acid has been extracted.

c. (1) Extract a new portion of propanoic acid solution with three portions of organic solvent. The total volume of organic solvent in this section should equal the volume used in part (b).

(2) Titrate the aqueous layer as in part (a).

(3) Calculate how much acid has been extracted.

d. (1) Extract a portion of propanoic acid in ether with aqueous base.

(2) Titrate the aqueous layer.

(3) Calculate the amount of base left, the amount of base neutralized by the acid extraction, the amount of acid left in the aqueous layer, and the percent of acid extracted.

QUESTIONS 1–5 ABOVE.

Q u e s t i o n s

6. Calculate the approximate distribution coefficient for propanoic acid between ether and water. Why would this probably not be the same as the theoretical value?

7. Find the boiling points of ether, methylene chloride, chloroform, and 1-octanol and their solubilities in water. Which solvent would be preferred for this experiment and why?

8. Explain why there is a buildup of pressure in the separatory funnel even when no gas is produced by a chemical reaction. Why is this pressure released frequently?

9. What advantages might 1-octanol have over ether, especially for multiple extractions with water? What precautions does this suggest for the extraction of substances from ether?

Disposal of Materials

aq. Propanoic acid (N)   Ether (OS)

aq. NaOH (N)   Ether solutions (OW, I)

aq. HCl (N)   Propanoic acid in organic solvents (OW)

Propanoic acid (N)   aq. NaHCO$_3$ (D)

Additions and Alternatives

Standardized HCl may be available for back titration if needed in (a), (b), or (c).

Recover the propanoic acid extracted into ether (part b or c) by evaporating the ether with a rotary evaporator (Figure 4.3). Determine the amount of recovered propanoic acid and compare it to the amount expected from the titration data. Your instructor may wish to tabulate class results for you to use in your report. Part or all of your class could reverse the procedures for parts (b) and (c) by extracting ether solutions of known concentration with water. Consider using methylene chloride or 1-octanol in place of or in addition to ether and ether solutions. The use of 1-octanol is particularly recommended because of its low volatility and apparent lesser toxicity. See also Umland, *J. Chem. Educ.*, 1983, *60*, 1081. One possibility would be to have members of your class use different solvents for extraction. You and your classmates could then compare your results.

A qualitative alternative to part (d) would consist of the extraction of benzaldehyde in ether with water and 10% aq. NaHSO$_3$. You could determine the relative amount of benzaldehyde extractable with NaHSO$_3$ by hydrolysis of the bisulfite addition compound and/or as a result of the formation of the 2,4-dinitrophenylhydrazine derivative (Experiments 25A and 55).

E X P E R I M E N T   8 B

Extraction of Propanoic Acid (M)
References: Sections 4.1 and 5.1.

a. *Determine the infrared spectrum.* of a 1N solution of propanoic acid in 1-octanol using a fixed thickness, sealed infrared cell with AgCl windows. Make several other more dilute samples of this solution by adding a known volume of the 1N solution (or a more dilute one of known concentration) to a known volume of 1-octanol. Have one sample with no acid, only solvent. Determine the infrared spectrum of each sample in the range of 1600–1800 cm^{-1}. The same piece of paper can be used if each run is labeled. Your goal is to collect 3–5 spectra on which the absorptions of

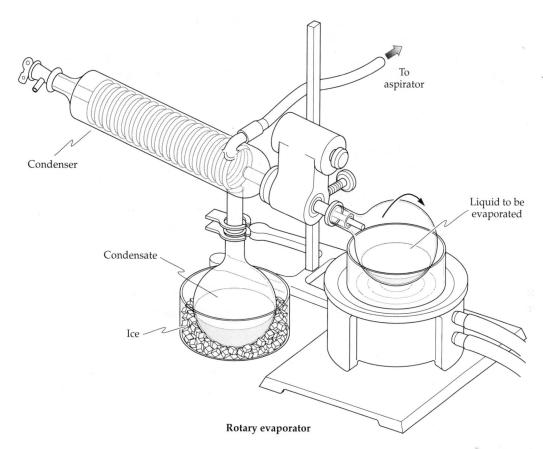

To
aspirator

Condenser

Liquid to be
evaporated

Condensate

Ice

Rotary evaporator

Figure 4.3 Rotary evaporation.

the major peaks in this area are in different ranges *between* complete transmission and complete absorption for solutions of known concentration. Graph the percent transmission or absorbance ($A = 2 - \log \%T$) for these peaks against their respective concentrations. Your goal is to find a linear range. If you use cells that must be assembled, take the average of three identical runs as the value for each concentration determination. You must have the same amount of solution in the cell for each determination.

b. *Introduce the reagents and initiate extraction.* Place a 0.5 mL sample of the original solution in a conical tube, flask, or vial and add 1.5 mL of water, seal, and mix or mix with a pipet.

Isolate the products. Remove the water layer with a pipet (Figure 4.2A or 4.2B). Carefully dry the organic layer by adding a few crystals of anhydrous magnesium or sodium sulfate and by removing the 1-octanol with a (plugged) pipet.

Interpret the results. With the aid of an infrared determination and with the graph from part (a), estimate the concentration of acid remaining in the organic layer. If it does not fall in the linear portion of the graph, repeat the extraction with a different amount of water that will lead to a result in the linear region of your graph.

c. *Repeat part (b) but with three successive extractions,* each using one-third of the volume of water used in part (b). Determine the concentration of the acid remaining in the organic layer, as before.

Summary

a. (1) Determine the infrared spectra of a series of acid solutions of different concentrations in 1-octanol.

(2) Prepare a graph of absorbance (A) versus concentration.

b. (1) Extract the organic solution of acid with one portion of water.

(2) Dry the remaining organic layer and determine its infrared spectrum.

(3) Estimate the concentration of acid remaining and extracted.

c. (1) Repeat part (b) but use three smaller portions (total volume equal to that used in b).

(2) Determine the acid remaining in the organic layer and the efficiency of each mode of extraction.

Questions

QUESTIONS 3, 4, 6, AND 7 FROM EXPERIMENT 8A (4 AND 6 SHOULD BE ANSWERED USING FIGURES FROM THIS EXPERIMENT IF AVAILABLE).

1. Could the extraction of benzaldehyde be studied in this manner (refer to Experiment 8A)?

2. What structural features of another solvent might interfere with the infrared determination?

Disposal of Materials

1-Octanol (OW)   Propanoic acid in 1-octanol (OW)

Propanoic acid in water (N)

Additions and Alternatives

You could use methylene chloride here as well as many other solvents that do not absorb in this region of the infrared spectrum. Comparisons of solvents could be made. For another quantitative use of infrared, see *J. Chem. Educ.*, 1966, *43*, 319.

E X P E R I M E N T   9 A

Caffeine from Tea

References: Sections 1.3l; 4.1, 3; 5.1 and:
Mayo, et al., *Microscale Organic Laboratory*, 2nd Ed.,
Wiley, New York, 1986, pp. 123–127.
Mitchell, et al., *J. Chem. Educ.*, 1974, *51*, 69.
Pavia, et al., *Introduction to Organic Laboratory Techniques,
A Microscale Approach*, 2nd Ed., Saunders, Philadelphia,
1990, p. 78.
Ault, *Techniques and Experiments for Organic Chemistry*,
5th Ed., Allyn & Bacon, Boston, 1987, p. 325.

Introduce the reagents. Weigh an empty tea bag with a staple; also weigh the tea bag to be used. Record the difference between these two weights as the weight of tea used. Place the tea bag in a 125-mL Erlenmeyer flask and *cover* it with 20 mL or so of a 15–20% aqueous solution of sodium carbonate. The base should convert any tannic acids present to their water-soluble sodium salts.

Initiate reaction. Warm the flask gently, and slowly boil the solution for 20–30 minutes. Do *not* allow water vapor to escape from the flask (place a glass funnel over its mouth, stem down). You don't want the solution to become too concentrated or the caffeine to steam distill. Cool the solution (to approx. 35–40°C) until it is comfortable to hold in the palm of your hand (BE CAREFUL WHEN TESTING!). *Decant* much of the liquid into two 12-mL centrifuge tubes through a funnel. Gently squeeze the bag with a spatula, and complete the liquid transfer using a Pasteur pipet. Use the pipet to make the volumes in the tubes the same (about 9–10 mL).

Isolate the products. To each tube, add exactly 2 mL of dichloromethane (d = 1.34); seal the tubes with rubber serum caps (or with other liquid-tight caps), and shake them vigorously. Centrifuge the two tubes (balanced) briefly to break any emulsion that forms. Put a cotton plug and a small amount of anhydrous magnesium or sodium sulfate (100–150 mg) into a Pasteur pipet (filter/drying pipet). Moisten the cotton plug with a bit of dichloromethane before transferring the organic layers, with a Pasteur pipet, through the filter pipet into a clean, dry 125-mL Erlenmeyer flask. This filter pipet will dry the organic extract.

Transfer the methylene chloride (= dichloromethane) layer from each centrifuge tube with a plugged pipet to the 125-mL Erlenmeyer flask, through the drying pipet prepared above. Extract each aqueous layer *three more times*. Rinse the drying agent with another small portion of organic solvent.

1.   Which layer is aqueous? How could you test?

Evaporate the solvent under the hood. Use a *warm* sand bath or aluminum block and a stream of clean, dry air or nitrogen, if available. If a

CAUTION

Figure 1.6A and Figure 4.5
(pages 19 and 148).

Question

See Figures 4.2A, B, C
(pages 131–133).

See Figure 1.9 (page 30).

stream of gas is not used, invert a funnel over the mouth of the flask, and connect the funnel to a water aspirator to speed the removal of the vapors. When the solvent has evaporated, purify the oily residue by sublimation. Dry the solid.

Interpret the results. Determine the yield, percent recovery, and the melting point (lit. value ca. 225–230°). This can be done at high temperatures or at reduced pressure and lower temperatures. It is best if done above 110°C with some reduction in pressure to avoid hydrate formation. (The residue can also be recrystallized from 95% ethanol if there is enough of it.)

Summary

1. Weigh the tea and steep it in aqueous sodium carbonate for 20–30 minutes.

2. Separate the tea from solution.

3. Extract the aqueous solution four times with methylene chloride.

4. Dry the combined organic layer and evaporate the solvent.

5. Purify by sublimation or recrystallization.

6. Check the product's melting point.

Questions

QUESTION 1 ABOVE.

2. Is caffeine a major component of tea according to your data? Explain.

3. Two extractions are performed here, one with aqueous base and one with dichloromethane. Explain the function of each.

4. Caffeine may sublime during the melting point determination. What precautions could be taken?

Disposal of Materials

Tea bags (S)   aq. Na_2CO_3 (D, N)

Dichloromethane (OS, HO) $MgSO_4$, Na_2SO_4 (S after evaporation)

Caffeine residue (S)

Additions and Alternatives

Try "brewing" the tea in water and then adding 2–4 g Na_2CO_3. Compare results.

Analyze the product by thin layer chromatography, Experiment 11.

For an isolation of caffeine that uses column chromatography, see *J. Chem. Educ.*, 1991, *68*, 73.

E X P E R I M E N T 9 B

Trimyristin from Nutmeg (M)

References: Sections 1.3H, I; 4.2C; Experiments 11 and 76; and: Beal, *Organic Synthesis*, Collective Vol. I, 2nd Ed. (Blatt, ed.), John Wiley & Sons, Inc., New York, 1964, p. 538.

NO FLAMES!!!

Trimyristin is an ester of glycerol coupled with three moles of myristic acid ($CH_3(CH_2)_{12}COOH$). Usually triglycerides are made from a variety of acids, but in the material extracted from nutmeg, it is primarily made from one acid. This experiment illustrates how some lipids can be isolated from natural sources.

Assemble the apparatus. Equip a 5-mL vial (or 10-mL flask) with a water condenser (Unit I, Figure 1.7 or IA, Figure 1.8).

Introduce the reagents. Crush and grind a nutmeg seed with mortar and pestle and weigh out 0.4 g of ground nutmeg. There should be enough for more than one student. Use ground nutmeg, as fresh as possible, if a seed is not available. Record the accurate weight. Place the nutmeg in the 5-mL vial (or 10-mL flask) with 3.5 mL (or 5 mL) of ethyl ether.

Initiate the extraction. Reflux the mixture *gently* for 10–20 minutes. Cool the vessel and then transfer the solution, with a Pasteur pipet (a plug may be helpful), to a small Erlenmeyer flask or test tube. If the solution will not all fit in this flask without exceeding two-thirds of its capacity, fill it only to that level. Evaporate the solution in order to make room for the rest. Evaporate (*No flames!!*) the ether until a viscous yellow oil or solid results. To do this, let the ether solution sit for a few minutes under an inverted funnel connected to a water aspirator (Figure 1.9). If this is not sufficient, support the vessel in a warm water bath (under the hood or with continued use of the funnel and aspirator). A boiling stick may be used to avoid bumping. If saponification is to be done, remove about half of the residue and place it into a clean 5-mL vial or 10-mL flask.

Purify the product and interpret the results. Recrystallize the remainder from acetone in a Craig tube. Allow the acetone solution to cool. It may take time (1 hour?) and scratching to get crystals. If you are doing the saponification, start it during this time. Completely cool the acetone solution in an ice bath. Collect the crystals by centrifuging the Craig tube, dry them, and determine their melting range (55–56°C). A test tube, Hirsch funnel, and suction can be used instead of the Craig tube.

CAUTION

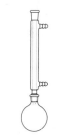

Figure 1.7, Unit I (page 24).

See Figure 1.9 (page 30).

See Figures 2.8A, B, C (pages 56–58).

Saponification

Assemble the apparatus. Equip the apparatus for reflux (Figure 1.7, Unit IA or Figure 1.8, IA).

Introduce the reagents. To the residue reserved for saponification, add 2 mL of 2.5N alcoholic potassium hydroxide.

Initiate reaction. Reflux the mixture for one hour. If you will be doing thin layer chromatography, withdraw a few drops of the solution with a long pipet midway through the reflux, and place these drops on a spot plate or in a small beaker to evaporate. If solvent is lost during reflux, replace it with ethanol. After refluxing, allow the vessel to cool a bit. With a Pasteur pipet, transfer *all* of the solution to a vessel containing 5–6 mL of water. Add concentrated hydrochloric acid dropwise (about 1 mL; count drops from a pipet calibrated to this solution by drops) to the diluted solution until no more precipitate forms.

Isolate the products and interpret the results. Remove most of the liquid with a (plugged) pipet, and transfer the solid and remaining solution to a Craig tube or a Hirsch funnel so that the crystals can be recovered by centrifugation or suction. Collect the myristic acid crystals, dry them, and determine their melting range (53–54°C).

Calculate the apparent **saponification equivalent** of trimyristin, the theoretical value, and the apparent error. The saponification equivalent is equal to the weight of the sample over the volume of the base that you started with multiplied by its molarity minus the volume of acid used to titrate the excess base multiplied by its molarity (= wt. of sample/{starting volume of base × its molarity − volume of acid used in the titration × its molarity}). The **theoretical saponification equivalent** is equal to the molecular mass of trimyristin divided by three.

Summary

1. Suggest sources of error.

2. Consider how this procedure might be improved.

Your accuracy should improve if you use larger quantities.

Thin Layer Chromatography

Assemble the apparatus. For thin layer chromatography (Section 4.2C), prepare a development chamber with a solvent consisting of absolute ethanol and benzene (Carcinogen) (1:9, v/v). Prepare a silica gel chromatographic sheet (ca. 4.5–6 × 10 cm) (e.g., Eastman Kodak 6060 or 13181 [UV indicator included]).

Introduce the reagents. Spot the sheet with four spots: the acid (isolated above), the ester, both, and the mixture isolated from the reflux (all as

CAUTION

ether solutions). Make sure that the alcohol has evaporated from the refluxed mixture, and then add a drop of concentrated hydrochloric acid to it. Only spot those substances that are soluble in ether.

Initiate chromatography. Allow the plates to develop as described in Experiment 11.

Analyze the plates. Use an iodine chamber for compound detection (or first use a short wavelength, UV lamp when #13181 is used). The acid should have the greater R_f value.

(How else could you demonstrate which spot is the acid? (3))

Question

The trail left by the refluxed mixture's spot may not show that there are different compounds in the mixture if the refluxing solution was too dilute.

(How could you investigate what might be done to correct this problem? (4))

Question

Interpret the results. Report the R_f value and the appearance of the spot for all compounds observed. What do you conclude?

Summary

1. Assemble the apparatus for reflux.

2. Introduce spice, ether, and a spin vane or boiling stone.

3. Reflux *gently* for twenty minutes.

4. Separate the solution and evaporate the solvent.

5. Recrystallize the material that will not be saponified.

6. Saponify by refluxing part of the product with alcoholic base.

7. Acidify the resulting mixture and collect the new product.

8. Do thin layer chromatographic analysis of the acid, ester, and intermediate fraction.

QUESTIONS 1–4 ABOVE.

Questions

5. Draw the structures of myristic acid and trimyristin.

6. One hour for saponification may not be the optimal length of time to reflux. How might thin layer chromatography be used to determine how long to reflux the mixture?

7. What other compound might be expected to show up in the thin layer chromatography trail left by the refluxed sample?

8. Some authors suggest that the ester should be recrystallized once or twice from hot ethanol. What problem might this create (especially if a trace of acid were present. See Experiment 47)?

Disposal of Materials

Nutmeg residue (S)   Ethyl ester (OS, OW)

Acetone (OS, OW)   Acetone solutions (D)

alc. KOH (D)   conc. HCl (N)

Ethanol (D)   Other aqueous solutions (D)

Additions and Alternatives

Compare the infrared spectra of the acid and the triglyceride. Determine a "mixed melting point" for these two.

The saponification may be done quantitatively to determine the saponification equivalent (see Experiment 70). For quantitative work, purify all of the material before saponification by recrystallization. Weigh out (and record the weight) 40–50 mg of the ester (duplicate samples if possible), add 2–2.5 mL of standardized alcoholic KOH (record the accurate volume and concentration) to it, and reflux this mixture for at least one hour. Supply ethanol as needed. Rinse the condenser with a bit of water and let it drain into the vessel; cool the vessel to room temperature. Transfer the solution into a 125-mL Erlenmeyer flask, rinse the vessel with water, and add the rinsings to the flask. Add about 20–25 mL of water and a couple of drops of phenolphthalein or methyl orange to the flask, and back titrate this solution using standardized 0.1M HCl. Calculate the apparent **saponification equivalent** of trimyristin, the theoretical value, and the apparent error. Your accuracy should improve if you use larger quantities.

You could do this experiment with 5 grams of nutmeg and the extraction scaled up. This should produce enough ester to use 0.4–0.5 g samples (scaled up proportionately but use only 150 mL of water in titration) in duplicate determinations. Beal suggests extracting 15–30 g of nutmeg in a Soxhlet extractor (Figure 4.4) with ether for two hours.

A more fanciful alternative would be to melt and resolidify the ester and/or the acid about a roll of filter paper or string to make a facsimile of a candle.

See also *J. Chem. Ed.*, 1971, *48*, 255; and 1990, *67*, 274 (M).

4.2 Chromatography

By name, chromatography could be misleading. Separation by chromatographic methods does not depend on color. The name probably persists because the earliest experiments using this technique separated colored plant pigments.

Chromatography utilizes a mobile phase that passes through a stationary phase with intimate contact. Separation can be achieved when the compounds to be separated differ in their relative affinities for the two

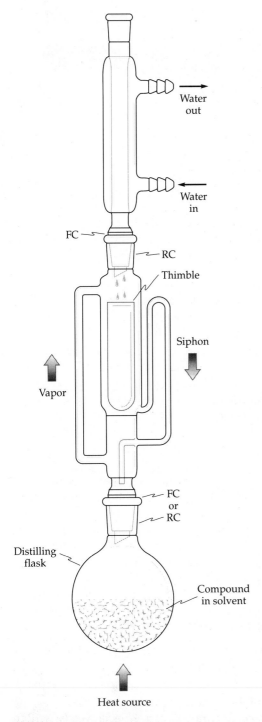

Figure 4.4 Using a Soxhlet extractor for continuous solid-liquid extraction.

Table 4.1 Some Suggested Separations
Types: C = Column, P = Paper, T = TLC

Type	Separation	Reference
P	Amino acids	B (5), M (5)
T	Amino acids	B (5)
C	Anthracene	D (76)
P	Anthracene	M (5)
C	Carvone/limonene	G (30), A (48)
C	Cholesterol acetate	F (14), K (10)
C	Colorless mixture	B (5)
T	Colorless mixture	M (3)
C	Dyes	D (76), M (2)
T	Dyes	M (3)
T	Essential oils	B (5)
C	Ferrocene/Acetylferrocene	W (41)
T	Ferrocene/Acetylferrocene	F (33), K (38)
P	Flower pigments	B (5)
C	Fluorene/Fluorenone	G (3N), F (14), K (10)
C	Nitroanilines	C (pp. 101–103)
C	Spinach leaves	M (2), B (5)
P	Spinach leaves	L (minilab 5)
T	Spinach leaves	M (3)
P	Sugars	B (5)
T	Terpenes	B (5)
C	Tomato paste, etc.	C (pp. 101–103), L (9), M (4)
C	2,4-Dinitrophenylhydrazones	V (sect. V, p. 21)
P	2,4-Dinitrophenylhydrazones	*J. Chem. Educ.*, 1966, *43*, 385.

phases. If a substance A has a greater affinity for the mobile phase than does substance B and B has a greater affinity for the stationary phase, they will be separated as the mobile phase moves through the system. In this case, A would progress by the stationary phase more rapidly than B. The stationary phase is most often a stationary liquid, a packed column, or a strip of adsorbent. The mobile phase is a gas or a liquid.

Over the years, many interesting chromatographic experiments have been described. Some of them are found in the following references. If you are assigned to develop a chromatographic project, you may want to consult them. They are referred to by letter in Table 4.1.

A. Ault, *Techniques and Experiments for Organic Chemistry*, 5th Ed., Allyn & Bacon, Boston, 1987, Sec. 14, Expt. 48.

B. Breiger, *A Laboratory Manual for Modern Organic Chemistry*, Harper & Row, New York, 1969, Chap. 5.

C. Cheronis and Entrikin, *Semimicro Qualitative Organic Analysis*, 2nd Ed., Interscience, New York, 1957, pp. 77–103.

D. Dannley and Crum, *Experimental Organic Chemistry*, Macmillan, New York, 1968, Expt. 76.

G. Durst and Gokel, *Experimental Organic Chemistry*, 2nd Ed., McGraw-Hill, New York, 1987, Sec. 3.6.

F. Fieser and Williamson, *Organic Experiments*, 6th Ed., D.C. Heath, Lexington, Mass., 1987, Chaps. 13–16 and 33.

K. Williamson, *Microscale Organic Experiments*, D.C. Heath, Lexington, Mass., 1987, Chaps. 9, 10, 12, and 38.

L. Lehman, *Operational Organic Chemistry*, Allyn & Bacon, Boston, 1988, Expt. 9, Minilab 5, pp. 16–18.

M. MacKenzie, *Experimental Organic Chemistry*, Prentice-Hall, Englewood Cliffs, N.J., 1971, Expt. 2–5.

V. Vogel, *Elementary Practical Organic Chemistry*, Part I, Longmans, Green, New York, 1957, Sect. V, 21.

W. Wilcox, *Experimental Organic Chemistry*, Macmillan, New York, 1988, Chap. 7 and Expt. 41.

4.2A Column Chromatography

(SM) The apparatus for column chromatography (Figure 4.5) consists of a glass tube constricted at one end and supported vertically. A plug of glass wool or cotton, or a filter plate and filter paper (sometimes also sand), is used as a foundation for the packing. A piece used as a column is included in some microscale kits. Some of the packing materials that have been used include alumina, calcium carbonate, calcium hydroxide, forms of silica, and sucrose. They may be used alone, mixed, in bands, or mixed with diatomaceous earth. The columns may be packed dry (in small portions with tapping) or with a slurry (poured in with a liquid). The former has advantages for preparing bands while the latter is often used to increase uniformity and to avoid trapped air.

1. How does slurry packing increase uniformity?

Question

A handy accessory is a long glass rod or a wooden dowel (for tamping or extruding) with a heavy rubber stopper on one end (for tapping the side of the column). In packing a column, the ultimate goal of any method is a uniformly packed solid.

(M) A convenient micro column can be made from a Pasteur pipet by placing a plug of cotton in its throat. This filter pipet (Section 1.3E) can be used for drying as well as for separation.

In column chromatography, the substances to be separated are dissolved in a suitable organic solvent (or mixture) and added to a column that is already filled with this solvent just to the top of the packed solid. The solvent passes down at a uniform rate and more solvent is added so that the top of the packing is never dry or disturbed during separation (a circular piece of filter paper on top will help). A change of solvent may be

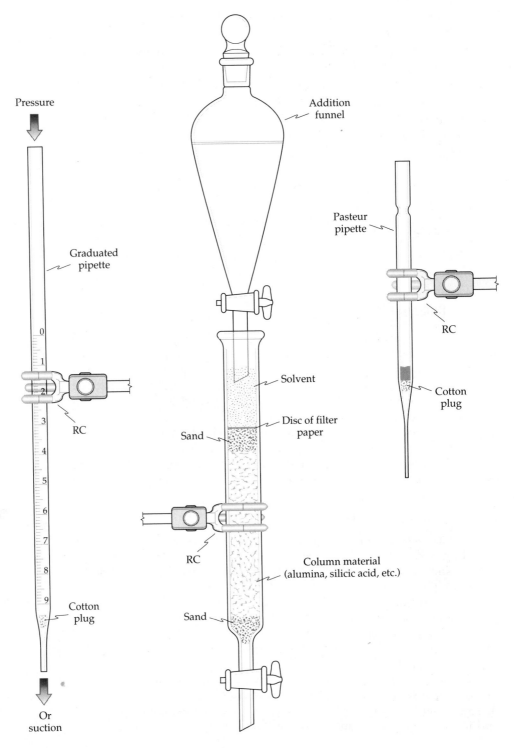

Pressure

Graduated
pipette

0
1
2
3
4
5
6
7
8
9

RC

Cotton
plug

Or
suction

Addition
funnel

Solvent

Sand

Disc of filter
paper

Column material
(alumina, silicic acid, etc.)

RC

Sand

Pasteur
pipette

RC

Cotton
plug

Figure 4.5 Apparatus for column chromatography.

Table 4.2 Elution Solvents: Some Solvents Used in Chromatography*
(By Increasing Polarity)

Compound	b.p. (°C)	Dielectric constant
m-Hexane	69	1.89
Cyclohexane	81	2.015
Dioxane	101	2.21
Carbon tetrachloride	77	2.23
Toluene	111	2.38
Carbon disulfide	46	2.64
Ether	34.6	4.34
Chloroform	61	4.8
Ethyl acetate	77	6.0
Acetic acid	118	6.15
Tetrahydrofuran	66	7.6
Methylene chloride	40	9.1
2-propanol	82	18.3
Acetone	56.3	20.7
Ethanol	78	24.3
Methanol	65	32.6
Acetonitrile	81.6	37.5
Dimethyl formamide	153	38
Dimethyl sulfoxide	189	48
Water	100	80.4

*Many of these solvents are too hazardous to be used for this purpose in the undergraduate laboratory.

needed to spread the materials sufficiently. A more polar solvent may be used to wash the substances from the column, or the column may be extruded and divided before dissolving off the substances. Tabulations of packing materials and eluents by polarity are available (see, for example, references A, B, L, and W in Section 4.2 above). Table 4.2 lists some common eluents, arranged by increasing polarity.

E X P E R I M E N T   1 0 A

Column Chromatography (SM)
Reference: Sections 4.2 and 4.2A.

Assemble the apparatus. Dry pack a 300-mm-long column (this may be just a piece of 8-mm or 10-mm tubing constricted at an end) with activated alumina or a uniform mixture of activated alumina and 5–10 percent of a filter aid such as Hyflo Supercel. Make the base with a plug of glass wool and just enough sand to make a smooth surface.

Introduce the solvent. When the column is uniformly packed, pass ethyl alcohol through the column at a slow, steady rate. Add alcohol as necessary to prevent the top of the column from drying out.

Initiate chromatography. At a time when the surface of the alcohol is only one mm above that of the packing, add 1 mL of a dye solution (a solution of 0.02 g of methyl orange and 0.10 g of methylene blue in 30 mL of alcohol for a class of 25) to the column all at once. Continue the alcohol passage at a moderately rapid rate; it will elute one dye. When the first dye is off the column, remove the second by passing water through the column. The solvent may be added automatically from a *closed* separatory funnel (as a reservoir) with its tip below the surface of the liquid waiting to enter the column (Figure 4.5).

Interpret the results. Record a diagram of the developed column including the size of the bands and the color appearing with both bands still on the column.

Discuss the behavior of the dyes in this separation as an indication of their solubilities in these solvents.

Summary

1. Pack the column and wet it with ethanol.
2. Add the dye mixture.
3. Elute the first component with ethanol.
4. Elute the second component with water.

Questions

QUESTION 1 IN SECTION 4.2A.

2. Explain the need for uniform packing in a column.
3. Why might suction be used with column chromatography?
4. Why is the column not allowed to run dry during separation?

Additions and Alternatives

Column packing (S)   Dye mixture (D)
Ethanol (D)   Dye solutions as eluted (D)

Additions and Alternatives

Try packing the column with an ethanol slurry. Develop directions for a separation from the suggestions here or in Table 4.1. Your instructor may prefer to use one or two specific separations and assign or allow variations in the choice of column packing, solvent, and developer; comparisons can then be made among the results. If the period is to be used for Experiment 10A only, you will be able to develop at least two columns. If you are doing both experiments, 10A and 10B, in the same period, 10B should be started first.

Water-soluble fluorescein may be used for methyl orange, and victoria blue B or methyl violet (0.1 g or less) for methylene blue. For other experiments you might try and details, see *J. Chem. Educ.*, **1951**, *28*, 39 and 546. Also see Table 4.1.

4.2B   Paper Chromatography

In paper chromatography, the strip of paper is both the column and the packing, and the solvent is drawn through it by capillary action. The rate of flow for the solute (R_f value) is usually calculated for each separated substance from the following equation:

$$R_f = \frac{\text{distance traveled by solute}}{\text{distance traveled by solvent}}$$

The solute distances are measured from the initial spot of the mixture to the center of the final, separated solute spots. The solvent distance is measured from the initial spot to the solvent front after development has finished. R_f values, then, are between zero and one.

E X P E R I M E N T   1 0 B

Paper Chromatography
References: Section 4.2B and those listed in Section 4.2.

Run paper chromatograms of 5% aqueous solutions of phenylalanine, glutamic acid, lysine, and an unknown. The unknown will contain one or more of these three amino acids. (Glycine, methionine, and leucine may also be used as the three amino acids.) This experiment may be done in pairs with four 200-mm boiling tube chromatographs, each containing a strip attached to the cork that seals the tube. Alternatively, a large beaker, covered with a large watch glass, may be used as a chromatograph chamber with four strips suspended from a splint or glass rod. A sheet of paper stapled into a cylinder and a wide-mouth jar can also be used. Be sure to record the number of the unknown sample issued to you.

Assemble the apparatus. Rolls of half-inch Eaton–Dikeman No. 613 and Whatman No. 1 paper are available for the chromatograms. Cut a strip long enough to be attached to the cork and just dip into the solvent. This length will be about 22 cm if a split cork is used (some extends a little above the cork) or 20 cm if it will be thumbtacked to the cork. Draw a light pencil line 0.6–0.8 cm from the bottom. Draw another 3–4 cm from the top. All handling and marking for identification MUST be above the top line. A staple or paper clip at the bottom of the strip will help to keep it extended. In the bottom of the test tube, place about 0.5 mL of the solvent so that it will not reach the bottom line. The solvent is a 1-butanol–acetic acid mixture (5:1) or a solution made by mixing 1-butanol, acetic acid, and water (4:1:5). Discard the aqueous layer.

Introduce the reagents. With a Pasteur pipet, place a very small spot of the solution to be tested in the middle of the bottom line, and allow it to dry. You may apply a little more to the spot after it has dried if a denser spot is desired. Arrange the paper in the tube, making sure that the spot is a few millimeters above the solvent.

Question

1. Why must the spot on the paper be above the solvent level?

Initiate chromatography. Allow the chromatogram to develop for two to two and one-fourth hours or until the solvent reaches the upper line. (If not enough time is available, go to the next step one-half to three-fourths of an hour before the end of the period.) Remove the strips; mark the leading edge of the solvent on each strip with a pencil before the strip dries. When the strip is almost dry (a heat lamp expedites drying), spray it uniformly until just moist with a 0.2% solution of ninhydrin in methanol, ethanol, or 1-butanol saturated with water. Dry the strip again by simple evaporation, in an oven or with a heat lamp. The heat lamp is preferred because it speeds up the process and permits the heating to be stopped as soon as the color appears.

Interpret the results. Calculate the R_f values for the three knowns. Measure the solvent distance from the starting line to the marked leading edge. Measure the solute distance from the starting line to the center of the spot.

Question

2. What appears to be the composition of your unknown? Explain.

Summary

1. Spot and develop the paper chromatograms.

2. Dry and spray them with ninhydrin.

3. Dry them with heat.

4. Calculate R_f values and characterize the unknown.

Questions

QUESTIONS 1 AND 2 ABOVE.

3. Explain the precaution for handling the paper strip.

4. Why is the paper chromatogram developed in an essentially closed system?

5. Look up the ninhydrin reaction in your text and write a specific example with one of the amino acids you used.

6. Why might proline give a different color in the ninhydrin reaction?

Disposal of Materials

Amino acid solutions (D)   Development solvent (N, D)

Paper chromatograms attached
in your notebook

Additions and Alternatives

You will find other possibilities in Table 4.1 and the references listed above. Try different development solvents to reduce the time required.

4.2C   Thin Layer Chromatography

See the references listed in Section 4.2.

Anal. Chem., 1961, *33*, 1138.

J. Chem. Educ., 1968, *45*, 510.

This method of chromatography employs a thin layer of adsorbent and/or a thin layer of support on a rigid surface such as a glass or plastic sheet. Although many chemists prefer to coat their own glass plates or microscope slides, prepared sheets (which can be cut with scissors) are available commercially with a wide variety of coatings. Some include an inorganic fluorescing agent that aids in solute detection.

Thin layer chromatography has many advantages compared with paper chromatography, for most samples. One, certainly, is the variety of possible stationary phases. It can be used to establish conditions for column chromatography. It is also more sensitive and certainly faster. Runs require 5 to 30 minutes and can be used to follow the progress of a reaction.

Development chambers that run small plates and large plates using a minimum of development solvent are available commercially, but simpler alternatives are often used in the laboratory. A beaker (400–600 mL), sealed with aluminum foil and a rubber band, will work for small plates. Jars with tight-fitting lids such as one-pint canning jars, some pickle jars, and even large baby food jars are good (see Figure 4.6A). A wick of filter paper should be included that extends from within the solvent pool to close to the top of the container. Always make sure that the vapor space of this container is saturated with development solvent vapor before beginning a run and that the plates do not touch the wick at any point.

For those who make their own plates, Jordan's method for making thin layer chromatographic vials (the *Journal of Chemical Education* reference above) has many advantages (see Figure 4.6B). The vials stand straight up, and it is easier to estimate where the origin should be with respect to the solvent. The coatings are fairly reproducible though a bit thicker at the

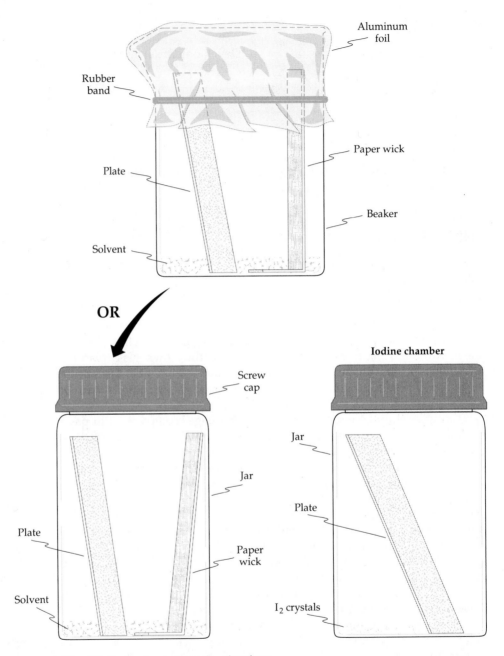

Figure 4.6A Thin layer chromatography chambers.

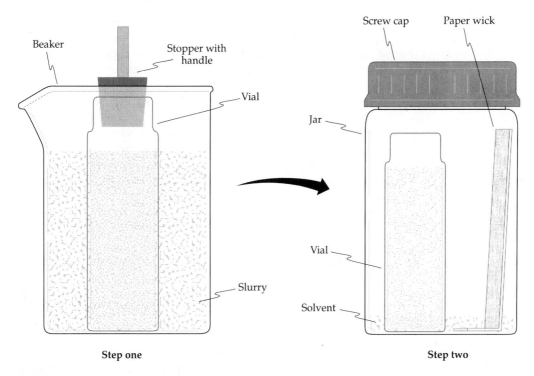

Figure 4.6B The preparation and use of student-made TLC vials.

bottom than at the top. The handles make handling easier, and there are no edges to give "edge effects."

Considerable care must be taken in spotting the substances to be analyzed. The substances must be dissolved in volatile solvents. They are spotted with their own microcapillaries (one for each substance!). These micropipets can be purchased or made by drawing melting point capillaries (open at both ends) down to a very fine diameter and cutting them to convenient lengths (open at both ends). If they are stored, they should be kept free of dust. The plate to be used is marked with a *light* pencil line (do not disturb the coating) a few millimeters above where the solvent will reach when the plate is placed in the development chamber. The spots will be placed at this line so that they are equidistant from each other (6–10 mm apart) and 6 mm from an edge, if there is one. The smaller end of the micropipet is placed in its solution and the liquid is allowed to rise in it by capillary action. It is moved to position over its place on the plate. The liquid, but not the glass of the pipet, should touch the coating, and a bit of it should drain into the coating. Do not linger. If the pipet is removed before the spot is 1 mm or more in diameter, the resulting spot will be less than 2 mm in diameter. Smaller spots run with less material "trailing" the main spot and are less likely to overlap neighboring spots. This spot is allowed to dry while the other spots are initiated with their own pipets. When a spot is dry, a bit more solution may be added. The spot should not be larger than 2 mm or too heavily loaded when analysis is begun. Overloaded spots have a tendency to "trail."

The spotted "plate" is placed (without touching the paper wick) in a development chamber saturated with the solvent vapor. The solvent climbs up to the origin and then through the rest of the plate, "carrying" parts of the mixture along with it at different rates. When the solvent front is 5–10 mm from the top or its rate of climb has just about stopped, the plate is removed from the chamber. The solvent front is marked immediately with a pencil. The solvent is then allowed to evaporate.

A number of methods are used to locate the spot after development and after the development solvent has evaporated. If all of the materials are colored, no other method may be needed. If this is not the case, reagents that react with the compounds may be sprayed on, streaked on, or used as dips (as in paper chromatography). Some are quite specific, but others react with almost everything except the adsorbent (as when streaking with concentrated H_2SO_4). If an inorganic fluorescing agent (insoluble in organic solvents) is included in the adsorbent, then illumination with short wavelength ultraviolet light will show dark spots against a light-emitting background (some compounds will also fluoresce). Do NOT LOOK INTO THE LAMP. ULTRAVIOLET RADIATION CAN CAUSE SEVERE DAMAGE TO THE EYES. With the lamp pointed at the plates and away from people, gently outline the spots in pencil to indicate where the greatest density occurs. Remember that the spots disappear again when the light is turned off.

CAUTION

Another method uses the iodine chamber. It depends upon the sublimation of I_2 and whether or not it reacts with or is adsorped by the compounds faster than it is adsorbed by the plate material. To use the iodine chamber, the "plates" are placed in the closed chamber, which also contains solid iodine and its vapors. *Do not touch the iodine.* Placing the chamber in *warm* water or on a *warm* surface will speed the process. When the brown of the spots is in maximum contrast with the background (which picks up I_2 after a while), the "plates" are removed and the spots are traced with a pencil. Again some of the spots may lose their iodine color upon standing, but some are permanently changed. Because of this fact, iodine development should be employed only after fluorescent analysis.

The R_f values are calculated in the usual manner in thin layer chromatography. The solute distance is from the original line to the center of density of the spot. The solvent distance is measured along the same path, from the origin to the light pencil line made at the solvent front as soon as the "plate" was removed from the development chamber, before the solvent evaporated. Whenever possible, knowns should be run alongside

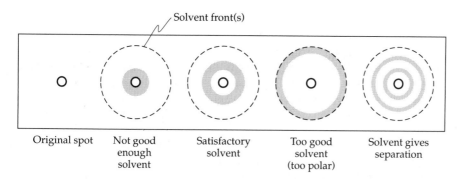

Figure 4.6C Thin layer chromatography solvent selection.

unknown mixtures rather than relying completely on published R_f values or even on values determined on separate thin layer chromatographic plates.

In the development of a thin layer chromatographic analysis, it is important to choose a workable combination of adsorbent and developing solvent. A quick and approximate method to do both at the same time is as follows. On a series of pencil dots, place spots of the materials to be separated on the different adsorbent materials available. They should *not* be too close to the edge or within 0.8–1.5 cm of each other. When a spot seems well dried, place a large drop (or two small drops quickly) of solvent precisely on the spot, and watch the solvent circle expand. Before it evaporates entirely, outline the circle with pencil. When the spot separation has been determined using the usual method of detection, outline the rings of the separated substances. If, for one substance, there is a ring intermediate between the dot and the solvent circle, this should mean that you have a workable combination. When separating two or more substances, this workable adsorbent–developing solvent combination should affect the materials to be separated differently (Figure 4.6C).

EXPERIMENT 11

Analgesic Drugs and Thin Layer Chromatographic Separation

References: Section 4.2C and:
    J. Chem. Educ., 1971, *48*, 478.
    J. Chem. Educ., 1979, *56*, 180.
    J. Chem. Educ., 1979, *56*, 834.

In this experiment, five potential components of analgesic drugs will be used for thin layer chromatographic analysis, and mixtures of them will be analyzed.

Assemble the apparatus. Prepare one or more of the development chambers with ethyl acetate with a trace of acetic acid (1–5%) as the developing solvent. Ethyl acetate or 1,2-dichloroethane with acetic acid has also been used. Introduce the solvent to a depth of 5–6 mm. Equip each chamber with a wick of filter paper (a strip about 2-cm wide) across the bottom and up the side and a vapor seal—a screw lid or aluminum foil or plastic wrap with a rubber band. A wick wet with solvent will saturate the vapor space faster.

Prepare 8–12 micropipets as described in the text above to be used for spotting. Obtain, from the stockroom or your instructor, two or three small thin layer chromatographic "plates" (4–6 × 10 cm), or prepare them as directed. Handle them carefully and only by their edges so that no coating is loosened. *Lightly* draw a *pencil* line (no ink!) across the shorter side of the plate 1.0–1.2 cm from the bottom on its coated face.

Question

1. Why should pencil and not pen be used?

Divide this line so that three to five spots will be equidistant from each other (6 mm or 8–10 mm apart) and 6 mm from an edge.

Prepare the plates. Six known solutions will be available containing acetaminophen, aspirin, caffeine, phenacetin, salicylamide, and a mixture of the five. They will be about 5% solute (W/V) in methylene chloride and absolute ethanol (chloroform or methanol have also been used). If you spot each solution once, you will need 2 or 3 plates. If you spot the solutions twice, you will need 4 or 5. As a compromise, do the mixture twice, and use 3–4 plates. Obtain two unknowns, and record their numbers. Make unknown solutions using 0.2–0.3 g of unknown or half a crushed tablet or caplet of analgesic in 4–5 mL of the same solvent as that used for the known. Mix the solid unknown with the solvent, and heat it gently to achieve dissolution. Cool this solution. Spot each single known at least once and the mixed known and the two unknowns twice (12–15 spots). Make sure that each plate has a spot in common with the others. Designate, where it will not interfere with the chromatography, the plate and what each spot is. Record a diagram of each plate in your notebook.

Initiate chromatography. When you have spotted a plate as described above, place it in a development chamber as upright as possible, so it does not touch the wick. The spots on the origin should be clearly above the top of the solvent so they can only be reached by capillary action.

Question

2. What would happen if the spots were below the solvent level?

Seal the chamber securely, and observe it regularly. Note the difference in the coating as the solvent rises.

Analyze the developed plates. When the solvent front has almost reached the top of the plate or when the rate of solvent rise has fallen significantly, remove the plate from the chamber, lightly trace the solvent front with a *pencil*, and allow the solvent to evaporate in the hood.

CAUTION

When the plate is dry, illuminate it with UV light. **Do NOT LOOK AT THE LIGHT**. Trace the spots lightly with a pencil; indicate areas of greatest intensity, and record the appearance of each spot. Make a diagram of this information in your notebook. Then place the plate in the iodine chamber until maximum contrast is achieved. Trace the spots, and record their appearances.

Interpret the results. Calculate and record the R_f value for each spot. Make a table in your notebook, similar to the table below, that includes the R_f, the spots' appearances in each detection system, and the identity of *each* spot. Be sure that you can detect and identify all five spots in the mixed known. Use this information to establish the composition of your unknowns. If your unknowns include commercial preparations, you may find that some materials like starch do not dissolve, that a component does not show because of low concentration, or that aspirin trails because of high concentration.

Sample table

Sample	Spot Identity	R_f	Appearance with UV	Appearance with I_2
1	Compound A	.38	dark purple	dark brown
2	Compound B	.62	glowed bluish	light brown
3	Compound C	.45	purple	light brown
3	Compound A	.38	dark purple	dark brown
3	Compound D	.12	dull purple	—

Summary

1. Set up development chambers and allow them to saturate with solvent vapors.
2. Make micropipets if they are not available.
3. Prepare solutions of unknowns and knowns, if needed.
4. Prepare and spot plates.
5. Develop the plates as far as practical.
6. Use UV and I_2 to locate the spots.
7. Calculate R_f values and record all data in your table.
8. Determine the composition of the unknown.

QUESTIONS 1 AND 2 ABOVE.

Questions

3. Why the insistence on a closed chamber?
4. If the development solvent front was allowed to reach the top of the plate and remain there for some time, what would be the effect on the R_f values calculated compared with those you should have found?

5. How does the solvent rise up the plate? What is the function of the solvent in thin layer chromatography?

6. Make a list of the ingredients of five different, common, commercial preparations from their labels. Compare this list with the composition of your unknowns.

Disposal of Materials

Development solvent (OW, I, HO)   Known and unknown
                                          solutions (HW)

TLC plates (notebook, S, I)

Additions and Alternatives

You may wish to omit phenacetin or acetaminophen or include ibuprofen.

4.2D   Gas Liquid Chromatography

DalNogare and Javet, *Gas Liquid Chromatography*, Interscience, New York, 1962.

Erskine, et al., *J. Chem. Educ.*, 1986, *63*, 1899.

Karasek, et al., *J. Chem. Educ.*, 1974, *51*, 816.

Miller and Harmon, *Elementary Theory of Gas Chromatography*, Gow-Mac Instrument Co., 1978.

Pacer, *J. Chem. Educ.*, 1976, *53*, 592.

Pardue, et al., *J. Chem. Educ.*, 1967, *44*, 695.

Sixma and Wynberg, *A Manual of Physical Methods in Organic Chemistry*, Wiley, New York, 1964, pp. 63–82.

Wiberg, *Laboratory Technique in Organic Chemistry*, McGraw-Hill, New York, 1960, pp. 165–177.

This versatile method of chromatography is based on the partition of components, originally in the gas phase, between a moving gas phase and a stationary liquid phase. An inert gas, such as helium or nitrogen, is the carrier of the moving phase. A variety of liquids that are usually viscous and high-boiling can make up the stationary phase. This stationary phase is usually coated on a solid support such as powdered fire brick and packed in a column. For some applications, the liquids may be coated directly on the walls of the capillary tubes used for the columns. Be careful of samples containing water at temperatures over 100°C because the steam may remove the stationary phase from the packing. Inject *dry* samples.

The gas liquid chromatography apparatus consists of four parts: the injector ports, the columns, the detectors, and the recorders. The gas flows in at the injection port, through the column, and out past the detector. The injection port is closed with a solid rubber septum, and gas enters as close to the septum as possible. The sample is introduced through the septum with a syringe as far into the port (and close to the column) as possible. The injection port is kept *hot enough* to volatilize any injected liquids instantly so that all gases are introduced together on to the column by the pressure of carrier gas. THE PORT IS *HOT.* AVOID TOUCHING IT WHEN MAKING INJECTIONS.

There are at least two standard ways of measuring the amount of sample injected. One consists of flushing and filling the syringe so that there is no air in the needle or syringe and then making the plunger advance only for the volume being introduced. This way can be quite accurate but may bleed some of the remaining liquid from the needle. The other approach is to draw up 0.5–1.0 μL of air with the needle out of the liquid, draw up some liquid, then draw up some more air (with the needle out of the liquid). This leaves a small volume (1–6 μL) in a 10-μL syringe surrounded by volumes of air. The total amount of sample between the two air pockets is estimated, and all of it is injected, including the air.

The introduction of the syringe and the injection require some care. The syringe is introduced with a straight, even, gentle push. Using the nondominant hand to guide the syringe during injection will reduce the chances of too strong a push. The thumb of the other hand should be positioned to prevent the blow-back of the plunger. To inject, the plunger is depressed with one gentle, firm stroke. The plunger is depressed without delay so that all of the sample to be introduced goes in at once without undue speed or pressure. The needle is then withdrawn, straight back without allowing gas to escape. The air comes through the column first and gives a reference peak near the injection point with some detectors.

The wide variety of columns in terms of size (diameter and length) and packing makes possible an even wider variety of analyses. These may vary from microanalytical to preparative commercial procedures and from monitoring gas mixtures to separating complex liquid mixtures.

Any description of a column should include the length, diameter, nature and state (how finely divided) of the support, the nature of the stationary phase, its percent coating, and the maximum operating temperatures. Temperatures that are too high can cause the stationary liquid to volatilize and bleed off of the column. There are many stationary phases, which vary widely in polarity and in what they will separate, as well as in other special functions. In the instrument, these columns are housed in an oven where the temperature of the separation can be controlled to minimize volatilization, bleed-off, and solubility changes for the compounds being tested.

Within the oven are two columns, which are part of the dual system of the instrument. In any analysis, one column is used for separation and the other for a reference system.

Separation is achieved in the column by differential solubilities. Different substances should have different solubilities in the stationary phase at the column temperature used. Those compounds that are less soluble in the liquid phase will be more soluble in the gaseous phase. Since the carrier gas is moving, these less-soluble substances will be moved through the column faster than those that are more soluble. Also, higher temperatures should reduce the time a substance stays on the column because the solubility of a gas decreases as the temperature increases.

Three types of detectors are most commonly employed: thermal conductivity (hot wire), flame ionization, and electron capture. Of these, the most common in student instruments is the "hot wire" detector. In this detector, a hot wire with a certain voltage across it is cooled by helium (or other carrier gas) as the gas flows through the detectors for both columns. When another gas (from the vaporized sample) is mixed in, less cooling occurs because as the molecules with less thermal conductivity are mixed with the carrier gas, the resistance of the wire increases and the current decreases in this detector compared with the reference detector. The conductivity of a wire decreases with increasing temperature. This difference can be converted to an electrical signal. The difference in cooling is proportional to the number of molecules (or moles) of the substance that was injected, though molecules of different compounds may have more or less of an effect (response factor) on the resistance/current difference. For this detector, the carrier gas must be flowing before the detector is turned on and for a while after it has been turned off. Oxygen will oxidize the hot wire and damage the detector.

The electrical signal from the detectors is carried to the recorder. The distance the recorder pen is driven across chart paper is proportional to this signal. If the paper is not moving, the pen will draw a line; if the paper is moving, the two movements should theoretically generate an isosceles triangle. The area under the triangle should be proportional to the moles of compound causing that signal. If two compounds are incompletely separated, they may generate one triangle, two overlapping triangles, or one triangle shouldered on the side(s). Ideally the pen should return to the base line before the next signal starts. Some signals are nonideal and don't look like triangles (e.g., water). The recorder is started before the injection to ensure that the base line is stable and to have one less thing to do during injection. A mark should be made on the base line when the injection is made. Some recorders have a marking device. The air peak may also be used to mark when the separation began, and useable retention times can be based on the air peak with a thermal conductivity detector.

For quantitative (or even semiquantitative) purposes, determination of the areas under the triangles is important. Some recorders have integrators that quantify this area; most do not. For very precise work without an integrator, the peaks are cut out and weighed. Another method involves planimeters, which are hand-held and calculate the area while tracing the perimeter. The area of the triangular peak should be equal to one-half of the base times the height. More often the width at half-peak height (which should be half of the base width) is multiplied by the height because the width of the base is sometimes difficult to determine. The area of one peak divided by the sum of all the areas equals the **mole fraction** of that compound in the sample. This assumes, of course, that all compounds in the mixture have the same effect per mole on the detector and that all areas are observable. For quantitative work, it is appropriate to test this assumption. One or more samples of known composition are injected to see if the ratio of the areas is as predicted. If not, one peak that will always be represented is taken as the reference peak, and **response factors** (correction factors) are calculated. For example, suppose a sample is made up of A, B, and C in a $2:1:1$ ratio, and observed areas are $8:5:3$. Choosing A as our standard, the response factors are B $(4/5)$ and C $(4/3)$. If A is 8, the expected B is 4 or $8 \times 1/2$. To get the response factor, divide the expected value by the observed value (here $8/2 \div 5$). Thus, the observed area times the response factor equals what "should have been observed" ($5 \times 4/5 = 4$). Note that the area units must all be the same and that the ratios can *now* represent moles, volume, etc. Early peaks tend to be very skinny triangles, whereas late peaks are very broad and rounded. The measurement of width then will be subject to quite a different percent error. One approach used to get around this is to measure peak heights (if they are all reasonable in height) and figure height response factors rather than the area response factors discussed above. (See Pacer above and Leber, *J. Chem. Educ.*, 1986, *63*, 550.)

The variables that should be observed, recorded, and controlled include: (A) sample size, (B) injector temperature, (C) column temperature, (D) column packing, (E) attenuation, (F) flow rate, and (G) chart speed. Sample size (A) should be consistent and not so large as to overload the column and cause incomplete resolution. The injector temperature (B) must be hot enough to instantly volatilize all components. The column temperature (C) will influence the solubility of the compounds in the stationary phase. Other things being equal, the higher this temperature the faster compounds will pass through the column. The column packing (D) also influences how fast things travel through the column. More-polar columns will have greater affinity for polar compounds, and so on. The attenuation (E) influences how high the peaks will be. It should be adjusted so that all peaks are of reasonable size without any going off of the scale. The flow rate (F) is controlled by the second valve at the tank or by a needle valve at the instrument. The greater the flow rate, the more pressure there is to push the compounds through the column. Some separations are very sensitive to flow rate. The device for measuring the flow rate usually consists of a soap bubble generator and a tube calibrated in milliliters. It is connected to the outlet port of the column used so that the exit gas lifts a bubble through the tube. The bubble is timed (as a passage of 10 mL), and the flow rate is calculated in mL/minute. The chart speed (G) does not affect the separation as such, but may aid one's ability to see and measure it. Too slow a speed makes skinny peaks and may make them appear overlapping. Too fast a speed makes them broad and wastes paper. If the *separation* is not satisfactory, changes in C, D, and F should be considered. When tailing occurs, try injecting so that all the sample enters the column at once and adjusting the flow rate before looking at C or D.

Retention times are measured for each significant peak and used the way R_f values are in paper and thin layer chromatography. The retention time of a peak is calculated from the *distance* the paper has traveled between injection and the time the peak reaches its maximum height (cm; e.g., to convert to time, t = distance/chart speed or (cm)/(cm/min)). If the conditions discussed above (A–G) are held constant, the retention times should be nearly reproducible and should be useful in identifying the compounds producing the peaks. This may be done by running known samples and mixtures. It is often helpful to add a pure sample of one component to the mixture in a *second* run to see which peak is proportionally enhanced. The retention times are a function of a particular set of conditions and often are not reported in the literature.

Once retention times have been determined, it is possible to *collect* the separated compounds to achieve the same purpose as distillation. Some kits come with devices for this purpose (Figure 4.7). These devices can be centrifuged to recover the liquid. Also, U- or V-shaped devices can be used to collect samples when the bottom part can be cooled. They should be designed so that the gas can pass by the collected liquid and the liquid can be reached with a Pasteur pipet. It has been suggested that tips for automatic pipets (1000-μL-size tip cooled with ice in part of a sandwich bag) or cold Teflon tubing can also be used. A collection device should be put in place just before or just as the peak emerges and removed before the base line is reached again or before any of the next compound starts to exit.

EXPERIMENT 12

Separation of an Unknown Mixture

References: Section 4.2D and its references (esp. Pacer) and:
Corcia, D. et al., *J. Chromatography*, 1979, *170*, 245
Corcia, D. et al., *J. Chromatography*, 1980, *198*, 347

Record the number and type of your unknown. Three types of unknowns may be given, which may contain two or three of the

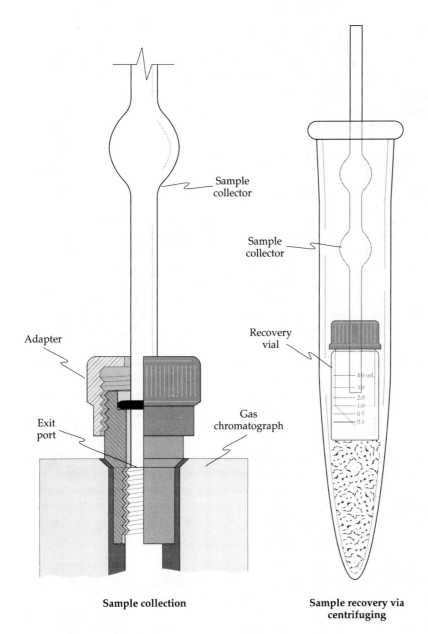

Sample collector

Sample collector

Adapter

Recovery vial

Exit port

Gas chromatograph

40 mL
3.0
2.0
1.0
0.5
0.1

Sample collection

Sample recovery via centrifuging

Figure 4.7 Gas liquid chromatography collection.

compounds in the group. Type A may contain methanol, 2-methyl-2-propanol, or 2-methyl-2-butanol; Type B may contain ethanol, 1-propanol, or 1-butanol; and Type C, ethyl acetate, ethyl propanoate, or ethyl butanoate. Determine if the types of columns available include a carbowax and/or silicone-type stationary phase.

Establish the conditions to be used. For the exploratory analysis, try a 70–80°C oven temperature (or lower for silicone-type columns). Stabilize the temperature of the column (based on the boiling point of the most volatile component) well in advance of the determination. If a change in column temperature is permitted, allow enough time to stabilize the temperature and the column at any new temperature. Your instructor may designate certain instruments with conditions partly worked out for each type of unknown. You may also be permitted, if you have the same type of unknown as another student, to work with that student to establish workable conditions.

Calibrate your analysis. Make a solution of known concentration using the three unknowns in your group, by mixing together 1 mL of each (accurately measured). Run individual knowns (ca. 0.5 μL) and a mixture (ca. 1.5 μL) to establish the retention times of each as they separate.

Inject 1.5–3.0 μL (depending on attenuation) of your known mixture into the gas chromatograph; identify the trace; record all variables, A–G. If all peaks are not separated and on scale, try another injection after modifying one or more variables (A, C, E–G; check with your instructor). If the results are still not satisfactory, discuss with your instructor other changes that could be made. Continue until a satisfactory trace is obtained. Using this trace, determine the response factor in height and/or the response factor in area.

Analyze the unknown mixture. Now inject the same size sample of your unknown into the chromatograph. If an early peak goes off scale, try a smaller sample; if the peaks are too small, try a slightly larger sample. These adjustments can be made with the attenuation if the multipliers are noted. DO NOT CHANGE ANY OTHER VARIABLE.

Question

1. Why must all the variables remain constant?

Calculate the composition of your mixture in volume percent, and tentatively identify the components by comparing their retention times with the retention times collected for your knowns.

In order to confirm or negate the identity of one component in the mixture, analyze a "spiked" mixture. Mix 8 μL of your unknown and 2 μL of a lesser component whose peak height is less than half scale to make a new mixture. Inject a sample of this "spiked" unknown mixture and analyze its peaks and composition. In your report, discuss the composition of your unknown and "spiked" mixture and your reasoning. Include in the report a list of successful conditions (A–G).

Summary

1. Establish separation conditions.

2. Determine retention times for knowns.

3. Analyze a known mixture to get correction factors.

4. Analyze your unknown.

5. Analyze a "spiked" sample of your unknown.

QUESTION 1 ABOVE.

2. Using the observed composition of your unknown, predict the percent composition for the "spiked" sample. Compare your prediction with your analysis of this sample.

3. How is spiking used to identify a peak?

4. If both types of response factors were calculated, how do the results using each compare?

5. What was the reasoning used to make changes in conditions for practical separation?

Disposal of Materials

Known and unknown liquids (OW, I)

Additions and Alternatives

You could request a more interesting Type A unknown, which could include a fourth potential unknown, 3-methyl-1-butanol (isoamyl alcohol). The trio of unknowns used by Pacer was benzene, toluene, and *p*-xylene.

You could make your own mixture to be separated by gas chromatography from compounds listed in Table 4.3.

E X P E R I M E N T   1 3 A

Free Radical Substitution (M)

References: Sections 1.3E, H; 4.2D; and Markgraf, *J. Chem. Educ.*, 1969, *46*, 610.

This experiment involves the chlorination of 2,3-dimethylbutane and utilizes gas liquid chromatography to analyze the product and explore the effect, if any, of the solvent on selectivity. The generalized reaction is:

$$RH_{(1)} + SO_2Cl_{2(1)} + \text{solvent} + \text{benzoyl peroxide} \longrightarrow HCl_{(g)} + SO_{2(g)} + RCl_{(1)}$$

Table 4.3 Some Possible Compounds That can be Separated by Gas Chromatography and Fractional Distillation

Compound	Boiling Point (°C)
n-Pentane	36.1
Dichloromethane	41
t-butyl chloride	51
Acetone	56
Methanol	64.7
Isopropyl formate	71
Ethanol	78.3
2-propanol	82.4
1-propanol	97.1
Methylcyclohexane	100.9
Toluene	110.6
1-butanol	116
1,3-dichloropropane	125
1-pentanol	138
m-xylene	139.1
Cumene (isopropyl benzene)	152.4
Cyclohexanol	161.1
Mesitylene (1,3,5-trimethyl benzene)	164.7

One convenience is that the *other* two expected major products are gases. The starting material, 2,3-dimethylbutane, has only two different kinds of hydrogens and, hence, can produce only two monosubstituted chlorination products (1-chloro-2,3-dimethylbutane (b.p. = 122°C) and 2-chloro-2,3-dimethylbutane (b.p. = 112°C)). The ratio of 3° hydrogens to 1° hydrogens is 2/12 or 1/6. If there was no selectivity and the products were formed by random abstraction of hydrogen, the ratio of products would be 1/6 (2-chloro 1-chloro). To calculate the selectivity, first divide the observed moles of each product by its **statistical value** (i.e., the number of equivalent hydrogens that form that product; this puts the results on a per hydrogen basis); then, use these results to formulate a ratio (e.g., 2-chloro/1-chloro). For example, if you collected 2 moles of the 2-chloro product and 4 moles of the 1-chloro product, the **product ratio** would be 2/4 or 1/2. Dividing each molar amount by each product's statistical value, you get 2/1 and 4/6; the ratio is thus (2/1)/(4/6) or a **selectivity** of 3. In this particular experiment, this process can be simplified to multiplying the observed reactivity product ratio of 2-chloro/1-chloro by six. Further chlorination is, of course, possible, but the ratio of reactants is designed to minimize this possibility.

1. Calculate the ratio and explain how this helps to minimize di- and polychlorination.

At the temperatures employed, benzoyl peroxide ($C_{14}H_{10}O_4$) breaks down to radicals ($Q\cdot$) that can attack sulfuryl chloride to produce a chlorine radical, which begins the chain reaction by abstracting one hydrogen.

2. Write the chain propagating and terminating steps. See your textbook for the mechanism of chlorination.

$$Q\cdot + SO_2Cl_2 \longrightarrow QCl + SO_2Cl\cdot SO_2Cl\cdot \longrightarrow SO_2 + Cl\cdot.$$

Assemble the apparatus. Equip a 5-mL vial with a condenser and gas trap (Figure 4.8). The rim of the funnel should touch the surface of the solution in the beaker.

3. Some such traps end in just a piece of glass tubing in the water of the trap. What could happen if the end of the tube got into the water when and if the volume in the trap increased? What advantages does the present setup have?

In the beaker, place a 5–10% Na_2CO_3 solution or water and a bit of solid NaOH or KOH. If you use the test tube trap, moisten the cotton in the tube with saturated sodium carbonate.

4. Why add base to the water in the trap?

Place a boiling stone or a magnetic stirring vane in the vessel.

Introduce the reagents. HANDLE THESE SOLVENTS IN THE HOOD AS THEY ARE TOXIC! DO NOT BREATHE THEM OR GET THEM ON YOUR SKIN! With care, introduce 2 mL of solvent (carbon tetrachloride, carbon disulfide, or benzene), 1 mL or 0.66 g of 2,3-dimethylbutane and 0.1 mL or 0.166 g of sulfuryl chloride. Transfer the liquids in this experiment with care; use automatic pipets if possible. You will be assigned or will choose the solvent from carbon tetrachloride, carbon disulfide, and benzene.

5. What are the health hazards of these solvents? Use chemical fact sheets, the Merck index, or a computer search.

Your instructor will then add approximately 2–5 mg of benzoyl peroxide to your reaction mixture.

Initiate reaction. Reassemble the apparatus, and regulate the stirrer if one is used. Reflux for about one hour.

Interpret the results. Set up a table of physical properties and determine the limiting reagent and the theoretical yield (or complete the first part of a report sheet). When the mixture has cooled to room temperature, analyze a sample by gas liquid chromatography as directed by your instructor. Try a 5-μL sample with the lowest possible attenuation. For most columns, the order of elution is: 2,3-dimethyl-1-butene = 2,3-

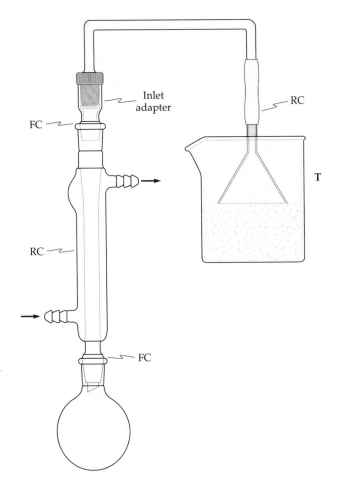

Figure 4.8 Simple reflux with fume trap (T).

dimethylbutane, carbon disulfide, 2,3-dimethyl-2-butene, chloroform, carbon tetrachloride, benzene, *2-chloro-2,3-dimethylbutane*, *1-chloro-2,3-dimethylbutane*, and then more chlorinated products. With no additional treatment, the starting material and the solvent will dominate the gas liquid chromatography trace. Their peaks will go off scale and, after the solvent peaks, two modest product peaks should appear. If authentic samples are available or collected, spiking may be informative. Calculate the areas for the monochloro products, the product ratio, and selectivity shown for your solvent. Dispose of the reaction/product mixture as directed by your instructor.

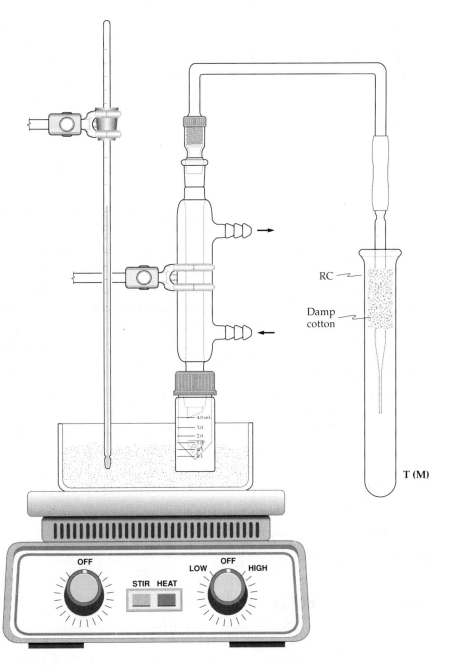

RC

Damp
cotton

T (M)

Figure 4.8 (*continued*)

Summary

1. Set up the apparatus and trap.
2. Transfer the liquids by pipet with *care*.
3. Reflux for one hour.
4. Cool.
5. Analyze the product by gas liquid chromatography.
6. Calculate the product ratio and selectivity.

Q u e s t i o n s

QUESTIONS 1–5 ABOVE.

6. Why is it not of concern in this experiment (as it usually is) that two major peaks go off scale?
7. Why can benzoyl peroxide be ignored in the calculation of theoretical yield?
8. Why do the reaction temperatures differ with different solvents?
9. Compare your selectivity with those found by students using the other solvents. Discuss what has been observed.
10. What would be the practical difficulties in trying to separate the components of the product mixture by distillation?
11. Why is it probably not necessary to determine and use response factors in your calculations of product ratio and selectivity?
12. Calculate the theoretical (statistical) ratio of monochlorinated products for 2,4-dimethylpentane. How would this compare with that observed experimentally?

Disposal of Materials

2,3-dimethylbutane (OW)	Solvent (HO, HW, I)
Sulfuryl chloride (HO, HW, I)	Benzoyl peroxide (HW, I)
Reaction/product mixtures (HO, I)	

Additions and Alternatives

If the gas chromatograph separation is good, the products can be collected and analyzed. It has been suggested that the starting material and solvent could be partially distilled off to enhance the observation of the product peaks. If you plan to do this distillation, *at least* double the quantities and use a semi-micro setup such as that shown in Figure 4.8. Any distillation should be attempted only after all analyses have been done. None of the alternatives should take the place of the primary analysis.

Hammons, in a private communication, has suggested the use of carbon tetrachloride as a principle solvent, with varying proportions of benzene. Chlorobenzene has also been used as a solvent and usually comes off after the monochloro products. With carbon disulfide, longer or more vigorous refluxing might be tried. You could also try mixing it with other solvents. Another possible solvent is hexachloroethane, which can be heated at 80–100°C rather than refluxed.

With a higher attenuation and a smaller sample, you could obtain a trace of reasonable peaks on scale for 2,3-dimethylbutane and the solvent. Since the molar amount of solvent should not change, you could estimate the amount of hydrocarbon consumed assuming equal response factors.

Other compounds that you could use in place of 2,3-dimethylbutane include 2,4-dimethyl pentane and 1-chlorobutane (see Reeves, *J. Chem. Educ.*, 1971, *48*, 636). These may be used without solvent if their amounts are doubled. Small amounts of solvents could also be added (carbon tetrachloride, hexachloroethane, chlorobenzene).

E X P E R I M E N T   1 3 B

Photohalogenation

This experiment demonstrates the differences in reactivity of different kinds of hydrogens and may be done in groups while other experiments, such as 12 or 13A, are being done.

Introduce the reagents. In five test tubes (18 × 150 mm), place 4 mL of carbon tetrachloride, 1 mL of 1M Br_2 or I_2 in carbon tetrachloride, and 2 mL of a different hydrocarbon in each tube. The hydrocarbons to be used are cumene (isopropylbenzene), toluene (methyl benzene), *t*-butyl benzene, cyclohexane, and methylcyclohexane. Try to add the hydrocarbons all at the same time.

Initiate reaction. Place the tubes in a beaker at room temperature where they can be illuminated *equally* by bright sun or a lamp. Provide each tube with a stirring rod and stir regularly. Record the times when the color of the halogen disappears. After a few minutes, when one or possibly more have decolorized, add a little water to the *beaker* and warm the water bath. Stop when four have reacted.

Interpret the results. For each hydrocarbon used, specify the types of hydrogens present. That is, 1°, 2°, 3°, aromatic, 1° α to an aromatic ring, etc. (e.g., cumene has 1°, 2° α to an aromatic ring, and aromatic hydrogens). From your observations, deduce the order of reactivity. Report in tabular form the types of hydrogens, in decreasing order of hydrogen reactivity; the compounds; and the times of reaction. Assume that the most reactive hydrogen in each compound is responsible for the decolorization. Include the logic used to deduce the order of reactivity. Compare with the synthesis in *J. Chem. Educ.*, 1971, *48*, 476. Dispose of the reaction mixtures as directed by your instructor.

Summary

1. Set up tubes for each of the five reactants.

2. Photolyse and record the time of decoloration.

3. Heat and photolyse the slowest ones.

4. Analyze your results.

Questions

1. What is the purpose of the light?

2. Write the steps in the reaction for one of these compounds.

3. Are the results consistent with prediction?

4. Are the results consistent with Experiment 13A?

Disposal of Materials

Reaction/product mixtures (HO, I)   Hydrocarbons (OW)

Carbon tetrachloride and solutions
   with it (HO, HW, I)

Additions and Alternatives

You could replace carbon tetrachloride with 1-octanol, glycerol, or methylene chloride (be careful heating).

4.3   Sublimation

Sublimation is a method of purification that may be used with some solids, sometimes at atmospheric pressure and often at reduced pressure. Consider Figure 4.9: if a solid is heated at temperature L, it will have a vapor pressure S and will not melt. If a cold surface is available, then the gas produced from the solid's heating will condense. If S-S is atmospheric pressure or the pressure of the system, the solid should sublime readily. Purification occurs as the solid volatilizes from the impurities and recondenses (solid to gas to solid). A variety of assemblies can be used for this process. A number of these are illustrated in Figure 4.10. See also Figure 1.10: use a 5-mL vial for the sublimator.

A few practical suggestions will assist this process. If reduced pressure is to be used, establish it before the coolant is introduced into the condenser. This will help to prevent the condensation of moisture on the

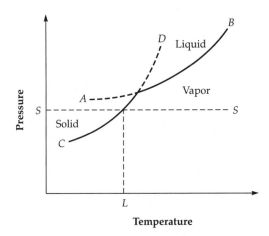

Figure 4.9 Curves showing the equilibrium between a pure solid and its pure liquid phase. (S is the sublimation pressure, L is the sublimation temperature.)

cold surface. The heating should be as gentle as possible, consistent with a reasonable rate of sublimation. The heat must be controlled so that the solid does not melt. When the apparatus is disassembled, care must be taken to ensure that none of the sublimate is shaken loose or shaken back into the original mixture. Sublimation is recommended for the purification of caffeine (Experiment 9A) and iodoform (Experiment 23). See also *J. Chem. Educ.*, 1990, *67*, 162.

4.4   Drying and Drying Agents

4.4A   Intensive Drying Agents

It is necessary to remove traces of water from nearly all liquid organic preparations before they are subjected to their final distillation. There is no universal drying agent for all compounds. The most effective desiccants are those that react chemically with water. Among these are some metals (SODIUM OR LITHIUM—DO NOT USE WITH COMPOUNDS CONTAINING HALOGENS OR ACTIVE HYDROGENS) and certain oxides that combine with water to form acids or bases. It is evident that no product can be dried by an agent that will react with it. Thus, organic acids cannot be dried by sodium or by calcium oxide, nor can amines or alcohols have contaminating moisture removed by sodium or phosphorus pentoxide. The really effective desiccants are limited in their usefulness.

CAUTION

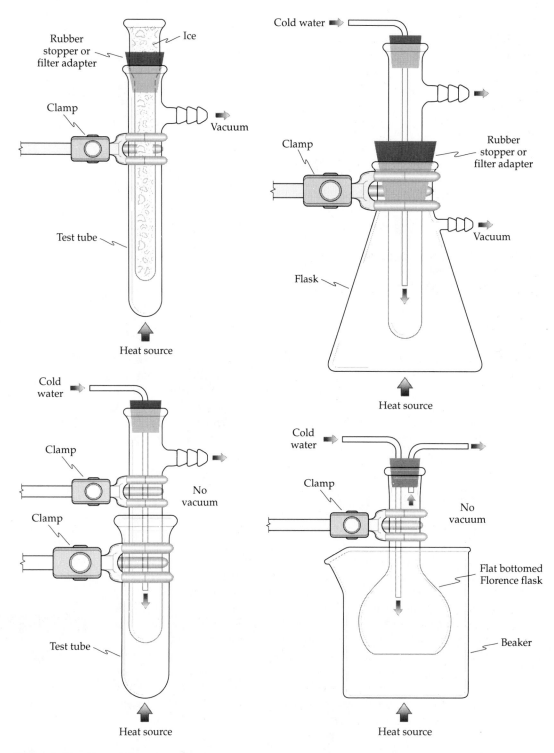

Figure 4.10 Assemblies for sublimation.

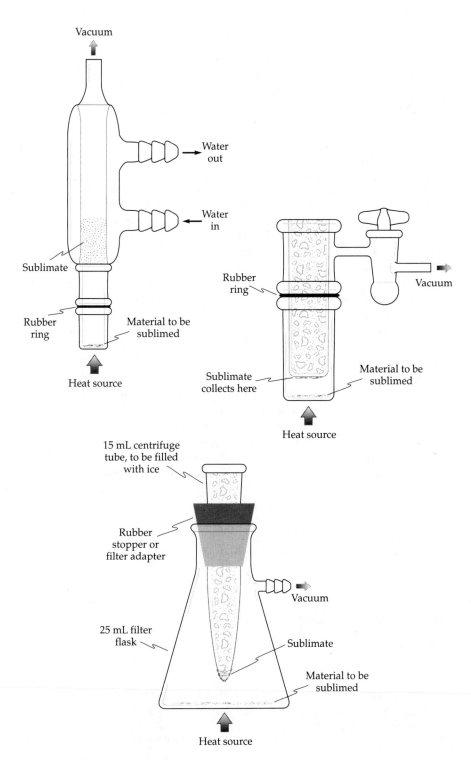

Figure 4.10 (*continued*)

4.4B   Molecular Sieves

Molecular sieves of a 4A pore size will adsorb water preferentially. Do not use them to dry methanol. They are reusable if they are washed with acetone and then with water to remove all traces of organic material and if they are stored in a drying oven (ca. 110°C).

4.4C   Hydrate-Forming Salts

The most generally useful drying agents are the anhydrous salts, which can bind water as "water of crystallization" to form hydrates of definite composition. These agents can reduce the water content of an organic liquid only to the point where the partial pressure of water vapor in the vapor phase is equal to the vapor pressure of the water in the hydrate–anhydrous salt system. This vapor pressure varies considerably from salt to salt. Some salts form a series of hydrates exhibiting higher vapor pressures when the extent of their hydration is greater. It is important in such cases to use enough of the salt so that the moisture present is insufficient to produce any of the higher hydrates. Alternatively, a saturated sodium chloride solution and/or a less efficient drying agent can remove enough water so these salts can then be used. The vapor pressure of calcium chloride dihydrate in contact with anhydrous salt is very low. Consequently, anhydrous calcium chloride should be an effective desiccant. It is, but the high solubility of this salt in water is a decided disadvantage in drying substances very soluble in water. This agent loses its effectiveness, too, upon frequent exposure to the air, even for short periods. This makes it unsatisfactory when many students are withdrawing small quantities from a common stock because the surface of the solid becomes glazed and unreceptive to moisture. Calcium chloride cannot be used to dry the lower alcohols because it forms "alcoholates" with them in the same way that it forms hydrates with water.

Basic salts such as anhydrous potassium carbonate are sometimes employed. They should not be used with acids but may be useful with basic substances.

4.4D   Anhydrous Sulfates

Anhydrous sodium sulfate and anhydrous magnesium sulfate are both satisfactory drying agents. The vapor pressures of their hydrate systems are not as low as that of calcium chloride, so they cannot be quite as effective as the latter. However, they leave little residual moisture in the product when properly used. Of the two, magnesium sulfate has certain advantages. Its hydrate is more stable than that of sodium sulfate, it removes water faster, and the anhydrous material clumps somewhat as it becomes hydrated. This last property makes it easier to remove this drying agent by filtration or to decant from it. Others have found that sodium sulfate forms clumps that are more easily separated by decantation. Some commercial products such as Drierite (with an indicator) and sodium sulfate beads are also useful.

4.4E Preparation

The two most commonly used sulfates are readily rendered anhydrous. If commercially made anhydrous salts are not available, the following procedure can be used for the preparation of either of the anhydrous sulfates. An evaporating dish is filled to two-thirds of its capacity with magnesium sulfate hydrate or sodium sulfate hydrate. The dish is placed over a low flame until a half-inch region around the edge of the material has become dehydrated and encrusted. A spatula is used to loosen the cake, and it is turned over in the dish in a single piece if possible. Heating is continued until, upon removal of the burner, a watch glass held over the mass no longer collects moisture. The cake is broken into pieces while still hot and stored in a dry, well-stoppered bottle or test tube.

4.4F Limitations

Drying agents of this kind, which act by the formation of hydrates, cannot be used to dry hot liquids. The hydrates give up their water at relatively low temperatures. The temperature at which the vapor pressure of magnesium sulfate heptahydrate in contact with its next lower hydrate becomes equal to one atmosphere is about 70°C; the corresponding temperature for sodium sulfate pentahydrate is about 32°C, and for the dihydrate considerably higher. For this reason it is necessary to remove such drying agents before the liquid is subjected to distillation.

Unless specifically directed to do otherwise, you will use anhydrous magnesium sulfate or sodium sulfate to dry your liquid products. Keep this desiccant tightly stoppered at all times except when actually withdrawing a small amount for use.

4.4G Drying Liquids

(SM) On the semi-micro scale, liquids are usually dried by adding them to a small amount of drying agent in a small flask (or by adding small amounts of drying agent to the liquid). A bit more drying agent is added to see if the water has been removed (if not, it will clump). When dry, the liquid is decanted through a filter paper into the next vessel. If the liquid is relatively dry, it may be poured through drying agent in a funnel.

(M) On the micro scale different approaches may be used. One would be a scaled-down version of that described above except that the liquid would be transferred away from the drying agent with a Pasteur or plugged pipet. A filter pipet or drying pipet might be set up (Section 1.3E) with a few grains of drying agent. The liquid to be dried is added at the top of the pipet and allowed to pass through or is pushed through using a bulb. A further modification would be to set up a micro column (Section 4.2A), with a column or a Pasteur pipet, where the packing in the column is a drying agent. This works well in extraction because each portion can be put on the column and worked through while other operations continue. Pouring a bit of an extracting solvent down the column first will improve its function; however, that requires evaporation of a bit more solvent (Section 1.3G).

4.4H   Drying Solids

The most convenient way to dry solids is to let them stand loosely covered for several days (on a watch glass or clay plate). If the solid can sublime, storage in a warm place could mean a loss. The clay plate can be used to accelerate the process (Section 2.2D). Crystals may be placed in a closed, side-arm test tube attached through a trap to a water aspirator or may be suspended in an open vial in a closed filter flask similarly connected to remove vapors.

Solids may be dried in an oven if the solid is reasonably dry, contains no organic solvents, melts 30° or more above the temperature of the oven, and does not sublime.

CHAPTER 5

Other Methods of Identification

Spectroscopy, the study of the interaction of electromagnetic energy with chemical substances, may be used to provide the scientist with information about the structure of a chemical substance. In this chapter, the interaction of infrared energy and of nuclear magnetic resonance energy with chemicals will be studied so that you can use the information obtained to determine what functional groups are present or absent in a molecule, how the atoms in the molecule are arranged relative to each other, and what the structure of the molecule is.

5.1 Infrared Spectroscopy

5.1A Background

Radiation, in the form of waves emitted from the sun, has many different energies. This radiation, or the electromagnetic spectrum as it is called, can be divided into several different regions depending upon the energy, wavelength, and frequency of the radiation.

Wavelength, λ, may be defined as the distance between any two similar points on a wave, i.e., from crest to crest or trough to trough.

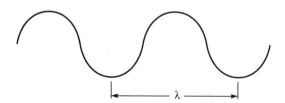

Frequency, ϑ, is the number of waves that pass a point within a certain period of time; for example, if a person counts the crests of ten ocean waves passing by within one minute, $\vartheta = 10$ waves/1 minute.

The three terms—energy, E, wavelength, λ, and frequency, ϑ—may be related to each other by the following equations:

$$\lambda\vartheta = c \tag{1}$$

where c = 2.998 $\times$ 10^{10} cm/sec, the velocity of light in a vacuum, and

$$E = h\vartheta \tag{2}$$

where h = 6.6256 $\times$ 10^{-27} erg · sec, Planck's constant. Solving equation (1) for ϑ gives:

$$\vartheta = c/\lambda \tag{3}$$

Substituting the result, equation (3), into equation (2) produces:

$$E = hc/\lambda \tag{4}$$

From the mathematical relationship in equation (2), it can be seen that when the value for E is large, the value of ϑ will be large; that is, when the energy of the radiation is large, the frequency of the radiation is large. Correspondingly, from equation (4) it can be seen that when E is large, the value of λ will be small; that is, when the energy of the radiation is large, the wavelength of the radiation will be small.

Electromagnetic Spectrum

$\longleftarrow$ Increasing frequency (ϑ)   Increasing wavelength (λ) $\longrightarrow$

Cosmic ray X ray Ultraviolet Visible Infrared Microwave Radio wave

$\longleftarrow$ Increasing energy (E)

5.1B   Recording Parameters

The infrared region of the electromagnetic spectrum consists of radiation having wavelengths of 2.5 μ to 15 μ (1 μ = 1 micron = 1 μm = 1 micrometer = 10^{-6} meters). Wavelength and frequency may be related by equation (3) above.

Infrared (IR) spectroscopy is the study of the interaction of infrared radiation with chemical substances. Infrared spectroscopy also uses the term *wave number*, $\bar{\vartheta}$, where $\bar{\vartheta}$ is defined by equation (5).

$$\bar{\vartheta} = 1/\lambda \tag{5}$$

The terms ϑ and $\bar{\vartheta}$ differ only by the constant c, the velocity of light. Therefore frequency, ϑ, and wave number, $\bar{\vartheta}$, are directly proportional. Wave numbers, $\bar{\vartheta}$, are reported in **reciprocal centimeters, cm^{-1}**

$$\textbf{1 meter (m)} = \textbf{10}^{2} \textbf{ centimeters (cm)} = \textbf{10}^{6} \textbf{ microns } (\boldsymbol{\mu}) \qquad (6)$$

Substituting in equation (5) and converting from microns to centimeters gives (1 cm = 10,000 μ):

$$\textbf{cm}^{-1} = \textbf{10,000}/\boldsymbol{\mu} \qquad (7)$$

Converting 2.5 μ and 15 μ into wave numbers gives 4000 cm^{-1} and 666 cm^{-1}, respectively, as the vibrational frequency units for the infrared range of radiation.

Visible light	Infrared radiation	Microwaves
2.5μ 4000 cm⁻¹	15μ 666 cm⁻¹	

When infrared radiation of successively varying wavelengths is directed at an organic compound, there will be absorption of those wavelengths that will bring about an increase in the vibrational–rotational motion of the molecule. This absorption can be recorded as transmittance, T, or absorbance, A, of the incident radiation. **Transmittance** may be defined as the ratio of the intensity of light that passes through a sample to the intensity of the incident light upon the sample.

$$T = \frac{I_{transmitted}}{I_{incident}} = \frac{I_T}{I_I} \qquad (8)$$

Percent transmittance is the value of T multiplied by one hundred.

$$\%T = T(100) \qquad (9)$$

Usually the infrared spectrum is recorded in percent T at all frequencies. **Absorbance** is defined as the logarithm, to the base ten, of the reciprocal of transmittance.

$$A = \log_{10} 1/T = \log_{10} I_I/I_T \qquad (10)$$

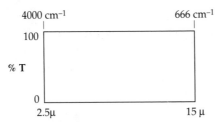

Parameters of chart paper for an infrared spectrum

5.1C Theory

A molecule may be considered to be masses or atoms joined together by springs or bonds. The molecules in a compound are in constant vibrational and rotational motion. When the energy of the incident radiation equals the transition energy, absorption of radiation occurs from one rotational or vibrational frequency to another.

A mathematical equation, **Hooke's law**, shows the relationship between the frequency of the vibration, the masses of atoms attached to a bond, and the bond strength.

$$\bar{\vartheta} = \frac{1}{2\pi c} \sqrt{\frac{k}{\frac{m_1 m_2}{m_1 + m_2}}} \tag{11}$$

where $\bar{\vartheta}$ = the vibrational "frequency" in cm^{-1}
 c = the velocity of light in cm/sec
 k = the force constant of the bond in dynes/cm
 m_1 = the mass of atom 1 in grams
 m_2 = the mass of atom 2 in grams

As an example, let us consider the triatomic molecule, H_2O, which can undergo symmetrical and unsymmetrical stretching as well as bending vibrations.

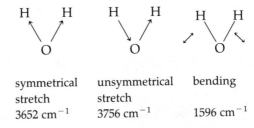

symmetrical   unsymmetrical   bending
stretch   stretch
3652 cm^{-1} 3756 cm^{-1} 1596 cm^{-1}

The vibrational frequency transitions of the O—H bond are produced by the absorption of energy at the "frequencies" shown above to produce absorption bands in the spectrum at these vibrational frequencies.

Since π and c in equation (11) are constants, the relationship may be stated as the following proportionality:

$$\bar{\vartheta} = \frac{1}{2\pi c}\sqrt{\frac{k}{\frac{m_1 m_2}{m_1 + m_2}}} = \frac{k'}{2\pi c}\sqrt{\frac{m_1 + m_2}{m_1 m_2}} = K\sqrt{\frac{m_1 + m_2}{m_1 m_2}} \quad (12)$$

From equation (12) it can be seen that atoms of large mass are related to smaller absorption "frequencies," whereas those with small mass show larger absorption "frequencies." For example, the C—H bond absorption is about 3000 cm^{-1} and the C—C bond absorption is about 1200 cm^{-1}. Similarly, with O—H and O—D stretching, where the k should be the same, the O—D absorption "frequency" is much lower than that for O—H. In addition, the larger the value of k, the strength of the bond or set of bonds, the larger will be the absorption "frequency." For example, the carbon-carbon single bond absorbs at lower "frequency" than the carbon-carbon double bond (the stronger bond) and than the carbon-carbon triple bond (the strongest bond).

C—C   C=C   C≡X

1200 cm^{-1} 1660 cm^{-1} 2200 cm^{-1}

In order for absorption of radiation to be observed in the infrared spectrum, there must be a change in the dipole moment of the bond during absorption of that radiant energy. The more polar a bond is, e.g., C—O, the greater is the intensity of the absorption band produced. The less polar the bond, e.g., C—C, the weaker the absorption band will be. In the case of a symmetrical bond, for example the C—C bond in $(CH_3)_3C—C(CH_3)_3$, there will be no change in the dipole moment during stretching, and the bond will be "infrared inactive."

5.1D Spectral Identification

Although the bond in a certain functional group absorbs at a certain vibrational "frequency," $\bar{\vartheta}$, the environment of the molecule in which the functional group is contained influences the specific vibrational frequencies observed. Consequently, the absorption frequency of a functional group in different molecules will generally fall within a range of values.

No two compounds produce the same infrared spectrum. The region of absorption of different compounds between 4000 and 1250 cm^{-1} may appear to be similar because of absorptions that are characteristic of the same functional groups. However, the region between 1250 and 666 cm^{-1}, the "fingerprint" region, is unique for every different compound because this region of absorption is characteristic of the molecule as a whole. Therefore, infrared spectra may be used to identify compounds by matching the spectrum of an unknown compound with that of a known compound, or the infrared spectrum may give information about the functional groups present in a molecule.

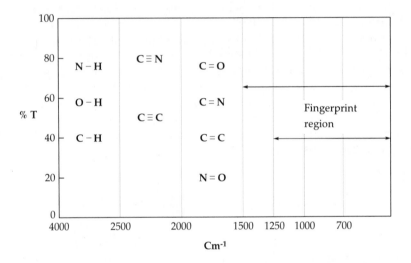

In the analysis of infrared spectra, the primary focus is on those regions of absorption that are characteristic of particular functional groups. Many of the most common of these regions are shown in Table 5.1. Tables of this type are very useful in spectral analysis. If a functional group is present in a compound, then there should be absorption in or near the regions characteristic of this group. The absence of an absorption in the appropriate region is very strong evidence for the absence of that functional group. However, there can be more than one reason for an absorption in a particular region. The molecular environment has considerable effect on the "frequency" of the absorption. Note the five C—H stretch regions shown in Table 5.1.

With this table in mind, examine the sample spectra in Figure 5.1. Look in the spectra of *o*-toluidine and diphenyl carbinol (diphenylmethanol, $C_6H_5CHOHC_6H_5$) for evidence of aromatic character in four regions. Two regions (1660–2000 and 660–920 cm^{-1}) suggest the type of substitution.

Table 5.1   Characteristic Infrared Absorptions

Type	Region (cm^{-1})	Intensity
Alkanes, Alkyl groups		
(C—H, stretch)	2850–2960	medium to strong
Alkenes		
(C—H, stretch)	3020–3100	medium
(C=C, stretch)	1650–1670	medium
Alkynes		
(C—H, stretch)	3300	strong
(C≡C, stretch)	2100–2260	medium
Aromatics		
(C—H, stretch)	3030	medium
(multiple C=C, stretch)	1660–2000	weak (multiple)
(ring breathing)	1500 and 1600	strong (2–4)
(C—H, bend)	660–920	medium to strong (1 or more)
Alkyl halides		
(C—Cl, stretch)	600–800	strong
(C—Br, stretch)	500–600	strong
(C—I, stretch)	500	strong
Carbonyl compounds		
(C=O, stretch)	1670–1780	strong
Carboxylic acids		
(O—H, stretch)	1500–3100	strong, very broad
Aldehydes		
(C—H, stretch)	2700–2775 and 2820–2900	medium to weak
Alcohols		
(O—H, stretch)	3400–3640	strong
(O—H, stretch, hydrogen bonding)	(3200) 3300–3640	strong, broad
(C—O, stretch)	1030–1150	strong
Amines		
(N—H, stretch) (1° and 2°)	3310–3500	medium to strong
(C—N, stretch)	1030 and 1230	medium
Nitriles		
(C≡N, stretch)	2210–2260	medium
Nitro		
(N=O, stretch)	1540	strong

Source: From McMurry, J. (1992). *Organic Chemistry, Third Edition*, p. 431, 433. Pacific Grove, CA: Brooks/Cole Publishing Company.

Note also the O—H and N—H stretch absorption above 3300 cm^{-1}. Find the suggestion of two types of C—H in *o*-toluidine. Compare 1-hexanol to these two. Note the absence of evidence for aromatic character in 1-hexanol and the presence of evidence for hydrogen-bonded OH.

5.1E Instrumentation

The double beam infrared spectrometer is commonly used to record spectra. This instrument may be divided into five parts: the source, sample area, photometer, monochromator, and detector. The instrument contains a source of infrared radiation that will irradiate the chemical sample and the reference. The photometer combines the beam passing through the sample and the beam passing through the reference into one beam made up of short, alternatingly equal segments of reference beam, sample beam, reference beam, etc., by means of mirrors and a rotating half mirror. The monochromator resolves or separates the beam into narrow ranges of absorption frequencies. The resolution is accomplished by slits, which are adjustable narrow openings through which the beam passes, and one or more gratings that separate the beam into various absorption frequencies. The detector "sees" the alternating segments of sample and reference beams and detects any difference in intensity between the two as a signal. The signal moves an attenuator in and out of the reference beam to keep the sample and reference beams of equal intensity. The attenuator is connected to the pen that records the infrared spectrum on the chart paper.

Some laboratories are now equipped with FT-IR (Fourier Transform). It is much faster than standard IR instrumentation and records multiple spectra to present one resultant spectra. The computer in the instrument that accomplishes the transformation often has standard spectra for comparison.

5.1F Sample Preparation

Infrared spectroscopy is a very sensitive method used for studying organic compounds. Samples as small as 1 mg in the hands of a beginning practitioner can give excellent spectra. However, just as the method can detect very small amounts of sample, it can also detect very small amounts of impurities, causing the resulting absorption bands to possibly be regarded as absorption bands in the sample's spectrum. It is important, therefore, that the sample be as pure as possible before a spectroscopic study is performed.

The materials used to prepare and hold samples in the path of the infrared radiation should be transparent to infrared radiation in the region

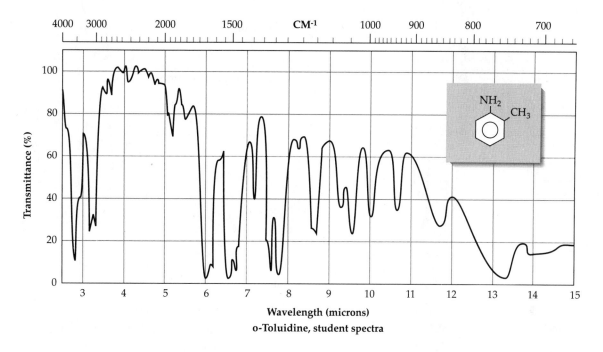

o-Toluidine, student spectra

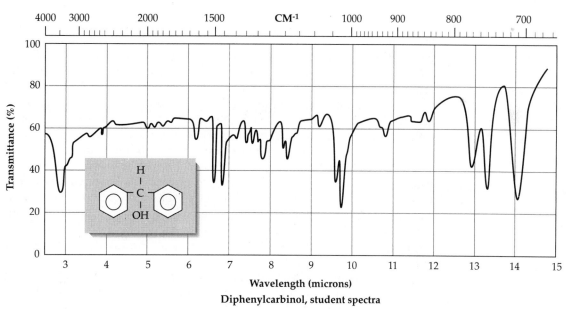

Diphenylcarbinol, student spectra

Figure 5.1 Sample infrared spectra.

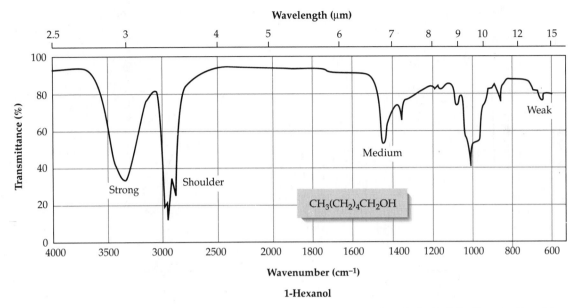

1-Hexanol

Figure 5.1 (*continued*) Source: From Fessenden, R. J., and Fessenden, J. S. (1990). *Organic Chemistry*, Fourth Edition, p. 328. Pacific Grove, CA: Brooks/Cole Publishing Company.

of interest; that is, they should not absorb infrared rays in the region where the sample is to be studied. If this is the case, then all the absorption bands observed should be attributable to absorption by the sample. This ideal is not always possible, as will be seen in the discussion below of the different methods of sample preparation. Glass generally absorbs in this region. Ionic substances, such as sodium chloride, potassium bromide, and silver chloride do not absorb the infrared radiation and can be used to hold the sample in front of the infrared beam. One major disadvantage to the use of sodium chloride, which is grown and cleaved from crystals into plates, and to the use of potassium bromide, which is frequently used as a powder, is their water solubility. All samples studied *must* be free from water, i.e., DRY. Silver chloride is not water soluble but is more expensive than sodium chloride or potassium bromide. It darkens with exposure to light, is softer and more fragile, and is attacked by some amines.

Gaseous, liquid, and solid samples may be studied using infrared spectroscopy. Gases may be studied in specially constructed gas cells, but since this method is generally not available to beginning students, no further information will be provided here. Liquid samples may be studied as liquid films between salt plates, solutions between salt plates, and solutions in fixed path length cells. Solids may be studied as potassium bromide pellets, melts, solutions in fixed path length cells, solutions between salt plates, and Nujol mulls. Each of these methods is briefly described below.

5.1G Liquid films between salt plates

The simplest method used to study a liquid sample is as a film between two salt plates. One to two drops of sample are placed on a salt plate. A second salt plate is gently placed on the first. (Hold the plates by the edges only, because moisture on your hands may cloud the surface of the plate.) The top plate is gently slid over the bottom plate to produce a thin, bubbleless film of liquid covering the entire surface of the plate that receives infrared radiation. The plates are placed in a metal holder, which is tightened with screws (Figure 5.2). The screws should not be tightened too much because the plates break easily. On the other hand, the screws should not be too loose because the plates may fall or the film of liquid may be too thick to register its strongest absorption bands on the scale. If the viscosity of the sample is too low, the film between the salt plates may be too thin to provide strong enough absorption peaks in the spectrum. (Sometimes a small aluminum foil wedge in the top of the space between the plates may help to hold more sample.) If the sample is quite volatile, it may evaporate from the heat of the infrared radiation, leaving nothing between the salt plates to register a spectrum. Air is used in the reference beam.

5.1H Solutions between salt plates

A solution of either a liquid or a solid solute dissolved in a liquid solvent can be placed between salt plates in the same fashion as described above. The weight ratio of solute to solvent should be about one to ten, respectively. Carbon tetrachloride is probably the most common solvent, but it has disadvantages: Carbon tetrachloride is strongly suspected to be a carcinogen, and absorption bands below approximately 1350 cm^{-1} may be blocked by its strong carbon-chlorine bond absorption. In this case, the reference beam passes through air.

Low-melting solids may be melted and a drop of the melt allowed to harden on one plate before assembly.

5.1I Silver chloride cells

The use of silver chloride cells differs because of the threaded plastic housing and because of the milling of the plates. Each plate has a flat side

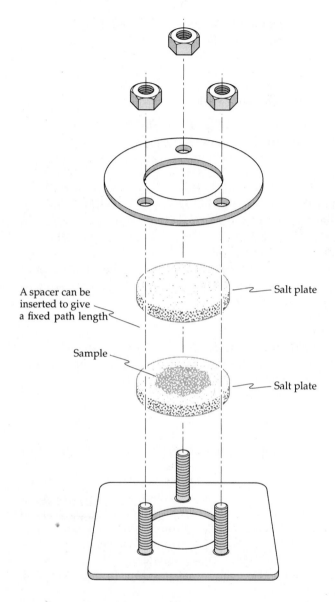

A spacer can be inserted to give a fixed path length

Sample

Salt plate

Salt plate

Figure 5.2 Salt plate assembly.

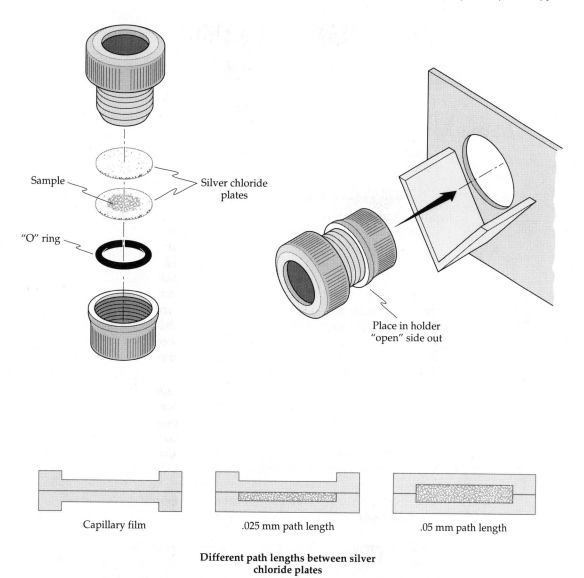

Different path lengths between silver
chloride plates

Figure 5.3 Silver chloride cell assembly.

and a side with a shallow cavity (.025 mm deep). Two such plates can be assembled in three ways to give two different cavity thicknesses or a capillary film (Figure 5.3). The plates are assembled with the sample in one half of the plastic housing. With a rubber O ring (washer) in place, the two parts of the housing are screwed together (Figure 5.3). They should be tight enough to prevent leakage but not so tight as to distort the plates.

Low-melting solids may be used with these cells in the manner described above. The temperature of the melt cannot be as high as with salt plates.

5.1J   Fixed path length cells

In using fixed path length cells, the neat liquid sample may be placed in the sample cell, and the spectrum may be taken using air in the reference beam path. Alternatively, the sample (solid or liquid) may be dissolved in solvent, and the spectrum may be taken using the solvent in the reference beam path. Two cells of the same path length are used, one for the sample plus solvent, the other for the solvent.

An example of this type of cell is shown in Figure 5.4. Two salt plates with a Teflon spacer of a specific thickness (0.1–1.0 mm, usually) between them produces the fixed path length. The salt plates are then held between metal plates. There is a small inlet and an outlet to the space between the plates for introducing and removing the sample using a Luer lock syringe without a needle. After the sample is introduced, the openings are fitted with Teflon plugs to seal the cell. Initially the two cells, filled only with solvent, should be run against each other, one in the sample and one in the reference beam. A relatively straight line with little to no absorption indicates that the cells are matched; that is, they have the same path length. Then the sample in the solvent, usually a 5–10% solution, is placed in the sample cell, and the spectrum is run. Common solvents used include carbon tetrachloride, chloroform, and carbon disulfide. Carbon tetrachloride is relatively free of absorption above 1350 cm^{-1}, whereas carbon disulfide has little absorption below 1350 cm^{-1}. Running a sample first in one solvent then in the other will allow study of the entire 4000–666 cm^{-1} range. Both carbon tetrachloride and carbon disulfide are nonpolar solvents. Not all substances dissolve in nonpolar solvents; sometimes it is necessary to dissolve the sample in a polar solvent such as chloroform. With chloroform, the regions between 1250–1200 cm^{-1} and 800–650 cm^{-1} cannot be studied because of the strong chloroform absorption in these regions. *All three solvents should be treated with care because they are all toxic.* Carbon disulfide is highly *flammable*.

5.1K   Potassium bromide pellets

A solid solution of the sample in potassium bromide powder is prepared by grinding and mixing a 1–2 mg sample of the solid to be analyzed in approximately 100 mg of potassium bromide. The potassium bromide

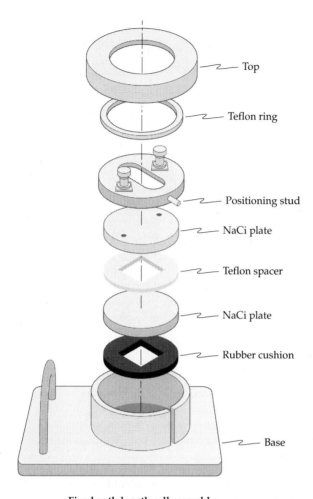

Top

Teflon ring

Positioning stud

NaCi plate

Teflon spacer

NaCi plate

Rubber cushion

Base

Fixed path length cell assembly

Figure 5.4 Fixed path length cell.

should be nitrate-free, purified, and carefully dried. It should be ground before the sample is added. If convenient, make twice the solid solution needed for one run so the half not used can be easily modified if you are dissatisfied with the first run. Prepare the solution by thoroughly grinding and mixing (three to five minutes) the KBr and the sample in an agate or glass mortar and pestle. A small watch glass or spot plate and rounded glass stirring rod may be substituted. The mortar and pestle should not have any cracks because these could harbor impurities that could be

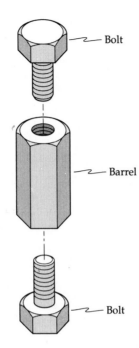

Bolt

Barrel

Bolt

Figure 5.5 KBr pellet maker.

added to the solution upon mixing. The potassium bromide should be dry and should have been stored in a desiccator. The sample mixture is then carefully transferred to a die (micro (Figure 5.5) or normal sized) that is subjected to high pressure (with wrenches or a press, respectively). (See Jordan, *J. Chem. Educ.*, 1977, *54*, 287.) The normal-sized die is partially evacuated while pressure is being applied in the press. The pressure is released from the die carefully and slowly in order to not crack the small pellet that has been formed. The pellet should be transparent. If it is cloudy, moisture may have been present, the preliminary mixing and grinding of the sample and the potassium bromide may have been insufficient, the ratio of sample to potassium bromide in the solid solution may have been too high, or possibly, the pressure applied may have been insufficient and/or may not have been applied long enough. The pellet, either in the micro die or in a sample holder, is placed in the path of the infrared radiation. In addition to the air in the reference beam, it may be necessary, especially in the case of the micro disc, to use a beam attenuator in the reference position. The beam attenuator may be positioned to reduce the reference beam's intensity so that the spectrum can be recorded on scale. After the spectrum is completed, the pellet holder and all parts that touched the sample and the potassium bromide should be washed, first with water, then with acetone, and the solvent allowed to evaporate.

5.1L   Nujol mull

A Nujol mull is prepared by grinding approximately 5 mg of solid to a very fine powder using a mortar and pestle. A few drops (1 or 2) of Nujol, a brand of white mineral oil, is added and mixed well into the solid sample. A small amount of the material is placed between two salt plates that are tightened to form a thin film of mull. Nujol, being a hydrocarbon, has absorptions in the CH, CH_2, and CH_3 regions of the spectrum, and consequently no sample absorptions may be studied in these regions (2800–3100 and about 1000–1600).

5.1M   Calibration

Calibration of the spectrum is necessary to be certain that the absorption bands recorded are actually at the frequencies shown. The sample and reference cells are removed from the sample area, and a card holding a thin film of polystyrene is inserted into the sample holder. The chart drum and the mechanically connected "frequency drive" are moved backward (Do not use force!) and the tips of certain polystyrene peaks, e.g., 3026, 3003, 2924, 1603, 1495, 906 cm^{-1}, are recorded on the sample's spectrum. This is done in order to compare the frequency of well-known peaks of polystyrene with the position in which they are recorded on the chart paper.

E X P E R I M E N T   1 4 A

An Infrared Spectrum

Analyze the unknown. Obtain an unknown with its molecular formula from your instructor. Calculate the degree of unsaturation* in that compound, and use that information to suggest what functional groups might be present.

Initiate infrared spectroscopy. Record a spectrum of the compound as your instructor directs.

Interpret the spectrum. Look at the spectrum to determine which functional groups are present and which are absent. Compare that information with spectral information on the possible functional groups for your unknown. Look through the spectrum for further information. Try to consider all possible structural formulas that are validated as possibilities by the data. Make a case or noncase, showing bands present or absent, for each possible structural formula. Negative evidence is particularly strong. For example, if there is no major absorption between 1600 and 1900 cm^{-1}, there is no carbonyl group in the compound. Try to select one formula as the one that best fits the data, but if it is not possible to differentiate among possible structural formulas, explain why not.

Degree of Unsaturation. This is a measure of the hydrogen deficiency of your compound compared with an acyclic, saturated hydrocarbon. The latter has the generic formula, C_nH_{2n+2}. The first step is to determine the number of hydrogens in the saturated compound by multiplying the number of carbons by 2 and adding 2. Call this answer **S**. Determine the number of hydrogens in the molecular formula of your compound and add one for every halogen. Call this answer **C**. The degree of unsaturation (**D**) is then $D = (S - C)/2$.

This value may be useful in the following way. Each double bond or ring represents one degree of unsaturation; a triple bond, two; and a six-carbon, aromatic ring, four. Thus, if you find $D = 2$, your compound might have a triple bond, two double bonds, or a ring and a double bond. The double bonds are *not* restricted to $C=C$. If your compound contains nitrogen, you must be careful in your interpretation. In calculating C, an $-NO_2$ group adds one, but for an $-NH_2$ group, you would subtract one.

Questions

1. Examine Spectra 1 and 2 in Appendix A. They are said to be the same compound made with a different catalyst (Experiment 16). Do the spectra bear this out? Account for any differences.

2. Repeat Question 1, but for Spectra 3 and 4.

3. Do the spectra (1 and 2 versus 3 and 4) show that they are different compounds? Give specific instances of difference. How do these spectra differ from the spectrum expected for the saturated compounds?

4. Compare Spectra 5 and 6. They are said to be the same compound made by different methods (Experiment 47). Do the spectra bear this out? Account for any differences. Are any due to the difference in methods?

5. What absorbances, present or absent, lead you to believe these spectra (5 and 6) are correctly identified? You can consider the method of preparation.

Additions and Alternatives

You might want to run the infrared spectrum of the product of one of the experiments in this text to verify its structure or the nature of impurities. This practice, however, should be routine.

Another alternative would be to use an infrared spectrum, an nmr spectrum, and molecular formula or mass spectra that have been taken and given to you to do this analysis.

5.2   Nuclear Magnetic Resonance

Radiation in the radio wave portion of the electromagnetic spectrum has low energy, low frequency, and long wavelengths. A study of the interaction of this radiation with the nuclei of certain atoms when placed in a strong magnetic field is called **nuclear magnetic resonance (nmr) spectroscopy**. Various nuclei, e.g., ^1H, ^{19}F, ^{13}C, and ^{31}P, among others, are capable of being studied in this way. In this section, only the hydrogen nucleus that contains one proton will be discussed, although the basic principles are the same for other nuclei able to be studied by nmr spectroscopy. The hydrogen nucleus has a uniform, spherical density. However, the nucleus spins about an axis, and a small nuclear magnetic moment, ε, is generated.

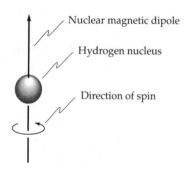

If the nucleus is subjected to an external magnetic field, H_0, the nuclear magnetic dipole of the nucleus traces a circular path or orbit, i.e., precesses about the axis of the external magnetic field. The magnetic dipole of the nucleus can align itself with, in the same direction as but not parallel to, the external magnetic field, H_0 (this is the lower of two possible energy states, E_1), or against, in the opposite direction from but not parallel to, the external magnetic field, H_0, (this is the higher of the two possible energy states, E_2).

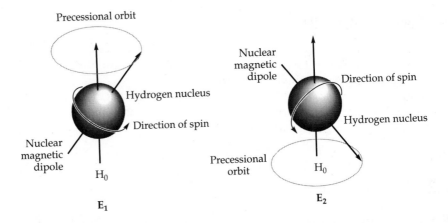

Since only a small amount of energy, on the order of 0.01 to 1.0 calorie, is necessary to enable the proton to "flip" from the lower energy state, E_1, to the higher energy state, E_2, hydrogen nuclei are present in both energy states with only a statistically slight excess of nuclei present in the lower energy state at room temperature. The absorption of radio wave radiation by this slight excess of nuclei in order to "flip" from the lower to the higher energy state is recorded as the nmr spectrum.

The energy needed to "flip" the nucleus from E_1 to E_2 is defined by equation (2).

$$E = h\vartheta \tag{2}$$

However, frequency, ϑ, may be related to the external applied magnetic field strength by equation (13).

$$\vartheta = \frac{\gamma_H H_0}{2\pi} \tag{13}$$

where H_0 = magnetic field strength
γ_H = magnetogyric constant for the hydrogen, H, nucleus
Substituting equation (13) into equation (2) gives

$$E = \frac{h\gamma_H H_0}{2\pi} \tag{14}$$

It can be seen in equation (14) that the amount of energy necessary to "flip" the nucleus is directly proportional to the value of H_0. The larger the value of H_0, the larger the value of E; that is, the stronger the magnetic field, the greater the difference in energy between E_1 and E_2.

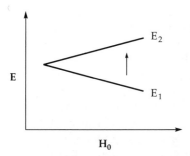

The energy necessary for the nucleus to "flip" from E_1 to E_2 is provided by an oscillating magnetic field, H_1. The field is produced by an oscillating coil aligned with its axis, perpendicular to the axis of the first magnetic field, H_0. One of the components of the oscillating magnetic field rotates in the same direction as the precessional orbit of the nucleus. When the angular velocity of the precessing nucleus and the angular velocity of the rotational component of H_1 are equal, the nucleus "flips" from E_1 to E_2, or in other words, **resonance** occurs.

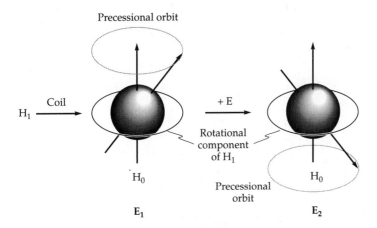

The energy supplied by the oscillator coil for absorption by the sample is related to the frequency of the radiation by equation (2).

$$E = h\vartheta \qquad\qquad (2)$$

As was shown and illustrated previously, the higher the external applied magnetic field strength, H_0, the greater the energy difference between E_1 and E_2. The greater the energy difference between the two states, the greater must be the amount of energy or the frequency supplied by the oscillating coil. Therefore, the higher the external applied magnetic field strength, H_0, the greater must be the frequency of the oscillating coil, ϑ, for resonance to occur.

Resonance of the nucleus may be induced two ways. The angular velocity of the rotational orbit may be kept constant by maintaining a constant oscillator frequency and the external magnetic field, H_0, varied until the angular velocity of the precessing proton equals the angular velocity of the rotational orbit. Alternatively, the external magnetic field, H_0, may be held constant and the oscillator frequency varied until the angular velocity of the rotational orbit equals the angular velocity of the precessing proton. The second approach has been the more common of the two due to instrumental considerations. When H_0 is held at 14,092 gauss, (G), and the oscillator frequency equals approximately 60 Mega-Hertz (MHz), resonance occurs. One Hertz (Hz) equals 1 cycle per second (cps). One MHz $= 10^6$ cps. The most common instruments are 60 MHz, but others may be encountered.

An nmr spectrum is a plot of absorption of radiation or peak intensity versus changing frequency of the oscillator coil. The conventional arrangement is shown here.

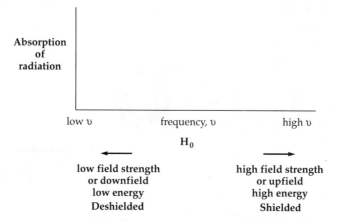

The left side of the spectrum corresponds to a low frequency, "downfield" or low field strength, and low energy. The right side of the spectrum corresponds to a high frequency, "upfield" or high field strength, and high energy.

In actuality the range usually recorded from low frequency to high frequency in the nmr spectrum is very small relative to the frequency of the oscillator coil. The range is generally 600 Hz compared with 60×10^6

Hz or 60 MHz for the oscillating coil.

$$\frac{600 \text{ Hz}}{60 \times 10^6 \text{ Hz}} \times 10^6 = 10 \text{ parts per million}$$

This amounts to a recorded range equivalent to a variation of 10 parts in one million parts, the range of the oscillator coil frequency. The nmr spectrum shows the 0 ppm to 8 ppm range,

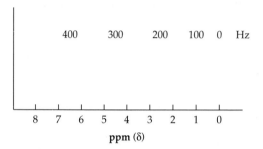

and may easily be extended below 8 ppm. Each ppm is given the unit δ (delta) (1 ppm = 1 δ).

The magnetic field strength experienced by the hydrogen nucleus varies very slightly from one kind of hydrogen nucleus to another. In other words, not all hydrogen atoms are equivalent; that is, they do not all absorb at the same frequency. For example, in isopropyl chloride, CH_3—$CHCl$—CH_3, there are two different types of hydrogen atoms—the six methyl hydrogens, —CH_3, which are equivalent and the methine hydrogen, —CH, which is different. In an nmr spectrum of isopropyl chloride, two absorption signals will be recorded, one for the methyl hydrogens and one for the methine hydrogens. In benzene all the hydrogens are equivalent. As a result, in the nmr spectrum of benzene, only one absorption signal is present.

The magnetic field experienced by the proton varies slightly depending upon the electron density around the nucleus being observed. The greater the electron density about the nucleus, the more the nucleus is protected, or **shielded**, from the external magnetic field and the stronger the magnetic field must be to bring about resonance. The signal is recorded "upfield." The less the electron density about the nucleus, the more **deshielded** the nucleus is and the weaker the magnetic field needed to induce resonance. The signal is recorded "downfield." The displacement of the signal from the reference signal of TMS, measured in delta, is known as the **chemical shift**.

$$H_{\text{nucleus}} + H_{\text{shielding}} = H_{\text{external}}$$

Electronegative elements nearby in the compound lower the electron density of the hydrogen atoms. For example, as mentioned earlier, isopropyl chloride exhibits two absorption signals, one for the methyl hydrogens and one for the methine hydrogen. Isopropyl bromide exhibits the same number of absorption signals, but they are "upfield" relative to those of isopropyl chloride because chlorine is a more electronegative element than is bromine. The chlorine removes more electron density from the hydrogen atoms than does bromine, and absorption consequently occurs at a lower frequency in isopropyl chloride than in isopropyl bromide.

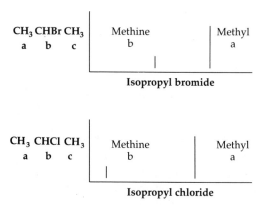

The signal for the methine hydrogen in isopropyl chloride is "downfield" relative to the signal for the methine hydrogen in isopropyl bromide. The signal for the methyl hydrogens in the chloride is "downfield" relative to the methyl hydrogen signal in the bromide. The downfield shift in the chloride relative to that in the bromide is greater for the methine hydrogen than for the methyl hydrogens because the electronegative chlorine is closer to the methine hydrogen than to the methyl hydrogens and consequently exerts a larger deshielding effect on the nearer neighbor. A brief table of some shift values is included in Table 5.2. For a more extensive table, see *J. Chem. Educ.*, 1964, *41*, 38.

Tetramethylsilane (TMS), $(CH_3)_4Si$, has a high electron density about its twelve equivalent hydrogen atoms. In fact, most hydrogens in organic compounds have lower electron densities than do those in TMS. TMS is used as a reference compound and is positioned on the "upfield" side of the spectrum at 0 ppm δ.

In the above discussion of the relative locations of signals for methyl and methine hydrogens in both isopropyl chloride and isopropyl bromide, no mention was made of what the signals looked like. One might initially assume that a single absorption is observed for each type of hydrogen atom, but this is not the case. In fact, the methyl absorption signal consists of two peaks while the methine absorption signal consists of seven peaks.

Table 5.2 Characteristic Chemical Shift Values

Type of Proton	Formula	Chemical shift (δ)
Reference peak	$(CH_3)_4Si$	0
Saturated primary	$-CH_3$	0.8–1.5
Saturated secondary	$-CH_2-$	1.0–1.8
Saturated tertiary	$\diagdown$C$-$H $\diagup$	1.4–1.7
Allylic primary	$\diagdown$C$=$C$-$CH$_3$ $\diagup$ $\vert$	1.6–1.9
Alkynyl	$-$C$\equiv$C$-$H	2.2–3.0
Methyl ketones	$\overset{\overset{\displaystyle O}{\|}}{-C}-CH_3$	2.1–2.4
Aromatic methyl	Ar$-$CH$_3$	2.5–2.7
Alkyl iodide	I$\diagdown$C$-$H $\diagup$	2.0–4.2
Alkyl bromide	Br$\diagdown$C$-$H $\diagup$	2.5–4.0
Alkyl chloride	Cl$\diagdown$C$-$H $\diagup$	3.0–4.0
Alkyl fluoride	F$\diagdown$C$-$H $\diagup$	4.2–4.8
Alcohol, ether	$-$O$\diagdown$C$-$H $\diagup$	3.3–4.0
Vinylic	$\diagdown$C$=$C$-$H $\diagup$ $\vert$	5.0–6.5
Aromatic	Ar$-$H	6.3–8.5
Aldehyde	$\overset{\overset{\displaystyle O}{\|}}{-C}-H$	9.7–10.0
Carboxylic acid	$\overset{\overset{\displaystyle O}{\|}}{-C}-O-H$	10.0–13.0
Alcohol	$\diagdown$C$-$O$-$H $\diagup$	Variable 1.5–5.0
Amine (1° and 2°)	$\diagdown$C$-$N$-$H $\diagup$	Variable 1.0–5.0

Source: From McMurry, J. (1992). *Organic Chemistry*, *Third Edition*, p. 466. Pacific Grove, CA: Brooks/Cole Publishing Company.

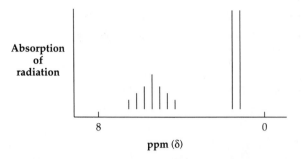

In order to explain the multiplicity of peaks for each of the different hydrogen atoms, **spin–spin coupling** or **splitting of peaks** must be considered. The multiplicity of the signal for equivalent hydrogen atoms absorbing radiation is determined by the neighboring hydrogen atoms. To illustrate this statement, consider a sample system containing both primary hydrogens, the methyl hydrogens, $—C—CH_3$, and a tertiary hydrogen, the methine hydrogen $C(3°)—H$. We can use the **n + 1 rule** to determine the multiplicity of peaks. The multiplicity of the absorption for the methyl hydrogens is determined by the hydrogen atom on the adjacent carbon atom. In the n + 1 rule, n equals the number of equivalent hydrogen atoms on the adjacent carbon atom. There is one methine hydrogen on the adjacent carbon atom. Therefore n = 1, 1 + 1 = 2, and the methyl hydrogens are split into two peaks of approximately equal intensity, a doublet. Correspondingly, there are three equivalent methyl hydrogens adjacent to the methine hydrogen, n = 3, 3 + 1 = 4, and the methine hydrogen is split into four peaks, a quartet, with the two outer peaks of approximately equal intensity and the two inner peaks of approximately equal intensity.

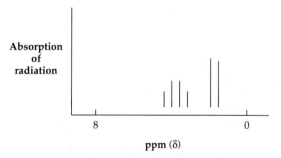

The methine hydrogen may be oriented either with or against the external applied magnetic field, H_0. When the methine hydrogen atom is aligned with the field, the external field that the methyl hydrogens require for resonance is slightly decreased relative to where the signal would be for the uncoupled protons because the methine hydrogen contributes to the field. On the other hand, when the methine hydrogen is aligned against the field, the external field that the methyl hydrogens require for resonance is slightly increased relative to where the signal would be for the uncoupled protons because the methine hydrogen field decreases the field experienced by the methyl protons

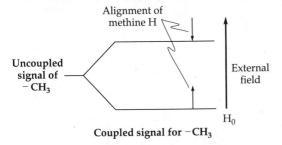

Coupled signal for $-CH_3$

There is close to an equal probability of finding the methine hydrogen in either alignment, and therefore the two peaks are of essentially equal intensity. The three hydrogens in the methyl group influence the multiplicity of the methine hydrogen. All three hydrogens may align with the field, all three against, or two may align with and one against the field, or vice versa.

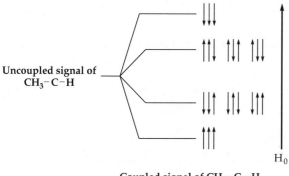

Coupled signal of $CH_3 - C - H$

There are eight equally probable alignments of the methyl hydrogens with two sets of three possible alignments being equivalent. The signal consists of a quartet where the two inner peaks are of approximately three times greater intensity or absorption than the two outer peaks. In practice, peaks nearest the chemical shift of the coupled protons will be enhanced. This comparison of two peaks within one signal that in theory should be equal will reveal a "slant" toward the coupled signal.

The n + 1 rule may be used to consider a few more simple examples. For ethyl bromide, CH_3-CH_2-Br, the spectrum consists of a triplet for the methyl hydrogens (n = the number of neighboring equivalent hydrogens = 2 on the methylene group, $-CH_2-$) and a quartet for the methylene hydrogens (n = 3 on the methyl group, $-CH_3$). Compare with Spectrum A, Figure 5.6. For $CH_3-CHBr-CH_3$, the two methyl groups are equivalent and are represented by a doublet (n = 1 on the methine group), while the methine hydrogen is represented by a septuplet (7

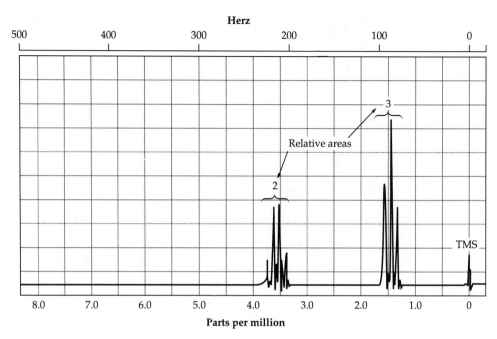

(A)
¹H nmr spectrum of chloroethane, CH₃CH₂Cl.

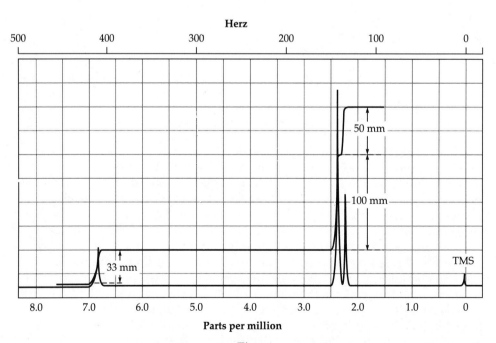

(B)
¹H nmr spectrum of 1-bromo-2,4,6-triemethylbenzene

Figure 5.6 Sample nmr spectra. Source: From Fessenden, R. J., and Fessenden, J. S. (1990). *Organic Chemistry*, Fourth Edition, p. 346, 351, 352, 355, 359, 360. Pacific Grove, CA: Brooks/Cole Publishing Company.

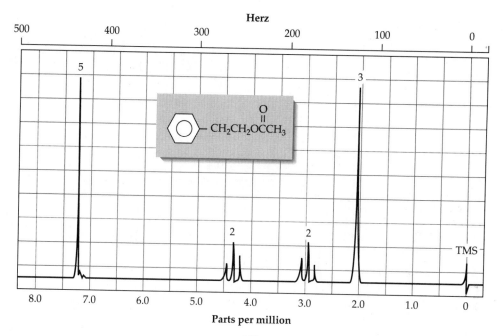

(C)
¹H nmr spectrum of 2-phenylethyl acetate.

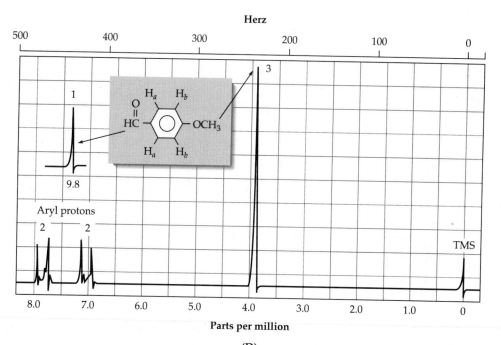

(D)
¹H nmr spectrum of *p*-methoxybenzaldehyde. (Note the offset signal for the aldehyde proton)

Figure 5.6 *(continued)*

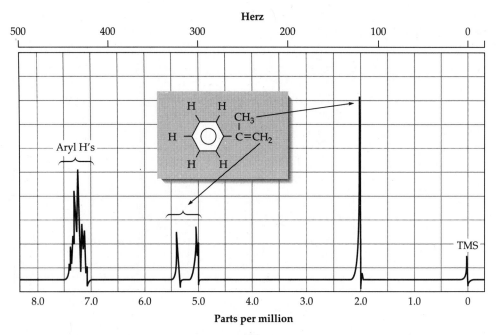

(E)
¹H nmr spectrum of 2-phenylpropene.

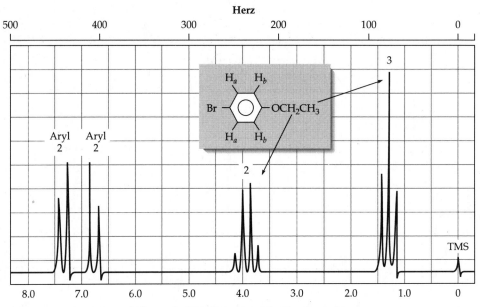

1-Bromo-4-ethoxybenzene of 4-ethoxybromobenzene

Figure 5.6 *(continued)*

peaks) (n = 6 for the methyl hydrogens on the two adjacent equivalent methyl groups). Generally, coupling does not occur between hydrogen atoms that are separated by more than two carbon atoms unless there are multiple bonds or rigid ring systems between the interacting hydrogen atoms.

The **coupling constant, J,** is a measure of the effectiveness of interaction of the spin–spin coupling between adjacent hydrogen atoms. The coupling constant may be determined by measuring the distance between peaks in a multiplet and is the same value in two multiplets that are coupled to each other. For example, in the spectrum of isopropyl bromide

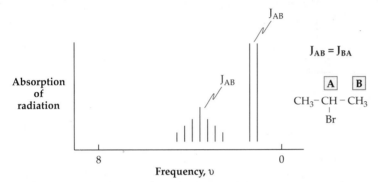

the distance between the peaks of the septuplet for the methine hydrogen must equal the distance between the peaks of the doublet for the methyl hydrogens because the hydrogen atoms of the two groups are coupled to each other. The coupling constant usually has a value of about 7 Hz but may have values ranging from about 0 to 18 Hz.

This very simple analysis of spectra is valid if two conditions are fulfilled. First, all protons in one group, e.g., $—CH_2—$ in ethyl iodide, $CH_3—CH_2—I$, must couple equally (i.e., J must be the same) with all protons in the other group (e.g., $—CH_3$ in $CH_3—CH_2—I$ and vice versa). Second, the coupling constant, J, must be small relative to the chemical shift difference between the two groups. When the value of J is large relative to the difference in chemical shift between two different types of hydrogen atoms, the multiplet becomes distorted, with the inner peaks becoming larger and the outer peaks becoming smaller than predicted. At such times it may be difficult to correctly identify the multiplet being observed.

One further aide in the analysis of spectra is the **integration area.** The area under an nmr multiplet is directly proportional to the number of hydrogen atoms represented by that multiplet. Integration areas are indicated by a stepped line over the multiplet. Measuring the distance vertically from the beginning of the multiplet to the end of the multiplet gives the area. For example, in the spectrum of isopropyl bromide, the area of the septuplet is about 2, whereas that of the doublet is about 12. Dividing both numbers by the smaller number, 2, gives a ratio of 1 to 6, the smallest whole number ratio. If the molecular formula is known, in this case C_3H_7Br for isopropyl bromide, it can be seen that the septuplet represents one hydrogen atom in a structure that has six equivalent neighboring hydrogen atoms, while the doublet represents six equivalent hydrogen atoms having one neighboring hydrogen atom. See Spectra B, Figure 5.6.

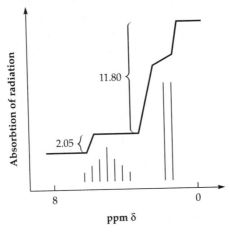

From the integration area, the splitting pattern, the equal spacing between peaks of the multiplets coupled to each other, the number of signals, and the shift values, it is sometimes possible to determine the structural formula of a compound from its molecular formula.

EXPERIMENT 14 B

Identification of an Unknown

Place the sample along with a few drops of tetramethylsilane, the internal standard, in an nmr tube, and insert it between the pole forces of a magnet of homogeneous field strength and inside the transmitter and receiver coils. The radio frequency transmitter produces the frequency, which is held constant while you vary the magnetic field strength, H_0, precisely and continuously over a small range of values, using the sweep generator. Rotate the sample freely to increase the homogeneity of the magnetic field in the sample tube.

Procedure

Choose a solvent. An nmr sample can be studied either as a pure liquid or as a solution. Some common solvents include carbon tetrachloride, CCl_4

(the solvent of choice if the sample is soluble in it since CCl_4 has no interfering proton absorption), and deuteriochloroform, $CDCl_3$, which is frequently used when the sample is not soluble in CCl_4. Since commercially available deuterated chloroform has a trace of undeuterated solvent, there will be a small peak present at 7.27 ppm due to the solvent. Deuterium oxide (heavy water) is often used when neither of the above solvents is appropriate.

Introduce the sample. Place approximately 0.5 mL of pure liquid or a solution of it (10–30% sample by weight or volume) along with 5–10 microliters of TMS (which must be refrigerated because of its high volatility) in an nmr tube approximately 5 mm in diameter and 15–20 cm in length. Carefully position the capped nmr tube to the appropriate depth within the magnet in the spectrometer. The tube is spun so that the sample can "feel" as uniform or homogeneous a magnetic field as possible.

Take the nuclear magnetic resonance spectrum. Record a spectrum of the compound as your instructor directs. Sharp, well-resolved peaks should result in the spectrum if this operation is completed correctly. If broad peaks occur, your sample may be too viscous and may need to be diluted; solid particles could be present and would need to be removed by filtration through a Pasteur pipet containing a small amount of glass wool; or ferromagnetic impurities may be present and can be removed by centrifugation, filtration, or by dipping a thin bar magnet into the nmr tube.

If only a very small sample (1–2 mg) is available, it is possible to obtain a spectrum by using a microcell nmr tube. This tube will require only 0.025 mL of liquid because the coils around the nmr tube "see" only a thin cross-sectional area of the sample. However, you must take greater than usual care that the vortex created in the sample by its spinning does not reach down into the area "seen" by the coils. Otherwise poor spectra will result.

Attend to other procedures. Because of limited resources, time on the instrument must be used wisely and care must be taken of the instrument.

Prepare samples before you come to the instrument. Do not work with chemicals or magnetizible material around the instrument. Spills should be avoided and any that do occur should be cleaned up immediately in an appropriate manner. Reserve your time on the log book ahead of time and sign out. Clean nmr tubes back in the lab. Use an nmr tube cleaner if available and make sure you leave the tubes clean and dry.

Unknown

Analyze the unknown. Obtain an unknown with its molecular formula from your instructor. Calculate the degree of unsaturation (see Experiment

14A) in the compound and use that information to determine which possible functional groups might be present.

Take its nuclear magnetic resonance spectrum. Record a spectrum of the compound as your instructor directs.

Interpret the spectrum. Look at the spectrum to determine the chemical shifts exhibited by the hydrogen atoms. Try to correlate this information with any functional groups that may be present in the molecule by considering the unknown's molecular formula. Examine the multiplicity of the peaks, the integration areas, and the coupling constants, along with the chemical shift values. Try to consider all structural formulas that are validated as possibilities by the data. Make a case or noncase, indicating peaks present or absent, for each possible structural formula. Try to select one formula as the one that best fits the data, but if it is not possible to differentiate among possible structural formulas, explain why not.

Questions

1. Consider Spectrum B (Figure 5.6). Find the simplest ratio of small whole numbers that fits the integration. Draw the structure of the compound. Does your ratio fit? Assign each signal (how many are there?) to a type of hydrogen. Explain your assignment.

2. Consider Spectrum C (Figure 5.6). Assign each of the two triplets to particular hydrogens. Explain your assignment.

3. Consider Spectra C, D, and E (Figure 5.6). Explain the differences in the signals for the aryl protons. Assign each signal for aryl protons in Spectra D. Explain. How do you know it is two doublets and not a quartet?

4. Sketch the nmr spectra of *p*-ethoxytoluene and *p*-methoxyethylbenzene in as much detail as possible. Comment on the similarities and differences. How could the nmr spectra of these isomers be used to distinguish between them?

Additions and Alternatives

You might want to run the nmr spectrum of the product of one of the experiments to verify its structure. To learn more about AB spin systems, see the discussion in *J. Chem. Educ.*, 1991, *68*, 83. See also *J. Chem. Educ.*, 1990, *67*, 438 or 513.

See the other alternatives given in Experiment 14A.

CHAPTER 6

Alcohols: Substitution and Elimination

6.1 Dehydration of Alcohols

6.1A Olefin Formation

All alcohols in which hydrogen is attached to the carbon atom adjacent to the carbon of the alcohol group can be dehydrated to yield water and the corresponding olefin. Although this reaction can be brought about by heat alone, the reaction proceeds more smoothly and with fewer by-products if dehydrating agents are used. The higher temperatures necessary for this reaction in the absence of a dehydrating agent cause extensive dehydrogenation to occur along with the dehydration.

6.1B Complicating Reactions

The required temperature and the yield (which is influenced by the extent of by-product formation) depend on the nature of the alcohol and the dehydrating agent used. By-products may result from side reactions involving the original alcohol or from those that consume a portion of the olefin. The most common side reaction of the first type is the oxidation of the alcohol to an aldehyde (or ketone), carboxylic acid, or free carbon when sulfuric acid is used as a dehydrating agent. Because sulfuric acid is an active oxidizing agent only when hot and concentrated, this difficulty arises most frequently in the dehydration of primary alcohols where these conditions are necessary. Another reaction that involves a loss of starting material when using a primary alcohol is ether formation. This process, induced by concentrated sulfuric acid, is rapid at temperatures around 130–150°C, whereas olefin formation does not become predominant much below 160°C for primary alcohols. Thus, it is evident that the composition of the product is often a matter of relative reaction velocities.

Product loss can result from an acid-catalyzed polymerization of the olefin. This can occur in almost any instance in which sulfuric acid is used. Easily polymerized branched-chain olefins are formed from easily dehydrated tertiary alcohols. The straight-chain olefins are less easily polymerized.

Experiments 15, 16, and 17 will illustrate the relative ease of dehydration for primary, secondary, and tertiary alcohols by sulfuric acid. Refer to your text for additional details. The reverse of this reaction is not illustrated, although the addition of water to an alkene proceeds rather readily in the presence of sulfuric acid. Experiment 18 illustrates the formation of alcohols from alkenes by means of hydroboration and oxidation.

6.1C   Vapor-Phase Dehydration

There are theoretical advantages to dehydrating alcohols in the vapor phase over hot alumina. The alumina (dehydrating agent) is not easily reduced, and losses by oxidation are low. Losses due to polymerization are slight for two reasons. First, alumina, being a neutral—or even slightly basic—contact agent, does not catalyze polymerization. Second, polymerization reactions are very sensitive to concentration effects. In the vapor-phase reaction at atmospheric pressure, the actual concentration of the olefin product in the heated zone is low. A low olefin concentration does not facilitate polymer formation. The olefin concentration in the liquid-phase reaction is probably at least one hundred times that in the vapor-phase reaction.

E X P E R I M E N T   1 5

Preparation of 1-Octene (SM)
References: Sections 1.3E, 3.4, 4.4G, 5.1, and 6.1B.

$$C_6H_{13}CH_2CH_2OH \xrightarrow[H_2SO_4]{-H_2O} C_6H_{13}CH=CH_2$$

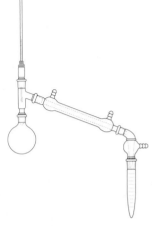

Figure 3.8 (page 87).

Assemble the apparatus and introduce the reagents. Place six milliliters of 1-octanol in a 25-mL flask, and add 1.5 mL of concentrated sulfuric acid, a few drops at a time, shaking vigorously after each addition. Add a boiling stone, and assemble the glassware for distillation as shown in Figure 3.8, with a graduated receiver or graduated centrifuge tube to collect the distillate. If a graduated receiver is not available, tare a small receiver. To **tare** a vessel, weigh it without the material that will be added later and record that weight.

Initiate the reaction. Cautiously begin heating. A rather vigorous and frothy reaction should ensue. During this process, control the heating so that no vapors or froth pass into the condenser. After about five minutes, when the froth subsides, increase the rate of heating so that slow distillation occurs. Distillation will cease abruptly when the thermometer reading is a little above 125°C. At this point, remove the source of heat.

Isolate the product. Shake the distillate with two pellets of sodium hydroxide or potassium hydroxide (if there are two distinct layers, the water should be separated first) to remove water and sulfur dioxide.

1. How were the water and sulfur dioxide formed?

Q u e s t i o n

Filter it through dry cotton in a drying pipet (without drying agent) into a 10-mL flask. Clean the distilling head, condenser, etc., by running a few drops of acetone through them and allowing complete evaporation of the acetone.

Reassemble the distillation unit used above (Unit III, Figure 1.7 or Figure 3.8) using the 10-mL flask, and start the distillation. Use a graduated receiver or a centrifuge tube to collect the fraction condensing between 120–125°C; this fraction is the product. If no graduated receiver is available, use a small tared flask as the receiver. In distillation, it is important that no vapor be permitted to pass to the condenser until the thermometer has come into equilibrium with the vapor.

2. Why must the bulb of the thermometer be bathed in vapor?

Q u e s t i o n

Record the volume and/or weight of product collected.

Interpret the results. Test these samples for unsaturation as directed in Exercises 1 and 2, Experiment 74. Determine the actual yield and the percentage yield. Report other pertinent information as directed. (The density of 1-octene is 0.716 g/mL; that of 1-octanol alcohol, 0.827. The refractive index for the alkene is 1.4088 and that for the alcohol is 1.4304 (20°C).) Record the infrared spectrum. Deliver the product to your instructor for credit.

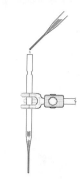

Summary

1. Mix the reactants and assemble an apparatus for distillation.

2. Cautiously distill out the product.

3. Separate it from the water and neutralize the distillate.

4. Dry and redistill the distillate.

5. Analyze the product.

Figure 1.6A and Figure 4.5 (pages 19 and 148).

Questions 1 and 2 above.

Q u e s t i o n s

3. What is the function of the H_2SO_4?

4. How does the boiling stone function?

5. What are the most probable by-products in this reaction?

6. Give the equations for the tests performed in Experiment 74.

Disposal of Materials

1-Octanol (OW, OS)	conc. H_2SO_4 (D, *Caution*: add acid to a large amount of water)
1-Octene (OW, I)	Boiling stone (Wash, S)
NaOH, KOH (D, N)	Drying agent (S)
Acetone (OS, OW)	conc. H_3PO_4 (D, see H_2SO_4 above)
Original residue (D, N)	Distillation residue (OW)

Additions and Alternatives

You can use 2.0–2.5 mL of conc. H_3PO_4 in place of H_2SO_4, which will avoid many problems.
See Experiment 16.

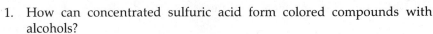

EXPERIMENT 16

Preparation of Cyclohexene (M)
References: Sections 1.3E, F; 3.4; 4.4G; 5.1; and 6.1B.

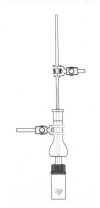

Figure 3.9A (page 91).

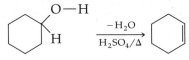

Preheat a sand bath or aluminum block to produce an internal temperature of 90–100°C (Section 1.3F).

Introduce the reagents. Place 2.5 mL of cyclohexanol and a stirring bar or spin vane in a 5-mL conical vial or equivalent vessel. With the stirring bar rotating slowly, add 0.4 mL of concentrated sulfuric acid. At this point you may notice a yellow to dark orange ring or strip around the stirring bar.

Question

1. How can concentrated sulfuric acid form colored compounds with alcohols?

One milliliter of concentrated phosphoric acid may be used in place of the sulfuric acid.

Assemble the apparatus and initiate heating. After the reagents have been thoroughly mixed, equip the vial with a Hickman still (or equivalent distillation apparatus, Figure 3.9A or 3.9B or Unit IIIA or B, Figure 1.8B), and place it in a preheated sand bath or aluminum block. Keep the sand bath or aluminum block hot enough to cause gentle distillation. Let the mixture distill until nothing more comes off below 90–95°C.

Figure 3.9B (page 95).

Isolate the product. With a Pasteur pipet or syringe, remove the liquid as it collects in the reservoir of the Hickman still. Transfer the liquid to a vial or a centrifuge tube reserved for the initial distillation. If you are not using a Hickman still, collect the distillate and go directly to the next step. Separate the layers if there are two; save the organic layer.

Q u e s t i o n s

2.   Which is it? How can you test for the water layer?

3.   Explain the formation of a cloudy mixture.

Place the organic mixture in a conical vial or capped centrifuge tube, and add some anhydrous NaOH (solid, ca. 150 mg). Place the cap on the vial; shake it, and let it stand for a few minutes. The organic layer should be clear. With a pipet, transfer the crude cyclohexene to a drying pipet containing 20–40 mg of anhydrous magnesium sulfate (Section 4.4G). Collect the dry product in a clean, dry small vial with a clean, dry spin vane.

Redistill the product. Attach the vial to a clean, dry Hickman still or short path head (Figure 3.9A or 3.9B or Unit IIIA or B, Figure 1.8B). Perform the distillation with care. If necessary, remove the product as it distills. It should distill between 80° and 90°C with the sand bath, or with the aluminum block well above that. When the distillation temperature falls, even with increased heating, or the vessel is dry, stop heating. Transfer the product to or collect it in a tared vial and weigh. (A tared vessel is one that has been preweighed.)

Figure 1.6A and Figure 4.5 (pages 19 and 148).

Interpret the results. Perform the analyses as directed, paying particular attention to the infrared spectrum. If directed to do so, carry out Exercises 1, 2, and 5 from Experiment 74 on half scale, using two drops each of product and cyclohexanol, to compare these two compounds.

4.   Compare the results of these tests.

Q u e s t i o n

Include your product, properly labeled, with your report. If gas liquid chromatographic analysis is to be done, try a silicone column at 65–70°C. The density of cyclohexene is 0.81 g/mL.

Summary

1.   Preheat the heating media.

2.   Mix the reactants and assemble an apparatus for distillation.

3.   Reflux and cautiously distill out the product.

4.   Separate the product from any water and neutralize the distillate.

5.   Dry and redistill the distillate.

6.   Analyze the product.

Questions

QUESTIONS 1–4 ABOVE.

5. How might the cyclohexanol have been made?
6. What other compound(s) would have the same molecular formula as cyclohexanol? How could they be distinguished?
7. Write the equations for the tests in Experiment 74.
8. What is the function of the sodium hydroxide? Why is it used as a solid?

Disposal of Materials

Cyclohexanol (OS, OW) H_2SO_4 or H_3PO_4 (D, *Caution*: acid to water)

Cyclohexene (OW, I)   NaOH (N)

Drying agent (evaporate, S)   Original residue (D, N)

Distillation residue (OW)

Additions and Alternatives

You can use a short path head to eliminate some transfers. This experiment can be done as well or better with 1 mL of 85% H_3PO_4 in place of the sulfuric acid. Try either acid with 2-methylcyclohexanol, which should lead to an interesting gas liquid chromatographic analysis (silicone at 70–80°C). See *J. Chem. Educ.*, 1967, *44*, 620. Try acidic ion exchange resins in place of the acid and basic resins in place of the sodium hydroxide (See *J. Chem. Educ.*, 1967, *44*, 620). Other secondary alcohols may be used and their isomeric products analyzed by gas liquid chromatography (e.g., *J. Chem. Educ.*, 1969, *46*, 765 and 849).

In the redistillation, you may employ a chaser such as toluene or xylene.

This experiment may be done on the semi-micro scale at 3–5 times the scale (Figure 1.7, Unit III). That scale should provide you with enough material for further experiments.

E X P E R I M E N T   1 7 A

Preparation of 2-Methyl-2-Butene (SM)
References: Sections 3.4, 4.4G, and 6.1B.

$$(CH_3)_2C(OH)CH_2CH_3 \xrightarrow[H_2SO_4]{-H_2O} (CH_3)_2C=CHCH_3$$

Introduce the reagents and initiate reaction. Slowly add 4 mL of concentrated sulfuric acid to 7 mL of water in a 25-mL or 50-mL flask, and cool it to room temperature. Add a boiling stone and 10 mL of tertiary amyl alcohol to the flask. Shake the flask a little to promote mixing. Attach the flask to an apparatus for fractional distillation (Figure 3.10A or Unit IV, Figure 1.7). Immerse the receiver in *ice* water to prevent loss of the *volatile* product.

Begin a *slow* distillation. Because the dehydration reaction is reversible, the yield is materially increased by this slow distillation, which ensures the removal of the olefin as soon as, and as long as, it is formed. Monitor the temperature, and discontinue distillation when the temperature reaches 40–45°C. Add a small amount of anhydrous magnesium sulfate to the distillate. Decant the liquid and determine its volume and/or weight.

Analyze the product. Test the product for unsaturation using the directions from Exercises 1 and 2, Experiment 74. Report the actual and percent yields, and turn in the product for credit. Tertiary amyl alcohol has a density of 0.809 g/mL; 2-methyl-2-butene, of 0.663.

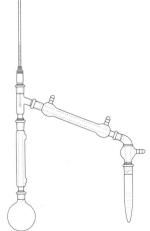

Figure 3.10A (page 96).

Summary

1. Mix the reactants and assemble an apparatus for fractional distillation.

2. Slowly distill out the product.

3. Dry the product.

4. Analyze the product.

Questions

1. How is most of the water produced in this reaction separated from the product? The rest?

2. Compare Experiments 15, 16, and 17. What differences are there in the acid used (molar ratio, concentration)? In the temperature needed for reaction?

3. What is the order of ease of dehydration of these three alcohols? Explain.

4. Why is 2-methyl-2-butene expected to be the major product? Show the E_1 mechanism.

5. Give the equations for the tests performed.

Disposal of Materials

t-Amyl alcohol (OW, OS)

2-Methyl-2-butene (OW)

Exercise 1 solution (HO, I)

Distillation residue (OW)

H_2SO_4 or H_3PO_4 (N)

Drying agent (evaporate, S)

Exercise 2 solution
  (thiosulfate solution, D, I)

Additions and Alternatives

Try 6–10 mL of concentrated H_3PO_4, with or without added water, in place of H_2SO_4. You should record the infrared spectrum; however, the volatility of the product makes the sample preparation difficult.

The gas liquid chromatographic analysis should work well. See the directions given in *J. Chem. Educ.*, 1962, *39*, 310. Note that the product must be kept in ice and that the minor product may be detected. Try the dehydration of 3,3-dimethyl-2-butanol, which also leads to an interesting gas liquid chromatographic analysis (*J. Chem. Educ.*, 1969, *46*, 849).

E X P E R I M E N T   1 7 B

Preparation of 2-methyl-2-butene (M)
References: Sections 3.4, 4.4G, 6.1B, and Experiment 17A.

$$(CH_3)_2C(OH)CH_2CH_3 \xrightarrow[H_2SO_4]{-H_2O} (CH_3)_2C{=}CHCH_3$$

Introduce the reagents and assemble the apparatus. In a 5-mL conical vial with a stir vane, place 0.7 mL of water and 0.5 mL of concentrated sulfuric acid. (Conc. H_3PO_4 (1.0–1.5 mL) without water may be used in place of H_2SO_4.) Some heat may evolve from this step, so cool the vial to room temperature or below before continuing. Carefully add 1.25 mL of tertiary amyl alcohol. Assemble the distillation apparatus with the Hickman still or the equivalent (Figure 3.9A, 3.10B, or 3.11B, Unit III). Add a water condenser and a Claisen adapter (or vented inlet adapter) above the Hickman still or use a short path head. Stir the solution to thoroughly mix it.

Initiate reaction. Place the vial in the sand bath or aluminum block, and turn the heat on low. Carefully monitor the external thermometer; distillation should be slow but steady, though the external temperature may go above 100°C. Collect about one mL of distillate or, with an internal thermometer, collect in the range of 38–44°C. Keep the *receiver* ice *cold*.

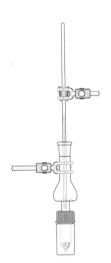

Figure 3.9A (page 91).

Monitor the vial at all times. Once the mixture starts to distill, it distills very rapidly. When the mixture stops boiling, when there is no condensation in the neck of the still, or before the vial goes dry, remove the heat.

Isolate the product. Transfer the distillate from the reservoir to a cold, dry vial with a (bent) pipet or syringe. Add to this vial, or to the receiver if you used a short path head, a small amount (20–50 mg) of anhydrous magnesium sulfate. Place a tight cap on the vial, and let it sit (cold) for several minutes. Remove the 2-methyl-2-butene from the vial with a pipet, and place it in a clean, dry, tared vial (cold). (Tared = preweighed.)

Interpret the results. Weigh the product. Do the analyses as directed. Gas liquid chromatography will be particularly interesting because, if the experiment was done with care, some 2-methyl-1-butene (b.p. = 31°C) may

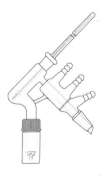

Figure 3.9B (page 95).

be detected (see Schimelpfenig, *J. Chem. Educ.*, 1962, *39*, 310). Test the product for unsaturation, using the directions from Exercises 1 and 2, Experiment 74 on half scale, if directed to do so.

Summary

1. Mix the reactants; cool and assemble a minimal apparatus for fractional distillation.

2. Slowly and carefully distill out the product.

3. Dry the product.

4. Analyze the product.

1. How is most of the water removed from the product? The rest?

2. Compare the reaction conditions (molar ratio, concentration, or temperature) for this experiment with those of Experiments 15 and 16.

3. What is the order of ease of dehydration of 1°, 2°, and 3° alcohols? Is this order observed in these experiments?

4. Why is 2-methyl-1-butene expected to be the *minor* product? Show the E_1 mechanism for the formation of the major product.

Disposal of Materials

t-Amyl alcohol (OW, OS)

2-Methyl-2-butene (OW)

Exercise 1 solution (HO, I)

Distillation residue (OW)

H_2SO_4 or H_3PO_4 (N)

Drying agent (evaporate, S)

Exercise 2 solution
  (this sulfate solution, D, I)

6.2   Alcohols from Olefins

The hydroboration–oxidation method of preparing alcohols from alkenes is of considerable interest because primary alcohols are formed in preference to secondary, whereas the usual method, involving the addition of water in the presence of sulfuric acid, produces secondary alcohols in preference to primary.

Briefly, this method involves reaction of an olefin with sodium borohydride and boron trifluoride etherate in a suitable solvent to give an organoborane, which is then oxidized by peroxide to give the alcohol.

The following experiments (18A and B), which describe two examples of the use of this method, were contributed by Nobel Laureates H. C. Brown and George Zweifel. The students at Purdue University have employed these preparations.

Preparation of Alcohols from Olefins

The preparation and handling of organoboranes require techniques and precautions similar to those used for the Grignard reaction. Although the necessity for a nitrogen atmosphere has not been definitely established, the hydroboration reactions are normally carried out under nitrogen.

CAUTION

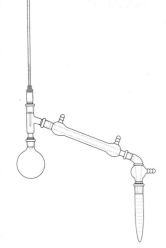

Figure 3.8 (page 87).
See Figure 3.11A (page 104).

Solvent Purification. ALL SOLVENTS MUST BE FREE OF PEROXIDES (SEE DISPOSAL OF MATERIALS). In order to avoid losses of active hydride in the hydroboration stage, it is important that the solvents be free of water and peroxides. Diglyme (dimethylether of diethylene glycol, b.p. = 162°C at 760 torr.), from Ansul Chemical Co., is purified in the following way. Store one liter of diglyme over 10 g of small pieces of calcium hydride for 12 hours. Then decant the diglyme into a distilling flask, and add sufficient lithium aluminum hydride to ensure an excess of active hydride. Distill the solvent at 62–63°C at 15 torr. Treat tetrahydrofuran (pure grade) with lithium aluminum hydride as described above; then distill it at atmospheric pressure; b.p. = 66–67°C. The excess hydride must be carefully destroyed.

Treat boron trifluoride ethyl etherate with a small amount of anhydrous ethyl ether (to ensure an excess of this component), and distill it in all glass apparatus at 46°C at 10 torr. from a small amount of granular calcium hydride. The latter serves to remove small quantities of volatile acids and greatly reduces bumping during the distillation. The excess hydride must be carefully destroyed.

Metal Hydride. Use sodium borohydride (98%) from Metal Hydrides, Inc. without purification.

General Procedure

$$RCH = CH_2 + NaBH_4 \xrightarrow{BF_3, Et_2O} (RCH_2CH_2)_4NaB$$

$$2NaOH + (RCH_2CH_2)_4NaB + 4H_2O_2 \longrightarrow 4\,RCH_2CH_2OH + Na_3BO_3 + 3H_2O$$

Assemble the apparatus. The apparatus consists of a three-neck flask equipped with a condenser and fitted with a calcium chloride tube, a pressure equalized dropping funnel, a thermometer, and a magnetic stirrer. Dry the apparatus in an oven, and assemble it under dry nitrogen. Alternatively, you can flame it dry in a stream of dry nitrogen.

Introduce the reagents and initiate reaction. To the flask, add the olefin and the hydride in an appropriate solvent. Use a 10–20 percent excess of the hydride over that calculated for the stoichiometric reaction. Add dropwise, in a nitrogen atmosphere, a quantity of boron trifluoride etherate, equivalent to that of the hydride used. After completion of addition, carefully decompose the excess hydride using water or ethylene glycol. You can oxidize the organoborane obtained in situ with alkaline hydrogen peroxide.

E X P E R I M E N T   1 8 A

Isopinocampheol (SM)

Introduce the reagents and initiate reaction. In a 50-mL flask, place 15.2 mL of a 1M solution of sodium borohydride in diglyme (15.2 mmoles excess of hydride) and 5.0 g of α-pinene (36.7 mmoles; $n^{20}_D = 1.4648$; $[\alpha]^{20}_D = +46.8°$) in 5 mL of diglyme. Immerse the flask in a water bath (set at 20°C). From the dropping funnel, add 2.6 mL of boron trifluoride etherate (20.2 mmoles; 2.9 g; $d^{20} = 1.125$ g/mL) dropwise to the stirred reaction mixture over a period of 30 minutes, keeping the temperature at 20–25°C. Permit the flask to remain for one hour at this temperature in a water bath. Decompose the excess hydride by carefully adding 5 mL of water dropwise. Oxidize the organoborane (30–50°C water bath) by adding 4 mL of 3N sodium hydroxide in one portion and then add dropwise 4 mL of 30% hydrogen peroxide (**CORROSIVE**). Permit the flask to remain at room temperature for an additional hour.

Isolate the product. Extract the reaction mixture with 50 mL of ether, and wash the ether extract five times with 50-mL portions of ice water to remove diglyme. Dry the ether extract over anhydrous magnesium sulfate. Crystallize the product, obtained after removal of the solvent, from a small amount of petroleum ether (35–37°C).

Interpret the results. Students may obtain 4.47 g of isopinocampheol (70% yield): m.p. = 55–57°C, $(\alpha)^{20}_D = -32.8°$ (in benzene).

CAUTION

Figure 4.1 (page 129).

Summary

1. Place the reagents in diglyme solution in the flask.

2. Cool to 20°C and add the etherate dropwise over 30 minutes, with stirring, at this temperature.

3. Allow the mixture to stand at 20°C for one hour.

4. Add water dropwise.

5. With the flask in a warm water bath, add base in one portion.

6. Add peroxide dropwise and allow to stand at room temperature for one hour.

7. Extract the product into ether and then extract the diglyme out of the ether solution.

8. Dry the ether solution and then evaporate the ether from the solution.

9. Recrystallize the residue from low-boiling petroleum ether.

Disposal of Materials

Excess $LiAlH_4$ solutions (add to water cautiously, empty hood, safety shield); after treatment: Diglyme (D, N, I), Tetrahydrofuran (N, OW, I), and boron trifluoride etherate (N, OW, I)

Boron trifluoride etherate (N with base, destroy peroxide, OW, I)

For the two solvents, destroy peroxides with care, before disposal, by reacting with an activated Al_2O_3 column or thiosulfate solution (OW, I). (To detect peroxide, mix equal amounts of the liquid and 10% (W:V) solution of NaI in glacial acetic acid. A brown color is positive.)

Sodium borohydride in diglyme (add water cautiously, hood, safety shield, N, D, I)

α-Pinene (OW, I)

aq. NaOH (N)

30% H_2O_2 (dilute, at least 10 water to 1 peroxide, treat with aq. $FeSO_4$ or thiosulfate solution, D)

Ethyl ether (destroy peroxides, OS, OW, I)

Petroleum ether (OW, OS, I)

Drying agent (evaporate, S)

Recrystallization residue (OW)

EXPERIMENT 18B

4-Methyl-1-pentanol (SM)

Introduce the reagents and initiate reaction. To a well-stirred suspension of 0.675 g of sodium borohydride (17.85 mmoles, pulverized) in 30 mL of tetrahydrofuran containing 5.0 g of 4-methyl-1-pentene (59.5 mmoles; $n^{20}_D = 1.3830$), add 3.0 mL of boron trifluoride etherate (23.8 mmoles) in 5 mL of tetrahydrofuran over a period of one hour, maintaining the temperature at 25°C. Keep the flask at 25°C for an additional hour. Decompose the excess hydride by adding water to the mixture. Oxidize the organoborane obtained at 30–40°C (water bath) by adding 6 mL of a 3N solution of sodium hydroxide, followed by a dropwise addition of 6 mL of 30% hydrogen peroxide. Saturate the reaction mixture with sodium chloride. Separate the tetrahydrofuran layer formed; then wash it with a saturated solution of sodium chloride. Dry the extract over anhydrous magnesium sulfate.

Isolate the product. Distill the product, obtained after removal of the tetrahydrofuran, at reduced pressure.

Interpret the results. Typically students obtained 4.86 g of 4-methyl-1-pentanol (80% yield): b.p. = 98–100°C at 23 mm Hg; n^{20}_D = 1.4295.

Summary

1. Stir the borohydride solution containing the alkene.

2. Add the etherate during a one-hour period at 25°C.

3. Add water until there is no further sign of the hydride decomposing.

4. Place the flask in a warm water bath and add base.

5. Add peroxide dropwise and then add sodium dichloride to saturate the aqueous layer.

6. Separate the organic layer and wash with saturated aqueous sodium chloride solution.

7. Dry the organic layer.

8. Distill off the tetrahydrofuran at atmospheric pressure.

9. Distill off the product at reduced pressure.

Disposal of Materials

Excess $LiAlH_4$ solutions (add to water cautiously, empty hood, safety shield); after treatment: Diglyme (D, N, I), Tetrahydrofuran (N, OW, I) and boron trifluoride etherate (N, OW, I)

Boron trifluoride etherate (N with base, destroy peroxide, OW, I)

For the two solvents, destroy peroxides with care, before disposal, by reacting with an activated Al_2O_3 column or thiosulfate solution (OW, I). (To detect peroxide, mix equal amounts of the liquid and 10% (W:V) solution of NaI in glacial acetic acid. A brown color is positive.)

See Figure 4.5 (page 148).

Sodium borohydride in diglyme (add water cautiously, hood, safety shield (N, D, I))

NaBH$_4$ (destroy very cautiously)

NaBH4 in tetrahydrofuran (cautiously destroy with water, N, OW)

aq. NaOH (N)

30% H$_2$O$_2$ (dilute, at least 10 water to 1 peroxide, treat with aq. FeSO$_4$ or a thiosulfate solution, D)

aq. NaCl (D)

MgSO$_4$ (evaporate, S)

Tetrahydrofuran (OW, see above)

Distillation residue (OW)

Questions

1. What is the "acid" added during the reaction? What is the purpose of its addition? Why is this particular one chosen? Explain.
2. What is the structural formula of diglyme? How does it compare with tetrahydrofuran?
3. Suggest the probable mechanism for the hydration of an alkene with water and sulfuric acid.
4. What influence(s) in the molecule govern(s) the direction of addition when water and sulfuric acid are used? When diborane (B$_2$H$_6$) or borane (BH$_3$, probably generated here in situ) is used? Why and how might the borane derivatives of an interior double bond rearrange?

Additions and Alternatives

Sheers and Chamberlain in their patent (U.S. 2,986,584) have suggested the preparation of 10-hydroxypinane from 2-pinene (α-pinene). They allowed externally generated B$_2$H$_6$ to react with the alkene (1 : 6) at room temperature. The product was then isomerized by refluxing for 15 minutes. The procedure was otherwise much the same as the above procedure. Consider using these two variations to give a preparation that would be interesting to compare with that of Experiment 18A. Other alternatives may be found in *J. Chem. Educ.*, 1971, *48*, 477 and 1990, *67*, 437.

You could adapt these experiments to employ the equipment usually available. A positive pressure of nitrogen may not be needed throughout the experiment. The temperature may be assumed to be the same as that of the water bath. An Erlenmeyer flask or a filtering flask could serve as the reaction vessel; the flat bottom of the flasks will facilitate the use of a magnetic stirrer. If a stirrer is not used, more or less constant shaking may be sufficient. An air condenser could be used, and additions made with a medicine dropper; or a macroscale Liebig condenser and a dropping funnel might be used. A semi-micro drying tube could be used, or wads of cotton inserted into the ends of open tubes may give some protection from atmospheric moisture. A small three-neck flask, to accommodate a thermometer, an addition funnel with a drying tube, and a condenser with a drying tube may be available. If an entry for nitrogen is desired, the Claisen adaptor and another inlet adaptor will be needed. An assembly like Unit II, Figure 1.7 could be tried.

6.3   Alkyl Halides from Alcohols

The alcohols constitute one of the most important classes of starting materials for syntheses in the aliphatic series. They are generally available and relatively inexpensive. From them a large variety of derivatives can be made by simple processes. These derivatives can, in turn, be converted into more elaborate compounds.

The alkyl halides are probably the most important and generally useful intermediates obtainable from the alcohols by direct synthesis. Among the classes of compounds into which they can be converted are organic acids, amines, hydrocarbons, and even other alcohols by means of the Grignard reagent. The conversion of alcohols into alkyl halides is most conveniently and cheaply brought about by the use of halogen acids or by reactions in which these acids are produced. The ease of their conversion depends upon the class of the alcohol and upon the particular halogen acid employed. The effectiveness of the acid diminishes in the order: HI > HBr > HCl $\gg$ HF; the last is of no value for this reaction. The susceptibility of the alcohol to reaction with these acids decreases in the order: tertiary > secondary > primary, with the secondary alcohol resembling the primary much more closely than it does the tertiary. In all of these reactions, the conversion is dependent upon the presence of an acidic substance. In fact, when aqueous solutions are used, the rate of reaction is very nearly proportional to the hydrogen-ion concentration of the acid used.

The relative ease of alkyl halide formation from the three classes of alcohols will be demonstrated in the following experiments. The use of sodium bromide and sulfuric acid (Experiments 19A and B) to produce hydrobromic acid, instead of employing constant-boiling (47%) hydrobromic acid, is simply a compromise. The presence of sulfuric acid in the reaction mixture causes side reactions, such as the production of dibutyl ether and butylene, which would be avoided by the use of the more costly halogen acid. These side reactions are minimized, though not avoided, by proper choice of the order in which the reagents are added to the reaction vessel. Such quantities of the ether, as are formed, do not greatly contaminate the final product, because butyl bromide and butyl ether form a minimum-boiling azeotrope that distills at a lower temperature than the product.

Experiments in which commercial HBr is used are included; the introduction of sulfuric acid into the reaction mixture could thus be avoided if lower yields are acceptable. Experiments 20A and B are written to use some sulfuric acid. These experiments may be done using only constant boiling HBr, but the yields are a bit lower. The hydrogen bromide method is especially recommended for the preparation of secondary butyl bromide (Experiment 20C) because, although ethers are somewhat more difficult to prepare from secondary alcohols than from primary, olefins are more easily prepared from secondary than from primary alcohols. Sulfuric acid is effective in causing this reaction, and so reduces the yield of the bromide.

Consideration of the conditions for the preparation of isoamylene (Experiment 17) will suggest that the presence of sulfuric acid is undesirable in a reaction mixture from which it is hoped that tertiary amyl chloride will be forthcoming. Fortunately, the relatively mild dehydrating action of the concentrated hydrochloric acid used is insufficient to promote olefin formation.

Alkyl halides can also be made from alcohols using PX_3 or PX_5 (where X = Cl or Br). Experiment 22 illustrates this type of reaction where it is not necessary to form the phosphorous halide prior to the reaction.

E X P E R I M E N T   1 9 A

Preparation of 1-Bromobutane (NaBr-H$_2$SO$_4$ Method) (SM)
References: Sections 1.3H; 3.4; 4.1C, 2D, 4G; and 5.1.

$$NaBr + H_2SO_4 + CH_3CH_2CH_2CH_2OH \longrightarrow$$

$$Na(HSO_4) + H_2O + CH_3CH_2CH_2CH_2Br$$

Assemble the apparatus. Set up a reflux apparatus with a trap as in Figure 6.1A. The trap should contain water to which a few sodium hydroxide pellets have been added. Adjust the water level so that the funnel just about reaches the water.

Introduce the reagents and initiate reaction. Add seven milliliters of water, 8.4 g of sodium bromide, and 5 mL of 1-butanol to a 50-mL round-bottom flask. Place this flask in an ice bath, and add 6 mL of concentrated sulfuric acid to it, 1 mL at a time with swirling and cooling. Because of the violence of this reaction, work under a hood and make sure to add the sulfuric acid *slowly*. Attach the flask to this apparatus, and reflux the mixture for 30 minutes. Allow the apparatus to cool and drain completely before continuing.

Isolate the product. Add 5 mL of water and a boiling stone to the flask, and arrange the apparatus for simple distillation (Unit III, Figure 1.7 or

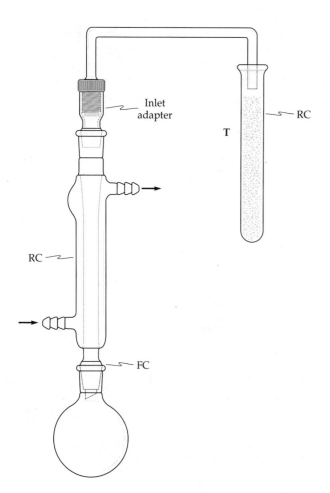

Figure 6.1A Simple reflux with fume trap.

Figure 3.8). Use a clean, dry receiver to collect the distillate. Support a beaker of ice water around the receiver during distillation. Distill the mixture as long as oily material condenses along with the water. If there is any doubt about this point, use a fresh receiver, and examine the next few drops of distillate for the presence of two layers.

Transfer the distillate to a 60-mL separatory funnel, and add an equal volume of a saturated sodium bicarbonate solution.

1. What is the function of this basic solution?

Insert a glass stopper into the neck of the separatory funnel, and quickly invert the funnel. (Hold the stopper and stopcock firmly in place!) Open the stopcock immediately to release any pressure that has built up. Then close it, and shake the funnel vigorously for a *few* seconds. Invert the

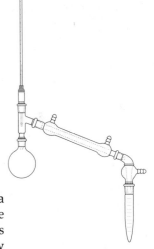

Figure 3.8 (page 87).

Q u e s t i o n

See Figure 4.1 (page 129).

funnel, and open the stopcock again before any pressure has built up to a point at which the stopper may be forced out. Repeat until the pressure buildup ceases, clamp the separatory funnel in a vertical position with the stopcock down, and allow the layers to separate.

Remove a few drops of the top layer with a Pasteur pipet, and test their solubility with an equal volume of water.

2. What is the expected result of this test?

The top layer should be the aqueous layer, which is to be discarded, but it should be tested before discarding.

3. Why would the aqueous layer be expected to be on top?

Draw off the lower layer.

Question

4. Why must the stopper be removed during this operation?

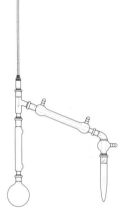

Collect the organic layer in a dry receiver, and shake it with a small amount of drying agent. Filter the crude organic layer through a small funnel, a glass wool plug, and a small amount of the drying agent into a clean, dry 25-mL round-bottom flask.

Purify the product. Add a boiling stone to the flask, and assemble a simple distillation apparatus (Figure 3.8) or a fractional distillation apparatus with an unpacked column (Figure 3.10A). Distill the mixture carefully. Collect the fraction distilling in the range of 92–98°C as the product.

Interpret the results. Determine the yield (the density of 1-bromobutane is 1.28 g/mL, of 1-butanol alcohol, 0.81). Calculate the theoretical yield. Report these values with the percent yield, and deliver the product to your instructor for credit. Determine the infrared spectrum and other physical properties as directed. Gas liquid chromatographic analysis would also be of interest.

Figure 3.10A (page 96).

Summary

1. Place the water, salt, and alcohol in a round-bottom flask.

2. With cooling and caution, add the sulfuric acid.

3. Reflux the mixture.

4. Add water and steam distill.

5. Extract the distillate with aqueous bicarbonate solution.

6. Separate and dry the organic layer with anhydrous $MgSO_4$ or Na_2SO_4.

7. Distill and analyze the product.

QUESTIONS 1–4 ABOVE.

5. Why is a trap used during reflux?

6. What are the principal by-products and how are they removed?

7. Explain the cause of the pressure buildup in the separatory funnel when the sodium bicarbonate solution is used.

8. Why does the first distillate contain "oil" and water? Why stop when only water distills?

Disposal of Materials

Solid NaOH (D, N)

1-Butanol (OW)

aq. $NaHCO_3$ and solutions (D)

Glass wool (S)

Product (I)

NaBr(S)

H_2SO_4 or H_3PO_4 (N)

Drying agent (S)

Boiling stones (S)

Distillation residue (HO)

E X P E R I M E N T   1 9 B

Preparation of 1-Bromobutane
(NaBr–H$_2$SO$_4$ Method) (M)
References: Sections 1.3H; 3.4; 4.1C, 2D, 4G; 5.1; and Experiment 19A.

$$NaBr + H_2SO_4 + 1\text{-}C_4H_9OH \longrightarrow Na(HSO_4) + H_2O + 1\text{-}C_4H_9Br$$

Introduce the reagents and initiate reaction. Place a spin vane in a 5 mL-conical vial or equivalent vessel, and add 800 μL of water, 1 g of NaBr, 600 μL of 1-butanol, and 750 μL of H_2SO_4 with constant stirring. Attach the vial to a water condenser and trap (Figure 6.1B), and start refluxing. It is important to thoroughly mix the reaction mixture during the reflux. If two distinct layers are present (a brown one and a clear one), your mixing is not efficient enough. To correct this, increase the rate of heating and agitation. After the layers are mixed and the mixture begins to reflux rapidly, continue the reflux for 10–15 more minutes. Allow the apparatus to cool and drain completely.

Isolate the product. Add, to the reaction vessel, 400 μL of water, and attach it to a Hickman still (or otherwise arrange for distillation as in Figure 1.8, Unit IIIA or B or Figure 3.9A or B). Carry out a careful distillation. Continue the distillation, and continuously remove the product with a (bent) pipet. Place it in a receiver until there is no more oily material in the distillate being collected. If there is a question about when this occurs, draw a few drops from the still or receiver into a Pasteur pipet.

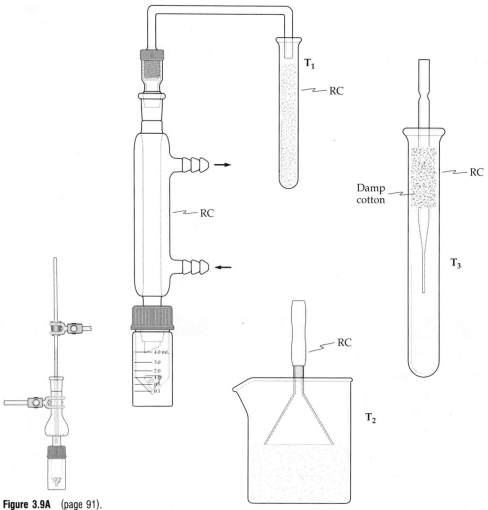

T_1

RC

RC

RC

4.0 mL
3.0
2.0
1.0
0.5
0.1

Damp
cotton

RC

T_3

RC

T_2

Figure 3.9A (page 91).

Figure 6.1B Simple reflux with trap.

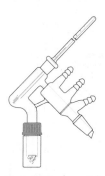

Figure 3.9B (page 95).

Examine these drops and decide whether or not you see any organic material. If you are using the short path head, collect the distillate directly in the receiver.

Add an equal volume of saturated sodium bicarbonate solution to the receiver containing the distillate. Stopper the receiver; mix it briefly, and immediately remove the stopper to allow the built-up pressure to be released. Stopper it again; shake it for a few seconds, and vent it. Continue this process until the buildup of pressure ceases. This mixing can also be done with a Pasteur pipet. Place the receiver completely vertical, and allow the layers to separate.

See Figures 4.2A, B, C (pages 131–133).

Remove a few drops of the top layer with a Pasteur pipet, and test their solubility with an equal volume of water. Remove the organic layer, and transfer it with a Pasteur pipet through a drying pipet with a small amount of drying agent into a 3-mL conical vial equipped with a spin vane. Attach the vial to a Hickman still or short path head, and carry out a slow distillation. Remember to include a thermometer in the throat of the still. Collect the product in the range of 93–101°C.

Interpret the results. Determine the theoretical and percent yields, and report them (the density of the product is 1.28 g/mL, that of 1-butanol, 0.81). Hand the product to your instructor for credit. Determine the infrared spectrum or do a gas liquid chromatographic analysis as directed.

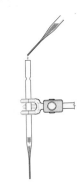

Figure 1.6A and Figure 4.5 (pages 19 and 148).

Summary

1. Mix the reactants and reflux.

2. Steam distill the alkyl bromide from the aqueous mixture.

3. Extract the distillate with aqueous base.

4. Separate and dry the product.

5. Distill and analyze the product.

Questions

1. Why are the by-products formed and how are they removed?

2. Where (how) is unreacted alcohol removed?

3. Why is an equal volume of saturated sodium bicarbonate solution added after the first distillation?

4. Explain the pressure buildup when sodium bicarbonate is used.

5. Why does the first distillate contain "oil" and water? Why stop when only water distills?

Disposal of Materials

NaBr(S)

H_2SO_4 or H_3PO_4 (N)

Drying pipet (S)

Distillation residue

1-Butanol (OW)

aq. $NaHCO_3$ and solutions (D)

Product (I)

E X P E R I M E N T   2 0 A

Figure 3.8 (page 87).

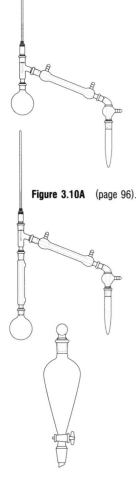

Figure 3.10A (page 96).

Figure 4.1 (page 129).

Figure 1.6A and Figure 4.5 (pages 19 and 148).

Question

Preparation of 1-Bromobutane (HBr Method) (SM)
References: Sections 3.4; 4.1C, 2D, 4G; and 5.1.

$$CH_3CH_2CH_2CH_2OH + HBr \longrightarrow CH_3CH_2CH_2CH_2Br + H_2O$$

Prepare one reagent. If the commercial reagent is less than 47%, this distillation of the azeotrope should result in a more concentrated acid. Place a boiling stone and 20 mL of commercial hydrobromic acid (40–47%) in a 50-mL flask. Attach this flask to an apparatus equipped for simple distillation (Unit III, Figure 1.7 or Figure 3.8). Use a 25-mL flask containing 5 mL of 1-butanol as the receiver. Begin heating, and allow the temperature to rise rapidly to about 122°C (the boiling point of constant-boiling HBr). Discontinue heating when one-half to two-thirds of the hydrobromic acid has been distilled. Allow the apparatus to cool and drain. Use a lot of water (to dilute the acid) when cleaning the apparatus because hydrobromic acid is a strong acid.

Introduce the reagents and initiate reaction. Add a boiling stone to the 25-mL flask containing the butanol–HBr mixture, and cautiously add, in small portions, 6 mL of concentrated sulfuric acid. Attach the flask to a fractional distillation apparatus (no packing in the fractionating column) (Unit IV, Figure 1.7 or Figure 3.10A) using a separatory funnel (stopcock closed) as the receiver. Begin heating slowly, allowing the mixture to reflux for 15–20 minutes before distillation begins. Increase the rate of heating, and collect the distillate until the temperature rises to 110°C or until no more of the organic layer distills with the water. Remove the heat, and allow the apparatus to cool and drain.

Isolate the product. Wash the distillate in the separatory funnel with 15 mL of 1N HCl. Wash the organic layer with 15-mL portions of 10% NaOH and cold water, and dry with anhydrous magnesium sulfate. Transfer the crude product through a drying pipet containing a few granules of anhydrous calcium chloride ($CaCl_2$) to a 10-mL flask containing a boiling stone, and attach it to an apparatus such as in Figure 6.2 (or Figure 3.10A).

1. What is the purpose of the Claisen adaptor?

Begin a slow distillation, and collect the fraction that distills in the region between 93° and 102°C as the product.

Interpret the results. Calculate the yields, and determine an infrared spectrum and other physical properties or perform gas liquid chromatographic analysis as directed.

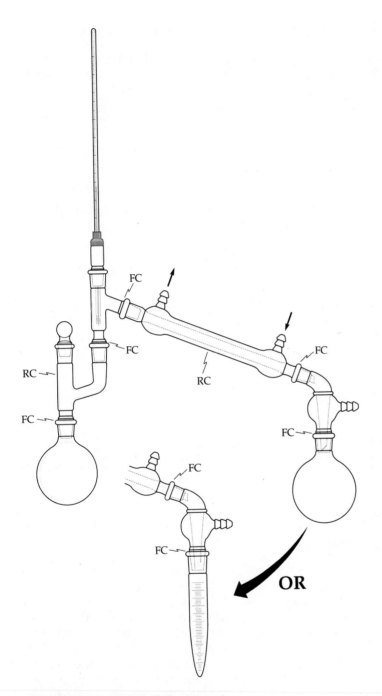

Figure 6.2 Modified distillation apparatus.

Summary

1. Concentrate the HBr solution by distillation and add it to the alcohol in the process of distillation.

2. Add H_2SO_4 and reflux.

3. Steam distill the mixture.

4. Wash, separate, and dry the product.

5. Distill and analyze the product.

Questions

QUESTION 1 ABOVE.

2. Why reflux the mixture before beginning the first distillation (into the separatory funnel)?

3. Which wash removed any unreacted alcohol from the product?

4. What is the purpose of each of the other two washing steps?

5. How could you test for the aqueous layer in each of the washings?

6. Compare the conditions used in Experiments 20 and 21. Does this reflect the reactivity of the alcohols (1° versus 2°)?

7. Analyze the infrared spectrum of the product. How would the spectrum of the alcohol differ?

Disposal of Materials

conc. HBr (D, N)	1-Butanol (OW)
H_2SO_4 or H_3PO_4 (N)	1 N HCl (N or D)
10% NaOH (N or D)	$MgSO_4$ or $CaCl_2$ and drying pipet (S)
Boiling stone (S)	Product (I)
Distillation residue (HO)	

Additions and Alternatives

You can do the final distillation with a regular fractionating apparatus (no packing) (Unit IV, Figure 1.7 or Figure 3.10A) using a chaser such as cumene, cymene, or tetralin.

The reaction can be done without preliminary distillation of the concentrated HBr if H_2SO_4 is added. The reaction also can be done with concentrated (47%) HBr (12–15 mL) and no H_2SO_4. The yield will be lower, and a bit of unreacted alcohol is usually present in the distilled product. This would make the analysis of the infrared spectrum and gas liquid chromatography trace more interesting. Record the infrared spectrum of the starting material. Try various amounts of H_2SO_4.

The experiment was originally designed and can be done with the production and distillation of HBr from 15 g of sodium bromide dihydrate and 20 mL of 85% phosphoric acid (with or without H_2SO_4 later). The 50-mL flask used in the generation of the HBr must be cleaned as soon as it can be handled, while the mixture is still hot, by pouring its contents into water and by flushing the flask with hot water. Otherwise, you will get a glasslike mass (mess) that is hard to remove.

For a different and quantitative look at nucleophilic substitution (HCl versus HBr and S_{N^1} vs. S_{N^2}), consider the suggestions of Warren and Newton, *J. Chem. Educ.*, 1980, *57*, 747.

E X P E R I M E N T 2 0 B

Preparation of 1-Bromobutane (HBr Method) (M)
References: Sections 2.4; 4.1C, 2D, 4G; 5.1; and Experiment 20A.

$$CH_3CH_2CH_2CH_2OH + HBr \longrightarrow CH_3CH_2CH_2CH_2Br + H_2O$$

Introduce the reagents. Place, in a 5-mL conical vial, a spin vane, 2 mL of concentrated hydrobromic acid (47%), 0.8 mL of 1-butanol, and, slowly in small portions, add 0.8 mL of concentrated sulfuric acid.

Initiate the reaction. Attach the vial to a Hickman still or short path head (Unit IIIA or B, Figure 1.8B or Figure 3.9A or B, or Figure 3.10B); include a thermometer (to monitor the temperature of the vapor), and begin heating it very gently in a sand bath or aluminum block, allowing the mixture to reflux for 5–10 minutes before distilling.

Collect the distillate as it forms, and transfer it to a 5-mL vial (with a short path head, it can be collected directly). Cease heating when the distillation temperature reaches 110°C or when the temperature falls off radically.

Isolate the product. Wash the distillate with 1 mL of 1N HCl. Repeat the washing process with 1-mL portions of 10% NaOH and cold water. Transfer the crude product through a drying pipet (with granular anhydrous calcium chloride) to a 3-mL vial equipped with a spin vane.

Set up the apparatus as in Unit IIIA or B, Figure 1.8B or Figure 3.9A or B, and slowly distill. Retain, as the product, the fraction that distills in the range of 93–102°C.

Interpret the results. Determine the yields. Record the infrared spectra of the product and the alcohol or the gas liquid chromatographic analysis, as directed by your instructor.

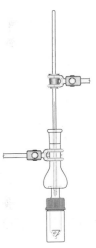

Figure 3.9A (page 91).

Figure 3.9B (page 95).

Summary

1. Mix the reagents and reflux the mixture.

2. Steam distill the mixture.

3. Wash, separate, and dry the product.

4. Distill and analyze the product.

Questions

ANSWER QUESTIONS 2–7 FROM EXPERIMENT 20A.

Disposal of Materials

HBr (D, N)   1-Butanol (OW)

H_2SO_4 (N)   10% NaOH (N or D)

1N HCl (N or D)   Drying pipet with $CaCl_2$ (S)

Product (I)   Distillation residue (HO)

Additions and Alternatives

See Experiment 20A.

To make this experiment analogous to the previous one (Experiment 20A), distill 2.5 mL of hydrobromic acid (5-mL vial), and remove the distillate as it forms (or let it collect directly in the next vial), transferring it to a 5-mL vial that contains 0.8 mL of 1-butanol. After the distillation is complete, add 0.8 mL of concentrated sulfuric acid to the vial. The rest of the experiment would be completed as above.

E X P E R I M E N T   2 0 C

Preparation of 2-Bromobutane (SM)
References: See Experiment 20A.

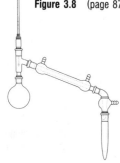

Figure 3.8   (page 87).

$$CH_3CH_2CH(OH)CH_3 + HBr \longrightarrow CH_3CH_2CH(Br)CH_3 + H_2O$$

Use the same procedure as in the preparation of 1-bromobutane (HBr method, Experiment 20A), except for the following modifications:

1. Use 2-butanol (density, 0.81 g/mL) in place of 1-butanol.

2. Distill two-thirds to three-fourths of the HBr into the reaction flask (or use 12–15 mL of concentrated HBr (47%) without distillation), and do *not* add any H_2SO_4.

3. Retain the fraction distilling in the range of 80–95°C as the product.

4. Take the density of the product to be 1.25 g/mL when calculating the percent yield.

Sᴇᴇ Exᴘᴇʀɪᴍᴇɴᴛ 20A.

Disposal of Materials

conc. HBr (D, N)

H_2SO_4 or H_3PO_4 (N)

10% NaOH (N or D)

Boiling stone (S)

Distillation residue (HO)

1-Butanol (OW)

1 N HCl (N or D)

$MgSO_4$ or $CaCl_2$ and drying pipet (S)

Product (I)

Additions and Alternatives

See Experiment 20A.

E X P E R I M E N T   2 1 A

Preparation of *tert*-Amyl Chloride (SM)
References: Sections 3.4; 4.1C, 4G; and 5.1.

$$CH_3CH_2C(OH)(CH_3)_2 + HCl \longrightarrow CH_3CH_2CCl(CH_3)_2 + H_2O$$

The quantities specified below are somewhat large for semi-micro work but are chosen to permit the building up of a stock of the chloride for use in the preparation of tertiary amyltoluene (Experiment 39). The product obtained in this exercise reacts more readily, after drying with anhydrous calcium chloride, than does commercial-grade tertiary amyl chloride. Apparently, some impurity remains that accelerates the reaction with amalgamated aluminum.

Introduce the reagents and initiate the reaction. Mix 15 mL of tertiary amyl alcohol and 40 mL of concentrated hydrochloric acid in a well-stoppered 125-mL Erlenmeyer flask, and shake it for ten minutes.

Isolate the products. Allow the mixture to separate into two layers (in the separatory funnel); draw off the aqueous layer (lower), and discard it.

Pour the upper layer into a dry, small Erlenmeyer flask containing a small amount of anhydrous sodium bicarbonate, and shake it to remove any residual acid.

Figure 4.1   (page 129).

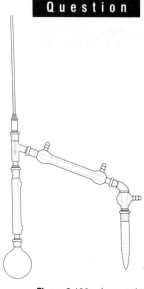

Figure 3.10A (page 96).

Question

1. How does shaking this mixture with anhydrous sodium bicarbonate remove residual acid?

After all effervescence has ceased, add a bit more sodium bicarbonate. Filter the liquid through a small, dry filter paper into another dry flask. Add a small amount of anhydrous magnesium sulfate and calcium chloride, and shake the flask for a minute or two to dry the crude product. Then add calcium chloride and repeat.

Purify the product. Filter the impure product directly into a 25-mL flask, which will be equipped with a fractionating column (Figure 3.10A). Collect the distillate in the range of 80–88°C as the product.

Analyze the product. Report the actual, theoretical, and percent yields (the density of the alcohol is 0.81 g/mL, that of the chloride, 0.87). Record the infrared spectrum of the product.

Summary

1. React the acid and alcohol.

2. Separate the organic layer from the aqueous layer.

3. Treat the organic layer with base and then dry it.

4. Distill and analyze the product.

Questions

QUESTION 1 ABOVE.

2. Where would any unreacted alcohol be removed in the above procedure?

3. Contrast the infrared spectrum with that expected for the alcohol.

Disposal of Materials

t-Amyl alcohol (OW)

Aqueous layer from reaction (N or D)

$MgSO_4$, $CaCl_2$, and filter (evaporate and S)

Product (I)

conc. HCl (N or D)

Solid $NaHCO_3(S)$

Boiling stone (S)

Distillation residue (HO)

Additions and Alternatives

Use gas liquid chromatography to check for impurities in the product.

Store the product over a few grains of anhydrous calcium chloride if it is to be used in Experiment 39.

You could use this reaction to prepare *t*-butyl chloride from *t*-butyl alcohol, but the product is much more volatile (b.p. = 50.7°C). The *t*-butyl chloride made in this manner could be used in Experiment 39 or 40.

EXPERIMENT 21 B

Preparation of *tert*-Amyl Chloride (M)
References: Sections 3.4; 4.1C, 4G; 5.1; and Experiment 21A.

$$CH_3CH_2C(OH)(CH_3)_2 + HCl \longrightarrow CH_3CH_2CCl(CH_3)_2 + H_2O$$

Introduce the reagents. Mix a 0.6 mL sample of tertiary amyl alcohol and 1.6 mL of concentrated hydrochloric acid in a graduated receiver, glass centrifuge tube, or 5-mL vial. Stopper tightly, and shake the vial vigorously for a few minutes. Unstopper and place the vessel in a flask or small beaker that contains some *warm* water (warm *not* hot). Adjust the water level so that it just extends above the height of the solution in the reaction vessel.

1. Why is the *warm* water used here? **Question**

Periodically shake the reaction vessel vigorously; vent; and then replace it in the water. Continue this activity for ten minutes. Let the mixture stand, and allow the layers to separate (two layers should be present).

Isolate the product. Use a Pasteur pipet to remove the lower layer. Applying some pressure to the bulb, lower the pipet to the bottom of the vial.

2. Why must you apply pressure? **Question**

Slowly lessen the pressure, and collect the aqueous layer. Be careful not to remove any of the top layer. It is better to leave behind a drop of the aqueous layer than to take a drop of the top layer. When working on the micro scale every drop is important!

Add very small amount of dry sodium bicarbonate to the organic layer, and shake to remove any residual acid.

3. How does shaking this mixture with anhydrous sodium bicarbonate remove residual acid? **Question**

After all effervescence has ceased, remove the liquid by means of a (plugged) pipet, and transfer it to another graduated receiver, centrifuge tube, or vial. Add a small amount of anhydrous magnesium sulfate and then calcium chloride to this liquid. Shake the vessel for a minute or two to dry the crude product. Employing a Pasteur pipet or a plugged pipet, collect and transfer the liquid to a 3-mL conical vial that contains a small spin vane.

Purify the product. Connect the vial to a Hickman still or short path head, and carry out a distillation. Save the portion of the distillate collected in the range of 80–88°C as the product. To remove the product from the still, use a syringe or a Pasteur pipet with or without a bent tip.

See Figure 3.9 (pages 91, 95).

To bend the pipet tip, rotate it in a gentle flame until a bend that corresponds to your Hickman still is achieved.

Analyze the product. The density of *t*-amyl alcohol is 0.81 g/mL, that of the chloride, 0.87. Report the actual, theoretical, and percent yields. Record the infrared spectrum and determine other properties as directed. Deliver the product to your instructor for credit.

Summary

1. React the acid and the alcohol.

2. Separate the organic layer.

3. Treat the organic layer with base and then dry it.

4. Distill and analyze the product.

Questions   QUESTIONS 1–3 ABOVE.

Answer questions 2 and 3 from Experiment 21A as appropriate.

Disposal of Materials

t-Amyl alcohol (OW)

conc. HCl (N or D)

Aqueous layer from reaction (D)

Solid NaHCO$_3$ (S)

MgSO$_4$ and CaCl$_2$ (S)

Product (I)

Distillation residue (HO)

EXPERIMENT 22

Preparation of 1-Iodopropane (SM)
References: Sections 1.3H, 3.4, 4.1C, and 4.4G.

$$6 \, CH_3CH_2CH_2OH + 3 \, I_2 + 2 \, P \longrightarrow 6 \, CH_3CH_2CH_2I + 2 \, P(OH)_3$$

CAUTION   RED PHOSPHOROUS IS EXTREMELY DANGEROUS. FRICTION MAY CAUSE IT TO IGNITE AND SUCH FIRES ARE DIFFICULT TO EXTINGUISH. THE ELEMENT ITSELF CAN CAUSE PAINFUL BURNS OF THE SKIN THAT ARE SLOW TO HEAL. HANDLE AND CLEAN UP SPILLS WITH CARE.

Introduce the reagents.   Place (use wide-mouth funnel) 400 mg of phosphorous, 3.0 mL of 1-propanol, and a boiling stone (or spin bar or vane) in a 10-mL or 25-mL flask equipped for reflux with a gas trap containing 10% sodium bicarbonate solution (Figure 6.1A). Have an ice water bath ready, and arrange the apparatus so that the change from heating to cooling may be done easily.

Initiate the reaction. Divide 5.1 g of I_2 into 5–8 small portions; add one portion through the top of the apparatus so that the I_2 reaches the flask, and reassemble the apparatus. Warm and mix (swirl or stir) the flask to start the reaction (cool if necessary). When the reaction starts to subside, continue the addition. Be sure that the reaction picks up each time, and allow 2–3 minutes for it to settle down between additions. When the additions are completed, gradually bring the reaction mixture to a boil, and reflux it for 30–45 minutes.

Isolate the product. When the flask has cooled, remove the condenser and trap; equip the flask for simple distillation (Figure 1.7, Unit III or Figure 3.8), but use a test tube or a 12-mL or 15-mL glass centrifuge tube containing 4 mL of 2% sodium hydroxide solution as the receiver. Put a new boiling stone (unless you use magnetic stirring) in the flask, and distill as much of the mixture as possible below 110–120°C without heating the flask to dryness. Thoroughly mix (cap and shake) the distillate, and decant or remove most of the water layer with a pipet.

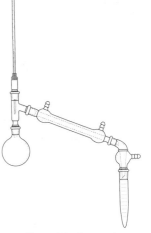

Figure 3.8   (page 87).

1.   Which layer is it? How do you test it?

Q u e s t i o n

Add 2 mL of water to the organic layer and repeat this procedure.

Remove the bottom layer with a Pasteur pipet, and dry it with 40–60 mg of drying agent (a filter pipet containing a column of drying agent may be used). Separate the product.

Interpret the results. Analyze the product by infrared spectroscopy and gas liquid chromatography, if directed to do so, before and/or after a final distillation. Check the density or refractive index of the product to see if distillation is needed. The density of the iodide is 1.743 g/mL and that of the alcohol, 0.804. The refractive indexes are 1.5055 and 1.3854, respectively, at 25°C. Distill the dry product if necessary, collecting a fraction in the vicinity of 100°C.

Summary

1.   Assemble the apparatus and add phosphorus and alcohol.

2.   Initiate the reaction by adding iodine in small portions.

3.   Reflux the mixture.

4.   Distill the product into base.

5.   Separate and wash the organic layer with water.

6.   Dry the product and distill if necessary.

7.   Analyze the product.

QUESTION 1 ABOVE.

2. What is the purpose of the sodium hydroxide wash?
3. How is unreacted alcohol removed from the product?
4. Discuss the major features of the product's infrared spectrum and contrast with those of the alcohol.

Disposal of Materials

Red phosphorous (HW, special instructions)   1-Propanol (OW, OS)

I_2 (S, special container) aq. $NaHCO_3$ (D)   aq. NaOH (N)

Boiling stones (S)   Distillation residue (D, N)

Drying agent (S)   Product (I)

2nd Distillation Residue (HO)   Chaser (HO)

Additions and Alternatives

You could use a chaser in the final distillation. You can do this preparation on one-tenth the scale (M).

CHAPTER 7

Organic Oxidation and Reduction Reactions

7.1 Balancing Oxidation–Reduction Equations

During an oxidation–reduction reaction, an atom in one compound or ion that gains one or more electron(s) is reduced, while another atom in the reaction loses one or more electron(s), and is oxidized. Many important organic reactions involve oxidation–reduction. Thus, the development of reactions that convert organic compounds from one oxidation level to another have received a great deal of attention. Table 7.1 suggests the different compounds that correspond to various oxidation levels. It is based on a construct that oxidation can be represented by the substitution of a hydrogen by an OH. Note that two hydroxides on one carbon are unstable in most cases and split out water to give a carbonyl function (OH and NH_2 on the same carbon split out ammonia). Substitution (Chapter 6) interchanges functional groups on the same oxidation level.

Table 7.1 Oxidation States in Organic Compounds

Number of OH	Example	Types of Compounds
4	CO_2	H_2CO_3 ($(HO)_2C{=}O$)
3	HCOOH	Acids and their derivatives
2	$H_2C{=}O$	Aldehydes, ketones, alkynes
1	CH_3OH	Alcohols, ethers, halides, and Alkenes
0	CH_4	Alkanes

7.1A Ion–Electron Method

It is possible to extend the ion–electron method learned in general chemistry to the balancing of organic oxidation–reduction equations. If you know how to balance this type of equation, you may skip this section and proceed to Experiment 23. The steps of this procedure (similarly designated for these following examples) are:

a. Separate into "half-reactions" those molecules or ions that change apparent oxidation state (only one kind of change in each).

b. Balance these "half-reactions" for principal elements, i.e., those elements whose apparent oxidation states change.

c. Balance for all other elements in the "half-reactions" except oxygen and hydrogen.

d. Balance oxygen by adding the proper number of water molecules to the oxygen-deficient side.

e. Balance hydrogen by adding the proper number of hydrogen ions (H^+) to the hydrogen-deficient side.

(Check to see that all elements balance.)

f. Balance charge by adding electrons (e^-) to the more positive (or less negative) side.

g. Multiply each "half-reaction" by coefficients that will balance the total number of electrons on the left-hand side and on the right-hand side.

h. Add and cancel to put the final equation in the simplest, net form.

i. If the reaction is to be carried out in basic solution, add a number of hydroxide ions (OH^-) to each side equal to the number of hydrogen ions. (Remember, $H^+ + OH^- \rightarrow HOH$.) Add and cancel as directed in (h).

Thus, in the reaction

$$CH_3CHOHCH_3 + Cr_2O_7{}^{-2} \longrightarrow CH_3COCH_3 + Cr^{+3}$$

these steps become:

a. $Cr_2O_7{}^{-2} \longrightarrow Cr^{+3}$

b. $Cr_2O_7{}^{-2} \longrightarrow 2\ Cr^{+3}$

c. (already balanced)

d. $Cr_2O_7{}^{-2} \longrightarrow 2\ Cr^{+3} + 7\ H_2O$

e. $14\ H^+ + Cr_2O_7{}^{-2} \longrightarrow 2\ Cr^{+3} + 7\ H_2O$

f. $6\ e^- + 14\ H^+ + Cr_2O_7{}^{-2} \longrightarrow 2\ Cr^{+3} + 7\ H_2O$

a′. $CH_3CHOHCH_3 \longrightarrow CH_3COCH_3$

b′. (Step not needed)

c′. (Step not needed)

d′. (Step not needed)

e′. $CH_3CHOHCH_3 \longrightarrow CH_3COCH_3 + 2\ H^+$

f′. $CH_3CHOHCH_3 \longrightarrow CH_3COCH_3 + 2\ H^+ + 2\ e^-$

g. $1(6\ e^- + 14\ H^+ + Cr_2O_7^{-2} \longrightarrow 2\ Cr^{+3} + 7\ H_2O)$

g′. $3\ (CH_3CHOHCH_3 \longrightarrow CH_3COCH_3 + 2H^+ + 2\ e^-)$

h. $14\ H^+ + Cr_2O_7^{-2} + 3\ CH_3CHOHCH_3 \longrightarrow$

$$2\ Cr^{+3} + 7\ H_2O + 3\ CH_3COCH_3 + 6\ H^+$$

or

$$8\ H^+ + Cr_2O_7^{-2} + 3\ CH_3CHOHCH_3 \longrightarrow$$

$$2\ Cr^{+3} + 7\ H_2O + 3\ CH_3COCH_3$$

For the reaction

$$C_6H_{12}O_6 \longrightarrow 2\ CO_2 + 2\ C_2H_5OH$$

following the designated steps gives

$$\overset{(g)\ \ (d)}{1(6\ H_2O + C_6H_{12}O_6} \longrightarrow \overset{(b)}{6CO_2} + \overset{(e)}{24H^+} \overset{(f)}{+24e^-})$$

$$\overset{(g)\ (f)}{2(12e^-} + \overset{(e)}{12H^+} + C_6H_{12}O_6 \longrightarrow \overset{(b)}{3C_2H_5OH} \overset{(d)}{+3\ H_2O})$$

(h) $6\ H_2O + 3\ C_6H_{12}O_6 + 24\ H^+ \longrightarrow$

$$6\ CO_2 + 24\ H^+ + 6\ C_2H_5OH + 6\ H_2O$$

or

$$3\ C_6H_{12}O_6 \longrightarrow 6\ CO_2 + 6\ C_2H_5OH$$

or

$$C_6H_{12}O_6 \longrightarrow 2\ CO_2 + 2\ C_2H_5OH$$

It is suggested that you practice on the following skeleton equations:

1. $C_6H_5CH_2OH + Na_2Cr_2O_7 \cdot 2H_2O + H_2SO_4 \longrightarrow$

$$C_6H_5COOH + Na_2SO_4 + Cr_2(SO_4)_3 + H_2O$$

2. $C_{12}H_{22}O_{11} + HNO_3 \longrightarrow H_2C_2O_4 + NO_2 + H_2O$

3. $CH_3CHO + HCHO + NaOH \longrightarrow CH_3CH_2OH + HCOONa + H_2O$

4. $KMnO_4 + H_2C_2O_4 + H_2SO_4 \longrightarrow MnSO_4 + K_2SO_4 + CO_2 + H_2O$

See also Burrell, *J. Chem. Educ.*, **1959**, *36*, 77.

E X P E R I M E N T   2 3

Preparation of Iodoform (M)
References: Sections 2.3 and 4.3.

Iodoform, a yellow powder with a "hospital smell," was once used as an antiseptic by sprinkling it on wounds that would suffer if iodine were applied directly. This was extremely important before the availability of sulfa drugs. It is formed by the haloform reaction, which is a general reaction for certain structural features in organic molecules and so is used as a test reaction (iodoform test, see Exercise 10, Chapter 17). Essentially, an alcohol can be oxidized to a carbonyl (or a carbonyl is used directly); the three hydrogens of a terminal methyl group next to the carbonyl are substituted by the halogen. The bond between the carbonyl carbon and the trihalogenated carbon is broken in strongly basic solution.

In this preparation, use is made of a commonly available oxidizing agent, commercial bleach. Not all brands have the same concentration of NaOCl (sodium hypochlorite), so the directions must be modified accordingly. No pure iodine is added, but it is, in effect, generated from the iodide ion as it is oxidized by the hypochlorite ion. ($2\,I^- + 2\,H^+ + OCl^- \rightarrow I_2 + Cl^- + H_2O$ or $I^- + OCl^- \rightarrow IO^- + Cl^-$) Thus any iodide produced will also be recycled. Use the balanced equation for making any calculations.

$$CH_3COCH_3 + 3\,I_2 + 4\,NaOH \longrightarrow CH_3COONa + HCI_3 + 3\,H_2O + 3NaI$$

Introduce the reagents. Dissolve 0.75 g of KI in 5–6 mL of water in a small beaker, Erlenmeyer flask, or large test tube, and add 0.2 mL of acetone. Add, in small increments, 12–15 mL of 3% bleach solution (see Additions and Alternatives) with shaking or stirring. Allow the mixture to stand for five minutes with occasional shaking or stirring. Cool and collect the product carefully using a Hirsch funnel, a small filter flask, and suction.

Wash the product with a little cold water, and dry it.

Purify by sublimation. Some or all of the product is purified by sublimation at reduced (water aspirator) pressure. Be sure *not* to *overheat* the product. Collect the crystals.

To compare your results with those for simple heating, place a few crystals of iodoform in a test tube. With a test tube holder or clamp, hold the tube at a slant, gently heat, and try to sublime the iodoform. Record your observations.

See Figure 4.10 (pages 176–7).

Question

1. What is the solid/gas that is produced? How is this connected to the old use of iodoform or its smell?

Alternative purification. If a Craig tube is available, dissolve 100 mg of product in 2 mL of hot methanol in the large Craig tube bottom and concentrate (in the hood) the solution to three-fourths of the original volume. METHANOL CAN BE TOXIC. Cool. Collect and analyze the crystals. A test tube and Hirsch funnel can also be used.

CAUTION

See Figures 2.8A, B, C (pages 56–58).

Summary

1. Mix the reactants and initiate the reaction by adding bleach.

2. After standing, crystallizing, and cooling, collect the product by suction filtration.

3. Purify by sublimation or recrystallization.

QUESTION 1 ABOVE.

Questions

2. What is the function of the bleach?
3. Why does reduced pressure make sublimation possible?
4. Why would the use of iodoform be "safer" than elemental iodine for treating open wounds?

Disposal of Materials

KI (S)	aq. KI (D)
Acetone (OS, OW)	Bleach solutions (D or thiosulfate and D)
Methanol (OS, OW)	Methanol solutions (HO)
Product (I)	

Additions and Alternatives

If the commercial bleach available is 5–6% NaOCl, use 7 mL of it. If a standard 6% solution is made up, use 6 mL. Other water-soluble substrates such as ethanol, 2-propanol, 2-butanone and, possibly, 2-butanol may be used. This preparation can also be done by adding bleach and the base to tincture of iodine (from the drugstore).

5. Try it, and explain why it should work.

Try the recrystallization with methanol containing a few percent water.

If you wish to try other commercial oxidants for mild oxidation, see *J. Chem. Educ.*, 1969, *46*, 755; 1981, *58*, 824; 1982, *59*, 862; 1984, *61*, 1118; and 1990, *67*, 172. Substituted acetophenones can be converted to benzoic acids in a haloform reaction with bleach.

Compare with Exercise 10, Chapter 17.

E X P E R I M E N T   2 4 A

Preparation of 2-butanone (SM)
References: Sections 1.3E–H; 3.3, 4, 7, 8; 4.2D; 5.1 and 7.1A.

$$CH_3CH_2CHOHCH_3 + Cr_2O_7^{-2} + H_3O^+ \longrightarrow CH_3CH_2COCH_3 + Cr^{+3} + H_2O$$

Assemble the apparatus used for fractional distillation (Figure 3.10A) with an unpacked column and a 25-mL flask. Use a graduated receiver or a tared flask as the receiver.

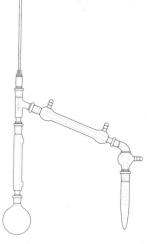

Figure 3.10A (page 96).

Introduce the reagents (HOOD). CHROMATE SALTS ARE PROBABLE CARCINO-GENS. In the 25-mL flask, dissolve 3 g of $NaCr_2O_7 \cdot 2\,H_2O$ (or $K_2Cr_2O_7$) in 5–6 mL of water (or 8 mL of a 36% aqueous solution). Add 2 mL of concentrated sulfuric acid a few drops at a time with cooling, if needed. Cool, add a boiling stone, and reassemble the apparatus except for the thermometer.

Initiate the reaction. Through the top of the apparatus, add dropwise 2.5 mL of 2-butanol with shaking and mixing. Use a Pasteur pipet, and make sure that the drops reach the flask. Put the thermometer in place. Heat the mixture quickly to a brisk boil. Collect the product when there is a steady distillation temperature in the range of 77–85°C.

Analyze the product. Use 0.81 g/mL as the density of the alcohol and the ketone. Determine the weight and volume of the product. Balance the equation above before calculating the theoretical yield; determine the percent yield, and turn in the properly labeled product with your report. As directed by your instructor, analyze the product using micro boiling point, density, refractive index, gas liquid chromatography, or infrared. The latter two are particularly interesting as they may reveal traces of alcohol or water.

Prepare a derivative. Many derivatives of hydrazine (NH_2NH_2) can be used to make crystalline compounds (derivatives), from carbonyl compounds, that are suitable for identification. HYDRAZINES ARE HAZARDOUS. AVOID BREATHING OR SKIN CONTACT. In this experiment, semicarbazine ($NH_2C(O)NHNH_2$) will be used. In a boiling tube, 5-mL or 10-mL flask, or in a 5-mL vial equipped for reflux, place 200 mg of semicarbazine hydro-chloride ($NH_2C(O)NHNH_2 \cdot HCl$), 300 mg of sodium acetate, and 2 mL of water. Warm the mixture with a sand or water bath to bring the solids into solution.

Figure 1.7, Unit I
(pages 24–27).

7. What is the function of the sodium acetate?

Question

Add 250 μL (0.25 mL or 6–8 drops) of product to the mixture with a calibrated pipet, and reassemble the apparatus. Heat this mixture in a bath at 70–75°C for ten minutes. Turn off the heat, and allow the apparatus to remain in the bath for another ten minutes. Cool the vessel in ice water, and filter the product using a small Hirsch funnel and suction. Save a few crystals, and recrystallize the rest from a minimum of hot water (or 20–50% aqueous ethanol). Use the large Craig tube if available, or filter as above. Dry, weigh, and determine the melting range (about 130–140°C) for the product collected.

See Figures 2.8A, B, C
(pages 56–58).

Summary

1. Assemble the apparatus with oxidizing agent and acid.

2. Add the alcohol and heat.

3. Collect the product by fractional distillation.

4. Analyze the product.

5. Prepare and recrystallize a derivative.

QUESTION 1 ABOVE.

Questions

2. Write the balanced chemical equations for the preparation of the ketone and its derivative.

3. Experimentally, what signs are there of impurities in the ketone? What are the likely impurities? What could be done to purify it?

4. Why does this distillation give a reasonably pure product for this preparation? What other liquids are present and what becomes of them?

5. Could acetaldehyde (b.p. = 20°) or butryaldehyde (b.p. = 76°) be effectively prepared by this method? Explain. Compare to subsequent experiments.

6. The preparation of derivatives followed by their recrystallization and decomposition has been used as a method of purification for liquids. Assuming that your derivative preparation gave an 80% yield, calculate back to determine the total yield of pure ketone and the new percent yield. Compare with your original percent yield. How does this compare with your gas liquid chromatographic analysis?

7. Analyze the infrared spectrum of the product and contrast it with that expected for the starting material. What impurities are suggested?

Disposal of Materials

Solid $Na_2Cr_2O_7 \cdot 2H_2O$ or $K_2Cr_2O_7$ (HW)

Distillation residue or dichromate solutions (D or N with thiosulfate or dithionate, D)

Recrystallization filtrate (add dilute HCl, evaporate, HW)

conc. H_2SO_4 (D or N)　　　　　2-Butanol (OS, OW)

Product (I)　　　　　Solid semicarbazine hydrochloride (HW)

Sodium Acetate (S)　　　　　Recrystallization solvent (D)

Additions and Alternatives

You could check both the alcohol and the ketone with the iodoform test (Chapter 17, Exercise 10). You could check the alcohol in an attempt to form the derivative. The 2,4-diphenylhydrazone derivative can be made (Experiments 25A and 53). You could develop a thin layer chromatographic analysis on silica and use it to analyze the reaction and product. Other secondary alcohols such as 2-propanol can be used.

Similar reactions have been used to measure alcohol levels (Breathalyzer, *J. Chem. Educ.*, 1986, *63*, 897) or as a spot test for 1° or 2° alcohols.

Question 8. What about the reaction of aldehydes?

See *J. Chem. Educ.*, 1962, *39*, 308. Bleach and related products can also be used as the oxidant. See *J. Chem. Educ.*, 1981, *58*, 824; 1982, *59*, 981; and other references in Experiment 23.

E X P E R I M E N T　　2 4 B

Preparation of 2-Butanone (M)
References: Sections 1.3E–H; 3.4, 7, 8; 4.2D; 5.1; 7.1A;
　　　　　　and Experiment 24A. (See hazards mentioned!)

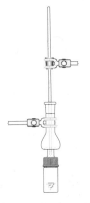

Assemble the apparatus for distillation (Figures 3.9A or B or Unit IIIA or B, Figure 1.8B).

Introduce the reagents (HOOD). In a 5-mL vial with a spin vane, dissolve 0.6 g of $Na_2Cr_2O_7 \cdot 2 H_2O$ or $K_2Cr_2O_7$ in 1 mL of water (or 1.5 mL of 40% aqueous solution), and slowly and carefully add 0.4 mL of concentrated sulfuric acid. Reassemble the apparatus except for the thermometer. If no stirring bar is used, add a boiling stone when assembled.

Initiate the reaction. Add 0.5 mL of 2-butanol dropwise through the top of the apparatus into the mixture with stirring or shaking. Replace the thermometer, and initiate brisk boiling (with stirring). Distill all the mate-

Figure 3.9A (page 91).

rial boiling below 90°C. Remove the distillate from the Hickman still as it is formed. A short path head may be used. Collect (or transfer) the distillate in (to) a tared receiver or vial.

Analyze the product and prepare a derivative. Determine the actual yield, percent yield, and other properties or analysis as directed. Prepare the derivative as directed in the preceding experiment.

Summary

See Experiment 24A.

SEE EXPERIMENT 24A.

Figure 3.9B (pages 91–95).

Questions

Disposal of Materials

Solid $Na_2Cr_2O_7 \cdot 2H_2O$ or $K_2Cr_2O_7$ (HW)

Distillation residue or dichromate solutions (D or N with thiosulfate or dithionate, D)

Recrystallization filtrate (add dilute HCl, evaporate, HW)

conc. H_2SO_4 (D or N)   2-Butanol (OS, OW)

Product (I)   Solid semicarbazine hydrochloride (HW)

Sodium acetate (S)   Recrystallization solvent (D)

Additions and Alternatives

See Experiment 24A.

EXPERIMENT 25A

Oxidation of 1-Propanol (SM)
References: Sections 2.4, 4.2D, 5.1, 7.1A, and Experiment 75.

$$CH_3CH_2CH_2OH + Cr_2O_7^{-2} + H_3O^+ \longrightarrow CH_3CH_2CHO + Cr^{+3} + H_2O$$

Primary alcohols may be converted to aldehydes by many different oxidizing agents. Chromic acid, produced from a mixture of sodium or potassium dichromate and sulfuric acid, is commonly used for this purpose. Care must be exercised to prevent further oxidation of the aldehyde to the corresponding carboxylic acid. In the production of the higher homologs, the carboxylic acid that is produced esterifies some of the original alcohol, and this introduces more complications. In the preparation of low-boiling aldehydes such as acetaldehyde, which boils at 20°C, it may be possible to distill the product from the reaction mixture as it is formed.

Assemble the apparatus and introduce the reagents. Place, in a 50-mL flask, a boiling stone and 12 mL of water. Add, with caution and cooling, 6 mL of concentrated sulfuric acid. When the flask no longer feels warm to your hand, cautiously add 6 mL of 1-propanol. Assemble one of the types of apparatus shown in Figure 7.1 (no packing; clamp it so that the heater can be moved in or out). Surround the graduated receiver used to collect the product with an ice water bath. In the separatory funnel, place 7.5 g of sodium dichromate dihydrate dissolved in 10 mL of water or 12 mL of a 60% aqueous solution. CHROMATE SALTS ARE PROBABLE CARCINOGENS.

Initiate the reaction. Heat the mixture in the flask until the condensate reaches well into the Claisen adaptor or column. Then withdraw the heat, and add the dichromate solution at a moderate rate. Control the rate so that, with only occasional use of heating, the reaction is maintained, and presently propanal (b.p. = 49–50°) will distill. Collect the product in the range of 40–70°C. If the temperature rises above 60–65°C, diminish the rate of addition and heating so that distillation is just barely continued. Once the addition is completed and the reaction subsides, heat the mixture cautiously to distill any remaining aldehyde. If the aldehyde is distilled as it is formed and fractionated carefully as it is distilled, a good yield of relatively pure product may be obtained.

Analyze the product. Redistill the product (Figure 3.10A, 10-mL flask) if analysis by infrared or gas liquid chromatography indicates it is appropriate. Test by Exercises 5, 7, and 8 or 9, Chapter 17 as directed and compare the results with those of 2-butanol.

Prepare a derivative. Add 1 mL of the distillate to 0.1 g of 2,4-dinitrophenylhydrazine in a test tube. AVOID BREATHING OR SKIN CONTACT. Add 5 mL of 95% ethanol, and heat the mixture just to boiling. Add three drops of concentrated hydrochloric acid, and maintain the mixture just below the boiling point for five minutes. Add water to the hot solution until a turbidity appears and persists upon shaking. Cool the mixture. Make a blank test using *n*-propyl alcohol instead of the distillate. This derivative is used in the routine identification of aldehydes and ketones. It is easily recrystallized in ethanol or aqueous ethanol and possesses a sharp melting point. Isolate and recrystallize the derivative as directed.

CAUTION

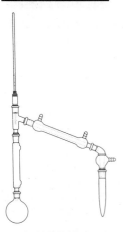

Figure 3.10A (page 96).

CAUTION

See Figures 2.6–2.8 (pages 52–58).

Summary

1. Assemble the apparatus, and add the reactants with care in the order given.

2. Begin the reflux.

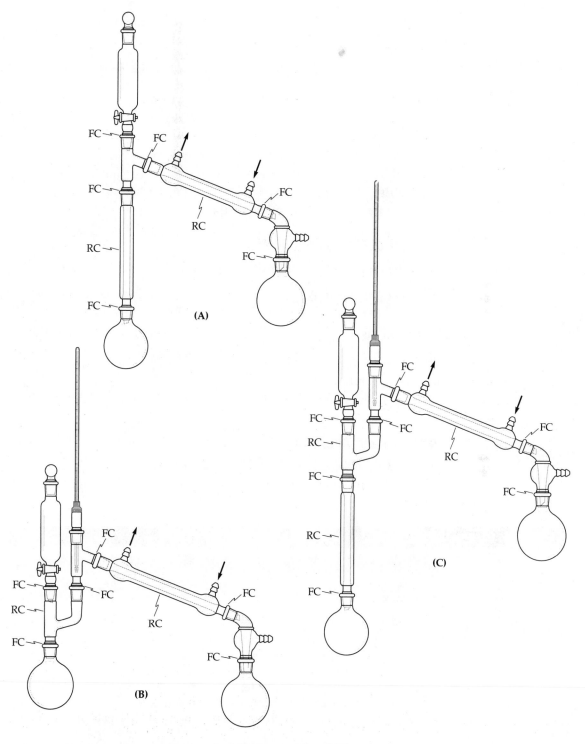

Figure 7.1 Simultaneous distillation and addition assemblies. (SM)

3. Slowly add the oxidant as the product is carefully distilled.

4. Analyze the product.

5. Prepare and recrystallize a derivative as directed.

1. Why test the product for acid? For easily oxidized compounds?

2. What is the difference between the boiling points of the reactant and the product? How does that make this procedure work?

3. Why is it important to add the oxidizing agent gradually?

4. Analyze the infrared spectrum of the product, and contrast those features with the features expected for the starting material.

5. Write the balanced equation for this preparation.

Disposal of Materials

conc. H_2SO_4 (D or N)   1-Propanol (OS)

Solid dichromates (HW)   Dichromate solutions (D or N, thiosulfate or dithionate, D)

Test reactions (I)   2,4-Dinitrophenylhydrazine (HW)

Ethanol (D)   conc. HCl (N or D)

Derivative solutions (I)

Product (I)

Additions and Alternatives

Modify the experiment to use ethanol or 1-butanol.

Compare, in your report, your observations and results with those of a classmate using a different experimental apparatus.

E X P E R I M E N T   2 5 B

Oxidation of 1-Propanol (M)
References: Sections 3.4, 4.2D, 5.1, 7.1A, and Experiments 25A and 75.

$$CH_3CH_2CH_2OH + Cr_2O_7{}^{-2} + H_3O^+ \longrightarrow CH_3CH_2CHO + Cr^{+3} + H_2O$$

Assemble the apparatus and introduce the reagents. Place, in a 5-mL vial, a spin vane and 0.6 mL of water. Cautiously add 0.3 mL of concentrated sulfuric acid with cooling as needed. When the acid has been added and the vessel is about at room temperature, add 0.3 mL of 1-propanol dropwise with stirring. Assemble as shown in Figure 7.2. Dissolve 425 mg of sodium dichromate dihydrate or potassium dichromate in 0.5 mL of

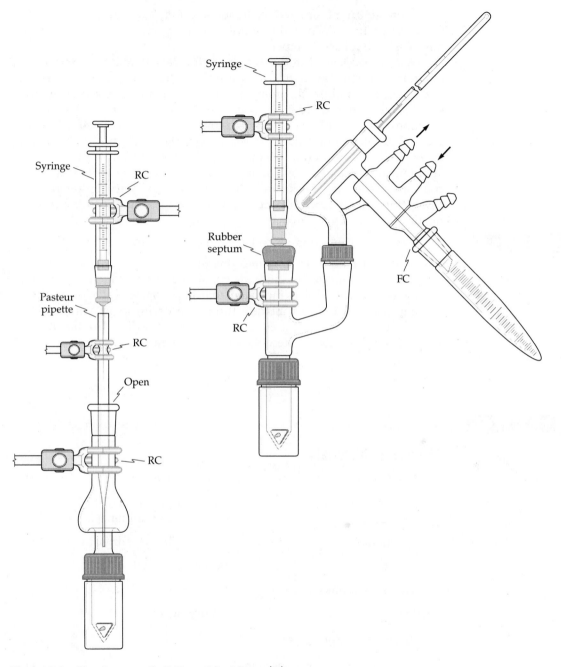

Figure 7.2 Simultaneous distillation and addition. (M)

water, or use 0.6 mL of a 60% aqueous solution. If a syringe is to be used (all glass preferred), load this solution into it, and put it into position; otherwise use a Pasteur pipet.

Initiate the reaction. Heat the mixture in the vessel until reflux has begun, but don't permit any liquid to distill. Withdraw the heat or raise the apparatus, and add the dichromate solution at a moderate rate. The drops should reach the solution in the vessel *directly.* Control the rate of addition so that, with occasional heating, reflux is maintained. Allow the gradual distillation of product to accompany the addition of the oxidizing agent. Collect any product in the range of 40–70°C, but try to keep the addition and heating slow enough so that the distillation just barely continues. With the Hickman still it will be necessary to remove the aldehyde as it is formed. Do not allow any of it to return to the reaction mixture.

Question

6. Why should the product be kept out of the reaction mixture?

Use of a short path head simplifies this process.

Analyze the distillate by infrared, gas liquid chromatography, or other means as directed. If it is indicated that appreciable water or alcohol is present, pass the product mixture through a drying pipet with about 30 mg of drying agent (Na_2SO_4 or $MgSO_4$ and $CaCl_2$). Do the tests or prepare the derivative as described in Experiment 25A, as directed.

Summary

See Experiment 25A.

Questions

ANSWER QUESTIONS 1–5 FROM EXPERIMENT 25A AND QUESTION 6 ABOVE.

Disposal of Materials

conc. H_2SO_4 (D or N)	1-Propanol (OS)
Solid dichromates (HW)	Dichromate solutions (D or N, thiosulfate or dithionate, D)
Test reactions (I)	2,4-Dinitrophenylhadryazine (HW)
Ethanol (D)	conc. HCl (N or D)
Derivative solutions (I)	
Product (I)	Solid drying agent (S)

Additions and Alternatives

See those suggested for Experiment 25A.

If a short path head is available, you can use Unit III, Figure 3.11B with a septum and syringe (put the needle of the syringe through the septum)

Figure 1.6A and Figure 4.5
(pages 19 and 48).

in the central arm of the Claisen adaptor (see Figure 7.2). This apparatus can be further modified by using the air condenser between the Claisen adaptor and the head. If you make this modification, use a 5-mL or 10-mL flask, and scale up by a factor of 2.

E X P E R I M E N T   2 6 A

Preparation of Benzoic Acid (SM)
References: Sections 1.3F–H; 2.3; 4.2C; 5.1; and 7.1.

$$\text{C}_6\text{H}_5\text{CH}_2\text{OH} + \text{HNO}_3 \longrightarrow \text{C}_6\text{H}_5\text{COOH} + \text{NO}_2$$

CAUTION

DURING THE OXIDATION OF A PRIMARY ALCOHOL TO AN ACID, CONSIDER WEARING A FACE SHIELD IN ADDITION TO GOGGLES BECAUSE OF THE POSSIBILITY OF SPLATTERING.

Assemble the apparatus and introduce the reagents. Place 1 mL of benzyl alcohol in a 10-mL or 25-mL flask equipped as shown in Figure 7.3. Arrange the apparatus so that heating and cooling can be interchanged. Support the test tube containing a 5% aqueous solution of sodium hydroxide so that the oxides of nitrogen will be delivered into the solution with the solution touching the tubing. Start the water flowing in the condenser. Place 3 mL of concentrated nitric acid in the addition funnel, and heat the flask to 95–105°C with a water or sand bath or an aluminum block. Have an ice bath ready to cool the reaction if needed. Add the nitric acid a few drops at a time with enough time between each addition to prevent the reaction from becoming too violent.

Question

1. Explain how this controls the reaction. What should you do if it threatens to become too lively?

This should take 4–10 minutes. When all of the acid has been added, continue heating until a separate alcohol layer has disappeared (15–20 minutes).

Isolate, purify, and analyze the product. Cool the reaction mixture, and filter (gravity, no paper) with a porcelain or glass frit Hirsch funnel.

See Figures 2.6A, B; 2.7 (pages 52–53, 55).

Question

2. Why use no suction or paper?

Use the filtrate and/or ice water to complete the transfer of the crystals. Recrystallize the product from a minimum amount of boiling water in the usual manner. When dry, determine the weight, yields, and melting range.

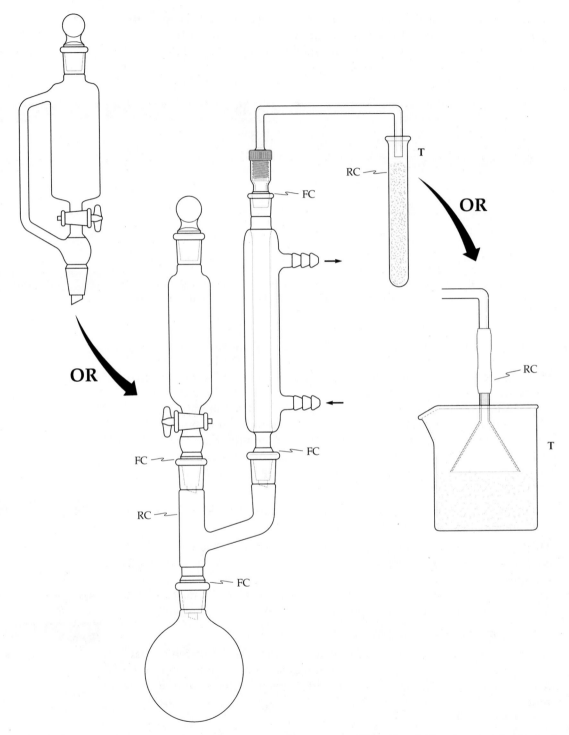

Figure 7.3 A Claisen tube equipped with a reflux condenser and addition funnel. The glassware is standard taper ground jointed. See also Fig. 6.1.

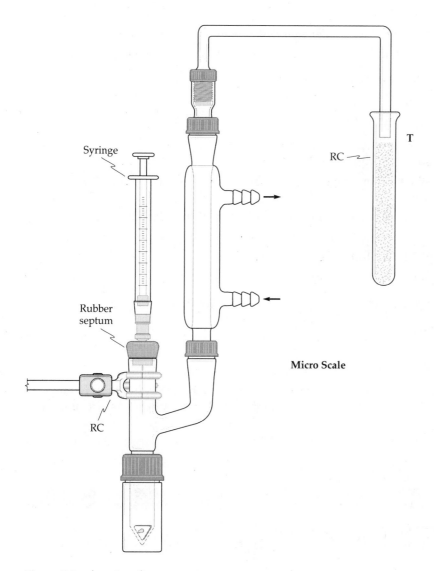

Figure 7.3 *(continued)*

Summary

1. Assemble the apparatus with alcohol in the flask and base in the trap.

2. Heat the flask, and add the nitric acid dropwise.

3. Heat or cool as needed.

4. Reflux briefly to react all of the alcohol.

5. Cool and collect the product without paper or suction.

6. Recrystallize from a minimum of hot water.

7. Analyze the product.

QUESTIONS 1 AND 2 ABOVE.

3. Is there any evidence of nitration? Why might there be? What would such by-products be?

4. What other reagents could be used to oxidize primary alcohols to acids?

5. What is the problem in the oxidation of this alcohol to benzaldehyde?

Disposal of Materials

Benzyl alcohol (OS, OW)	aq. NaOH (N or D)
conc. HNO$_3$ (N, D)	Product (I)
Filtrate (N, D)	

Additions and Alternatives

An infrared spectrum (KBr pellet) would be interesting, and thin layer chromatographic analysis should show by-products. As an alternative method you could use acetophenone and bleach (*J. Chem. Educ.*, 1984, *61*, 551) or the method in Experiment 23.

Q u e s t i o n 6. Two solids are formed. How could they be separated?

E X P E R I M E N T 2 6 B

Preparation of Benzoic Acid (M)
References: Sections 1.3F–H, 2.3, 4.2C, 5.1, 7.1, and Experiment 26A.

Assemble the apparatus and introduce the reagents. Place 0.2 mL of benzyl alcohol in a vessel suitable for 3-mL to 5-mL volumes, equipped for addition with reflux (Unit II, Figure 1.8B) and modified for a gas trap (Figure 7.3) using the equipment available. Support the test tube or beaker containing 5% sodium hydroxide solution so that the solution will not be sucked back into the reaction mixture. Cotton, wet with aqueous base, can surround the tube in a test tube as the trap.

Initiate the reaction. Heat the vessel in a water or sand bath or aluminum block at 95–105°C. Activate the condenser, and add 0.6 mL of concentrated nitric acid a few drops at a time so that the reaction does not become too violent. When the addition is complete, continue heating until the alcohol layer is gone (5–20 minutes).

See Figures 2.8A, B, C (pages 56–58).

Isolate, purify, and analyze the product. Cool the vessel, and collect the crystals that form by gravity (no paper) filtration with a Hirsch funnel. Complete the transfer with the filtrate or ice water. Recrystallize with a minimum of hot water (Craig tube or test tube and Hirsch funnel). Dry, weigh, and determine the yields and the melting range.

Summary

See Experiment 26A.

SEE EXPERIMENT 26A.

Questions

Disposal of Materials

Benzyl alcohol (OS, OW)

conc. HNO_3 (N, D)

Product (I)

aq. NaOH (N or D)

Filtrate (N, D)

Additions and Alternatives

There is room for some creative engineering in setting up an apparatus for this experiment. Also see Experiment 26A.

Try acetophenone as the substrate (half the volume of the alcohol) and 2 mL of 10–12% NaOCl solution (or 4 mL of 5–6% household bleach) and 2–3 drops of 3N NaOH. Look at Figures 7.2 and 7.3. Use an apparatus for simple reflux, with or without a trap, or a large test tube. Reflux a bit longer. Destroy the excess oxidant by adding 2–5 drops of acetone (Experiment 23) or a small amount of $NaHSO_3$. Transfer to or continue in a test tube. Heat in a boiling water bath for five to ten minutes (hood or suction). Acidify with dilute or concentrated HCl. Cool and continue as above.

EXPERIMENT 27

Preparation of Hydrocinnamic Acid (SM)
References: Sections 1.3F–H, 2.3, 3.4, 4.2C, 5.1, and 7.1.

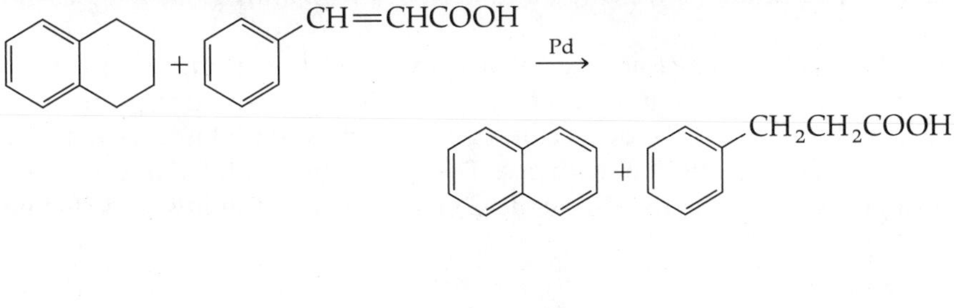

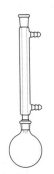

Figure 1.7, Unit I
(page 24).

Catalytic hydrogenation is one of the most important methods of reduction and perhaps the one most easily visualized. Although this method finds considerable use in industry and research, it is largely neglected in the undergraduate laboratory. This neglect is due primarily to the special equipment usually needed and to the difficulties and possible dangers inherent in the handling of gaseous hydrogen. These problems may be overcome in large measure through the generation of the hydrogen in solution by means of the aromatization of the easily dehydrogenated tetralin. Tetralin also serves as the solvent. This particular combination was discussed in detail by Wilen and Kremer, *J. Chem. Educ.*, 1962, *39*, 209. For an alternative, see McGinley, *Chemistry*, 1974, *47*, 28.

Assemble the apparatus and introduce the reagents. Place 2.0–2.5 g of cinnamic acid (accurately weighed) and 13 mL of tetrahydronaphthalene (tetralin) in a 25-mL flask equipped for reflux (Unit I, Figure 1.7) (no water, but open as an air condenser). Use a stir bar or boiling stone. Disperse 0.1 g of 15% palladium on carbon (0.05 g of 30%; 0.25 g of 5%) throughout the solution with stirring or swirling.

Initiate the reaction. Reflux the mixture for 80–90 minutes.

Question

1.   At what temperature is this reaction being run?

Cool the reaction mixture to room temperature or below.

Question

2.   Why cooled? (See the next step.)

Isolate the acid. NO FLAMES! Add 10–12 mL of ether, and separate the carbon and the catalyst by gravity filtration; place a watch glass on top.

Question

3.   Why not suction?

The catalyst may be deposited in a jar under ethanol or under another suitable solvent for future recovery or may be disposed of as instructed. Instructions will also be given for destroying the filter paper (burn).

Question

4.   What might happen as the solvent evaporates from the catalyst and carbon on the filter paper?

Extract the filtrate twice with 6-mL portions of 3M NaOH. If the other product is to be isolated, dry the nonaqueous layer; otherwise, dispose of this solution as directed by your instructor.

Acidify the aqueous layer containing the product with careful addition of concentrated hydrochloric acid. Extract the hydrocinnamic acid with two 15-mL portions of ethyl ether or methylene chloride. Wash the combined nonaqueous extracts with water and then dry with anhydrous magnesium sulfate.

Concentrate this dried organic solution by distillation using an assembly like those shown in Figure 7.1, but attach a tube to the vacuum outlet that reaches the bottom of a deep sink or the floor. Use a 50-mL flask as the receiver, and surround it with ice. Decant about 7 mL of this solution through a filter into the 10-mL flask used as the distilling vessel, and

Figure 4.1   (page 129).

decant the rest into the addition funnel. Start the water in the condenser, and heat the flask in a water bath. When 2–3 mL of solvent has been distilled, start the addition from the funnel as fast as possible without the volume in the pot increasing above 6 mL and without retarding the distillation. Distill until the solvent has been removed and a thick syrup is left.

Isolate the product. Cool the syrup intensely to get crystallization, and use seed crystals as needed. To get a seed crystal, take a few drops in an evaporating dish on a beaker of boiling water, and boil the syrup to dryness in the hood. Collect the solidified material with suction.

See Figures 2.6A, B; 2.7 (pages 52–53, 55).

Recrystallize the product. Equip the flask with a water condenser, and dissolve the product in petroleum ether with reflux. If it does not all go in, decant and reflux again with a fresh portion of solvent.

5.   What else should be done if the solution is not clear and colorless?

Question

Thoroughly cool the solution, and add a seed crystal if needed. If crystals do not form, evaporate a bit of the solvent, cool the solution, and try again. Collect the crystals with a Büchner or Hirsch funnel and *gentle* suction. Dry the crystals; determine the yield and melting range.

Summary

1.   Assemble the apparatus for reflux, introducing all of the reactants.

2.   Reflux for one and one-half hours.

3.   *Cool* and then add ether and filter by gravity.

4.   Dispose of the catalyst properly.

5.   Extract the ether layer twice with base and dry it.

6.   Acidify the basic aqueous layer with concentrated HCl.

7.   Extract the product into an organic solvent and dry this solution.

8.   Remove the solvent by distillation, and crystallize the product.

9.   Recrystallize the product.

10.   Analyze the product.

QUESTIONS 1–5 ABOVE.

Questions

6.   Why all the precautions (ice, long tube) when distilling the ether? How do they work?

7.   Why must the ether or methylene chloride be dry before distillation?

8.   Prepare a detailed flow chart showing how all of the materials are affected by each step. Where (how) are the major products separated?

9.   Discuss the differences in properties between cinnamic acid and hydrocinnamic acid.

Disposal of Materials

Cinnamic acid (S)

Pd on C (special recovery directions)

Filter paper with Pd on C (I, destroyed)

aq. NaOH (N)

Petroleum ether solutions
(OW or careful evaporation,
residue, S)

Ethyl ether (OS, OW)

$MgSO_4$ (S)

Petroleum ether (OS, OW)

Products (I)

Tetralin (OS)

Tetralin solutions

conc. HCl (N)

Methylene Chloride (OS, HO)

Ethyl ether or methylene
chloride solution (careful
evaporation, residue, S)

Additions and Alternatives

Determine the pKa of starting and product acids by titrating them with a pH meter.

An alternate method of recrystallization is given in the Wilen and Kremer article cited above. The isolation and identification of naphthalene is detailed as well. Try one or both of these. They also suggest converting this product into 1-indanone by the method of Koo (*J. Am. Chem. Soc.*, 1953, *75*, 1891). The cinnamic acid prepared in Experiment 56 may be used as the starting material. Compare the product and starting material with Exercises 1 and 2, Experiment 74. Their infrared spectra should be of interest, and thin layer chromatographic analysis might show the four compounds and how much reflux is needed.

E X P E R I M E N T   2 8 A

Preparation of Cyclohexanol (SM)
References: Sections 3.2, 4; 4.1, 2C, 2G; 5.1; and 7.1.

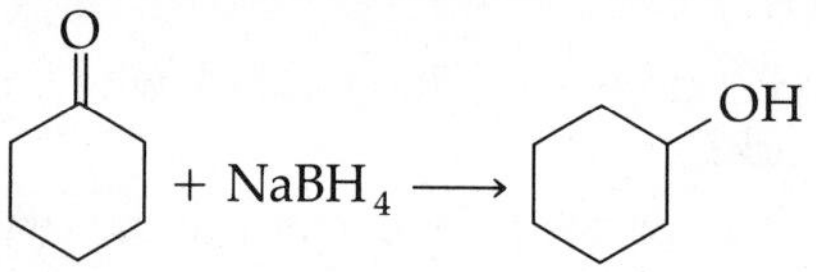

The introduction of the alkali metal alumino- and borohydrides as commercially available reagents has provided a wide variety of possibilities for the reduction of organic compounds. Lithium aluminum hydride is a powerful general reducing agent that may not be used in water or alcohol, whereas sodium borohydride is a more selective reducing agent that may be used in aqueous or alcoholic solution. For a discussion of these reagents and the development of a variety of selective reducing agents, see Brown, *J. Chem. Educ.*, 1961, *38*, 173. In most cases, these reagents will not reduce carbon-to-carbon double bonds or alkyl halides.

Introduce the reagents and initiate reaction. Add 3 mL of cyclohexanone (d = 0.95) to 6 mL of 1% NaOH in a 50-mL Erlenmeyer flask. Place the flask in a 150-mL beaker containing about 40 mL of water. If a magnetic stirrer is available, gently stir to disperse or dissolve the ketone in the aqueous solution. If not, swirl to achieve this purpose. Add dropwise 4 mL of 2.5M sodium borohydride solution in very dilute base. Monitor the temperature of the water in the beaker. Keep it close to room temperature (below 30°C) by adding ice chips as needed. *Do not overcool* or allow any water or ice into the flask. Swirl or stir the mixture for five to ten minutes more. Then take the flask out of the beaker and continue mixing gently for 10–15 minutes. **No FLAMES!** In the *hood*, add 3.0–3.5 mL of 3N HCl to the reaction mixture.

CAUTION

1. Why are these two precautions needed?

Question

Add the acid until the reaction mixture is clearly acidic. Test a drop of the mixture on external indicator paper. Be sure the mixture is cooled to room temperature.

Isolate the product. Transfer the aqueous mixture to a separatory funnel. Extract the aqueous layer three times with 8–10 mL portions of methylene chloride (or two portions, 12–15 mL each). Combine the organic layers and wash with 10 mL of water. Draw the organic layer into a flask containing some anhydrous magnesium sulfate.

Purify the product. Assemble an apparatus for simultaneous addition and distillation (Figure 7.2) with a 10-mL flask as the distilling vessel. Place, in this flask, 7 mL of the dried organic solution (no solid!) with a boiling stone. Put the rest of the nonaqueous solution in the addition funnel. Rinse the drying agent with a very little methylene chloride and add this rinse solution (no solid) to the funnel. Start the water in the condenser and initiate distillation. You may want an ice water bath around the receiver. A 50-mL flask is recommended as a receiver. Once distillation has well begun, add solution from the funnel, keeping the flask about half full. Distill off all of the material that boils below 100°C. Cool, change to a tared or graduated receiver, add a fresh boiling stone or two, and, if directed to do so by your instructor, 1–2 mL of a high-boiling chaser such as tetralin. Drain the water from the condenser and remove the hoses. Carefully distill the alcohol, collecting the fraction that boils near 160°C. This can be done with or without a chaser and at reduced pressure. The density of the alcohol is about 0.96 g/mL.

Figure 4.1 (page 129).

Interpret the results. Record the infrared spectra of the starting ketone and the product alcohol. Determine the refractive indexes.

Summary

1. Disperse the ketone in aqueous base.

2. Carefully add the basic solution of the reducing agent dropwise.

3. Allow the reaction to progress.

4. Acidify the mixture with dilute acid solution.

5. Extract the product into methylene chloride.

6. Dry the extract.

7. Strip the solvent from the product.

8. Collect the product by distillation.

Questions

QUESTION 1 ABOVE.

2. Give the balanced reaction steps for this experiment.

3. Why is the solvent stripped off by simultaneous addition and distillation?

4. How does a chaser work and how is it selected?

5. Why can the final atmospheric distillation be done with an air condenser?

6. What are the major differences in the two infrared spectra?

7. If the starting ketone and product alcohol were tested as in Exercises 5 and 7 (Chapter 17), what should be observed?

Disposal of Materials

Cyclohexanone (OS, OW)

aq. HCl (N or D)

solid $MgSO_4$ (S)

Rinse glassware with very dilute HCl before cleaning.

aq. $NaBH_4$ (treat with acid as described in third paragraph, I)

aq. NaOH (D or N)

Methylene Chloride (OS, HO)

Tetralin (OS, OW) and solutions (OW, HO, I)

Product (I)

Additions and Alternatives

Use exercises 5 and 7 (Chapter 17) to compare the starting material and the product. Ether (a bit more with suitable precautions) could be used to extract the aqueous layer after a saturated salt solution is added. You could use thin layer chromatography to follow the reaction. An interesting extension would be to reduce 2-methylcyclohexanone.

This experiment was originally developed to prepare 2,2,2-trichloroethanol from chloral hydrate. In that preparation, the $NaBH_4$ solution was made with methanol rather than with water. You may wish to try this reduction and/or the methanolic reducing agent with less-soluble aldehydes, ketones, and alcohols. For example, benzil could be used. It needs twice as much reagent. It also requires standing, heating, and adding water to cause the meso-product (m.p. < .140°C) to crystallize.

With careful planning, it may be possible for you to complete two such reductions in one period. They may be carried out in much the same fashion for many or all of the compounds reduced by Chaikin and Brown (*J. Am. Chem. Soc.*, 1949, *71*, 122). Some interesting compounds to reduce would be cinnamaldehyde, benzil, methyl ethyl ketone, and acetophenone. Some instructors prefer to mix the reactants in methanol solution, omitting the water. Heptanol works well. When possible, it is desirable to work with smaller quantities of organic reactant and relatively more reducing agent. With some starting materials, such as cinnamaldehyde, the reduction should be conducted at slightly higher temperatures, and the reaction time should be somewhat extended. In this case, the product is a low-melting solid, and it is more convenient to evaporate the solvent from an evaporating dish set on a steam bath or beaker of boiling water in the hood. Use *no flames*. A stream of dry nitrogen or air can be used with a vial or test tube. Complete cooling and rubbing should cause crystallization. (The addition of a little alcohol and another evaporation for a longer time may help.)

For the identification of smaller amounts of liquid product, you could convert it to a solid derivative, such as the 3,5-dinitrobenzoate (Section 17.2B), rather than attempt to purify it by distillation. With cinnamyl alcohol, the preparation of this derivative (m.p. = 121°) is probably the best method of demonstrating that the double bond has not been reduced.

8. What tests would usually be used? Discuss their probable success or reliability in this case.

Question

Vogel (*Practical Organic Chemistry*, 4th Ed., Longman, New York, 1978, p. 357) has suggested the reduction of *m*-nitrobenzaldehyde by much the same method but using the addition of sulfuric acid to demonstrate when an excess of reducing agent has been used. This interesting preparation illustrates the reduction of the aldehyde without effect upon the nitro group. For more details on hydride reductions, see *Chem. Eng. News*, 1954, *32*, 2424; *Organic Reactions*, Vol. VI, Ch. 10; *J. Am. Chem. Soc.*, 1951, *73*, 242; ibid., 1960, *82*, 681; *J. Chem. Educ.*, 1957, *34*, 367; *Chem. Eng. News*, 1980, Nov. 3, 19; and *J. Am. Chem. Soc.*, 1986, *108*, 6761.

Preparation of Vanillyl Alcohol (M)
References: Sections 2.3, 4.2C, 5.1, 7.1, and Experiment 28A.

Introduce the reagents. Add 0.3 g (or 305 mg) of vanillin to 2 mL of 5% aqueous sodium hydroxide in a 25-mL Erlenmeyer flask (if a magnetic stirrer is available, add a spin vane or bar) sitting in a 150-mL beaker containing 35–40 mL of water. Stir or swirl to dissolve most or all of the solid.

Initiate reaction. Add 2.5 mL of 2.5M $NaBH_4$ solution dropwise (for 3–5 minutes) with stirring or occasional swirling. The external water temperature should be kept at 20–25°C or a bit below. Allow the mixture to stand with occasional gentle agitation (15–30 minutes). In the *hood*, add 3N HCl dropwise until the evolution of hydrogen ceases. **NO FLAMES!!**

CAUTION

Question

1. Why are these two precautions appropriate?

Test the solution with external indicator paper to be sure that it is acidic.

Isolate the product. When the solution is clearly acidic, allow it to cool and then scratch the vessel to form crystals. Chill the vessel in an ice bath. Collect the crystals by suction using a Hirsch funnel. Draw air through them to dry. Weigh the crude product. Recrystallize the product from a minimum amount (about 4–5 mL/g) of hot ethyl acetate using a Craig tube. If the crude product exceeds 150–175 mg, use a test tube and a Hirsch funnel. Wash all of the glassware that contacts sodium borohydride with a preliminary rinse of dilute hydrochloric acid.

See Figures 2.8A, B, C (pages 56–58).

Analyze the results. Determine the weight, percent yield, and melting point range for the dry product. The melting point of the product should be between 110° and 115°C. Record the infrared spectra of vanillin and your product; comment on their comparison. Test the two compounds according to Exercises 7, 8, or 9 in Chapter 17, as directed by your instructor.

Summary

1. Dissolve vanillin in dilute base.

2. Add the slightly basic solution of reducing agent dropwise and with care.

3. Allow the reaction to progress.

4. Neutralize and react the mixture with acid.

5. Maximize crystallization and collect the crystals.

6. Recrystallize from a minimum of hot ethyl acetate.

7. Analyze the product.

QUESTION 1 ABOVE.

Q u e s t i o n s

2. Give the balanced reaction steps for this preparation of vanillyl alcohol.

3. What is the major difference between the infrared spectra?

4. Discuss the results of the test reactions.

Disposal of Materials

Vanillin (S)

aq. NaOH (D or N)

aq. $NaBH_4$ (treat with acid as described in the second paragraph, I)

aq. HCl (N or D)

Ethyl acetate (OS, OW) and solutions (OW, I)

Rinse glassware with very dilute acid before cleaning.

Product (I)

Additions and Alternatives

Compare with the experiment suggested by Fowler, *J. Chem. Educ.*, 1992, *69*, A43. Fowler has suggested four more reactions with vanillin on a micro scale: oxidation to an acid, condensation, bromination, and acetylation.

Try thin layer chromatography of the starting material, recrystallized alcohol, and crude alcohol. Dissolve the solids in methylene chloride, ether, or chloroform for spotting. The development solvent can be an equal mixture of toluene and ethyl acetate. You could try other mixtures of these solvents to maximize R_f values or separation. Compare with the separation methods in *J. Chem. Educ.*, 1986, *63*, 638.

This reaction can be done on five or ten times the scale with more attention to temperature control. Try this reaction on less than half of the scale in the bottom of the Craig tube (which is used for the first collection) and use 6N HCl instead of 3N.

Try the suggestions for Experiment 28A on a reduced scale by this method.

Preparation of *m*-Nitroaniline (SM)
References: Sections 2.3, 4.2C, 5.1, and 7.1.

$$C_6H_4(NO_2)_2 + 3(NH_4)_2S \longrightarrow C_6H_4(NO_2)NH_2 + 6\,NH_3 + H_2O + S_x$$

The presence of one nitro group renders another nitro group meta to it on the same aromatic ring more easily reduced than if it were alone on the ring. The amino group resulting from such a reduction does not have a comparable effect on the remaining nitro group. This situation makes possible a considerable degree of selectivity in the reduction of one of the nitro groups in *meta*-dinitrobenzene and of one of the ortho nitro groups in picric acid. Moderate excesses of the reducing agent are permissible in both cases and do not seem to have an adverse effect on either the quantity or quality of the product. The reducing agents most frequently used are the soluble sulfides. They may be used as such or prepared during the course of the reaction by passing hydrogen sulfide into an alcoholic solution of the nitro compound and a base, such as sodium hydroxide or ammonium hydroxide. If the latter procedure is used, the hydrogen sulfide may be added until the calculated increase in weight results.

Preparation of Reagent

The light ammonium sulfide solution is available commercially or may be prepared by saturating concentrated ammonium hydroxide with hydrogen sulfide gas. In the latter process, provision should be made to assure a positive pressure of the gas in order to prevent the base being drawn back into the generator; an undue increase in pressure should be avoided. A considerable amount of *heat* may be evolved in the saturation. THIS GAS IS VERY TOXIC; THIS SATURATION SHOULD BE DONE IN A HOOD, WITH CARE, AND A BASIC TRAP.

CAUTION

Experiment

Introduce the reagents and initiate reaction. **Note:** This portion of the experiment should be done in a hood. Dissolve 2 g of *m*-dinitrobenzene in 20 mL of hot 95% ethyl alcohol in a 125-mL or 250-mL Erlenmeyer flask. Add 10 mL of concentrated light ammonium sulfide solution in 1-mL portions with shaking after each addition. Heat the mixture to just below its boiling point. THE SOLVENT IS VERY FLAMMABLE; USE A WATER OR STEAM BATH OR PROTECTED SAND BATH for five minutes; cool it to room tempera-

CAUTION

ture, and pour it into 50 mL of cold water to precipitate the crude *m*-nitroaniline. Collect the crude product with suction filtration, and wash it in the approved manner. It also contains unreduced dinitrobenzene and sulfur.

Purify the product. In the purification of the amine, you will take advantage of the ability of amines to form water-soluble salts. Separate the impure product from the filter paper, and place it in 5 mL of concentrated hydrochloric acid. Add 20 mL of water. Heat the mixture to its boiling point in a hood, and filter it hot without suction, using a fluted paper. Cool the filtrate, and liberate the amine by the addition of 10 mL of concentrated ammonia. Cool the mixture again, and filter with suction. Wash the product with the minimum amount of dilute ammonia (2 drops of concentrated ammonia per milliliter of wash water). The product should be reasonably pure. If an especially pure amine is desired, recrystallize the product by dissolving it in the smallest possible amount of boiling water, filtering it without suction, cooling it, and refiltering with suction. Allow the silky crystals to dry on filter paper until the next laboratory period; determine their melting range and weight. *Use care* in handling the product; most aromatic amines are considered toxic. Avoid skin contact.

See Figures 2.6A, B; 2.7 (pages 52–53, 55).

Summary

1. Mix two of the reactants.

2. Add the reducing agent in 1 mL-portions.

3. Heat the mixture.

4. Cool and pour into cold water.

5. Collect the crude product.

6. Dissolve the product in dilute HCl with heat.

7. Filter hot and then cool the filtrate.

8. Liberate the amine with ammonium hydroxide.

9. Cool, filter, and wash the product.

10. Recrystallize if needed or directed to do so.

11. Analyze the product.

Questions

1. Why does the product dissolve in acid? Write the equation.

2. What does the addition of ammonia do after the addition of acid (question 1)? Write the equation.

3. How are dinitrobenzene and sulfur separated from the product?

4. How and why might this selective reduction be used in the preparation of *m*-iodobromobenzene? Write the equations for these reactions.

5. What products might form with the use of this method for picric acid? Where would this production stop?

6. During the five-minute heating, at what temperature should the heating bath be held for this reaction?

Disposal of Materials

m-Dinitrobenzene (OW, S)

aq. Ammonium sulfide (I)

conc. HCl (N, D, I)

conc. Ammonia (N, D)

Rinse glassware with very dilute acid before cleaning (Hood)

Product (I)

Residual Dinitrobenzene and sulfur (I, S)

Treat solution that may contain sulfide ion with $CuSO_4$ solution.

Additions and Alternatives

Consider thin layer chromatographic analysis of the product before and after recrystallization, as well as an infrared spectrum of the final product (KBr pellet). The product could be converted to *m*-nitroiodobenzene (Experiment 42A) or *m*-nitrobromobenzene and compared with the product collected by nitrating bromobenzene. You could determine the purity of the product by quantitative diazotization (Experiment 71) (two ways?). It could be diazotized and coupled in a dye experiment (Experiment 43).

Compare with the selective reductions discussed in *J. Chem. Educ.*, 1989, *66*, 611.

EXPERIMENT 29 B

Preparation of *m*-nitroaniline (M)
References: Sections 2.3, 4.2C, 5.1, 7.1, and Experiment 29A.

CAUTION

Introduce the reagents and initiate reaction. **WORK IN A HOOD.** In a 10-mL or 25-mL Erlenmeyer flask, dissolve 200 mg of *m*-dinitrobenzene in 2 mL of hot 95% ethanol. Add 1 mL of light ammonium sulfide solution a few drops at a time with thorough mixing and swirling (light shaking) after each addition. Heat the mixture to just below its boiling point for five minutes with occasional swirling or stirring. Cool the mixture to room temperature, and pour it into 5 mL of cold water.

Isolate the product. Carefully collect the crude solid on a Hirsch funnel (paper, suction), and wash it with a few drops of ice water. The product will be contaminated with unreduced dinitrobenzene and sulfur.

Purify the product. Gently remove the impure product from the filter paper, and add it to 0.5 mL of concentrated hydrochloric acid and 2 mL of water. Heat the mixture to its boiling point, and gravity filter the hot solution. Cool the mixture, and add 1 mL of concentrated ammonium hydroxide. Mix this solution thoroughly, and cool it to ice temperature. Suction filter the product with a Hirsch funnel, and wash it with a few drops of very dilute ammonium hydroxide (2 drops of concentrated NH_4OH to 1 mL of water). This product should be reasonably pure. To further purify it, recrystallize from a minimum amount of boiling water. Allow the silky crystals to dry; weigh them and determine yields and melting range.

See Figures 2.8A, B, C (pages 56–58).

Summary

See Experiment 29A.

SEE EXPERIMENT 29A.

Question

Disposal of Materials

m-Dinitrobenzene (OW, S)

aq. Ammonium sulfide (I)

conc. HCl (N, D, I)

conc. Ammonia (N, D)

Rinse glassware with very dilute acid before cleaning (Hood)

Product (I)

Residual Dinitrobenzene and sulfur (I, S)

Additions and Alternatives

See Experiment 29A.

CHAPTER 8

Aromatic Substitution Reactions

8.1 Electrophilic Aromatic Substitution Reactions

Electrophilic aromatic substitution reactions involve the reaction between an aromatic ring, which serves as a source of electrons, and a positive or partially positive electrophile, usually in the presence of an acidic catalyst. Contrast this reaction with the nucleophilic substitution illustrated in Chapter 6.

A benzene carbocation is believed to be formed and is subsequently stabilized by resonant dispersal of charge.

The ring then loses a proton in order to reform a stable aromatic ring rather than adding a nucleophile and forming the less-stable cyclohexadiene ring.

Aromatic compounds tend to react in a stepwise fashion. If a deactivating meta-directing group is present in the starting material (as in Experiment 35), the second substitution will generally require strong reaction conditions. The third substitution will take place only if the reaction conditions are made sufficiently more rigorous. If an activating, ortho-, para-directing group is present in the starting material (as in Experiment 32), the second or even the third substitution generally occurs with ease, and it may be necessary to closely control the reaction conditions to prevent polysubstitution.

For an interesting example of nucleophilic aromatic substitution, see *J. Chem. Educ.*, 1970, *47*, 41.

8.2 Halogenation

Halogens react with aromatic hydrocarbons both by substitution and by addition. The rate of the substitution reaction varies with the halogen involved. The violence of a fluorine substitution reaction makes the process very difficult to control; other methods are used to produce fluorobenzenes. Iodine reacts so slowly and with such difficulty that it is not practical to make iodobenzene by direct iodination of benzene. (See Chapter 9, Experiment 42A.) This leaves chlorine and bromine as halogens that replace hydrogen from a benzene molecule at a reasonable rate. Both of these halogens react so slowly that a catalyst, such as aluminum chloride or ferric bromide (Lewis acids), is often used.

The addition reaction, however, produces the hexahalocyclohexane after a prolonged period of time. If $X = Cl$, the product has been used as an insecticide.

E X P E R I M E N T   3 0 A

Preparation of 2,4,6-Tribromoaniline (SM)
References: Sections 1.3F–H, 2.3, 4.2C, and 8.2.

Prepare the brominating solution. Check to see if this solution has been made up. **BROMINE CAUSES EXTREMELY DANGEROUS BURNS.** Add 35.0 mL of glacial acetic acid (d., 1.05 g/mL) with care to 14.0 mL of bromine (d., 3.12 g/mL). This solution will be sufficient for ten students.

CAUTION

In the hood, introduce the reagents. Place 0.6 mL of aniline (d., 1.022 g/mL) (handle with care in the *hood*) in a 25-mL or 50-mL Erlenmeyer flask. With stirring add 4 mL of the brominating solution. After the mixture has stood for five minutes, add 8 mL of water. The procedure to this point should be carried out in the hood. If the solution is *highly* colored, you may use a little 20% sodium bisulfite to discharge the color. Collect the solid product by suction filtration, and wash it with 2.5 mL of 50% ethyl alcohol.

Purify the product. In a 50-mL round-bottom flask equipped with a reflux condenser, dissolve the solid product by warming it with 20 mL of 95% ethyl alcohol. Filter the resulting hot solution by gravity; reheat it to give a clear solution (a little alcohol may be added if needed to dissolve the product in hot solvent), and then allow it to cool to form the solid product. Collect the solid product by suction filtration; wash it with several portions of cold alcohol, dry, and weigh it. Determine its melting range, which should be between 115° and 125°C. Calculate the percent yield assuming aniline is the limiting reagent.

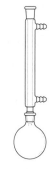

Figure 1.7, Unit I (page 24).

Summary

1. (Hood) Place one reagent in the reaction vessel and add the brominating reagent with stirring.

2. After standing with stirring, add water.

3. Decolorize if necessary.

4. Collect and wash the crude product.

5. Recrystallize from 95% ethanol.

6. Analyze the product.

1. Account for the difference in conditions between this experiment and the other two brominations (Experiments 31 and 32).

Questions

2. What type of director is the —NH$_2$ group?

3. Why could sulfuric acid not be used in place of acetic acid?

4. The color in the reaction mixture may be due to oxidation products of aniline or excess bromine. What is the function of the bisulfite? What does it become?

5. How are insoluble impurities removed?

6. How could *p*-bromoaniline be prepared?

Disposal of Materials

Bromine (HW, I)   Glacial acetic acid (D or D, N)

Aniline (HW, I)   Bromine solutions (D, N
                                      with bisulfite
                                      or thiosulfate)

aq. Ethanol solutions (add   Product (I)
   diluted HCl to N, I)

Additions and Alternatives

You could use thin layer chromatographic analysis to investigate the question of other products being present before or after recrystallization or the question of the length of time needed for reaction. You could test for the evolution of HBr.

These conditions may be used for other aromatic compounds that have ring activation.

See *J. Chem. Educ.*, 1971, *48*, 405.

E X P E R I M E N T   3 0 B

Preparation of 2,4,6-Tribromoaniline (M)
References: Sections 1.3F–H, 2.3, 4.2C, 8.2, and Experiment 30A.

Prepare the brominating solution. If it has not already been done, prepare the solution described above (30A), which should serve 30 or more students.

Initiate the reaction. (*Hood*) In a 5-mL reaction vial fitted with a cap and a stir vane, place 0.10 mL of aniline (d., 1.022 g/mL). Use a calibrated Pasteur pipet to slowly add 0.8 mL of the brominating solution to the reaction vial in the hood. Cap the reaction vial, and magnetically stir the mixture for five minutes. WHEN UNCAPPED, **HBr** WILL BE RELEASED. If the stir vane cannot stir the mixture, open the vessel under the hood from time to time, and mix it with a stirring rod. Add 2 mL of water to the reaction mixture using a calibrated pipet. If the solution is highly colored, you may use a little 20% sodium bisulfite to discharge the color. Collect the solid product by suction filtration in a Hirsch funnel, and wash it twice, using a few drops of cold 50% ethyl alcohol each time.

Recrystallize the crude product from 1.5–2.5 mL of 95% ethyl alcohol in a Craig tube. A test tube and Hirsch funnel can also be used. Dry and weigh the product.

Analyze the product. Determine the melting range, which should be 115–125°C. Calculate the percent yield assuming aniline is the limiting reagent.

CAUTION

See Figures 2.8A, B, C
(pages 56–58).

Summary

See Experiment 30A.

ANSWER THE QUESTIONS AS DIRECTED FOR EXPERIMENT 30A.

Disposal of Materials

Bromine (HW, I)

Aniline (HW, I)

aq. Ethanol solutions (add
diluted HCl to N, I)

Glacial acetic acid (D or D, N)

Bromine solutions (D, N with bisulfite
or thiosulfate)

Product (I)

Additions and Alternatives

See Experiment 30A.

If reaction vials and Craig tubes are not available, you can carry out the reaction in test tubes. The reaction mixture can stand in the hood with occasional shaking instead of stirring. In both cases, collection of crystals can be done with a Hirsch funnel.

E X P E R I M E N T   3 1 A

Preparation of *p*-Bromoacetanilide (SM)
References: Sections 1.3E, 2.3, 4.2C, and Experiment 30.

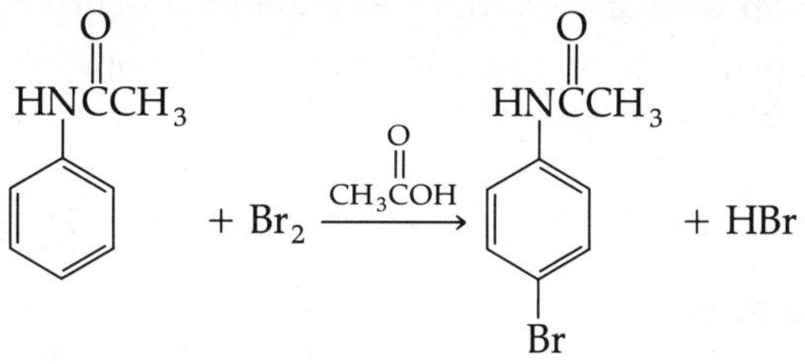

Prepare the brominating solution. Check to see if this solution has been made already. To prepare the brominating solution, add 10.5 mL of bromine (d., 3.12 g/mL) to 100 mL of glacial acetic acid. This should be enough for 20 students.

Initiate the reaction. This procedure should be carried out in the *hood*. Place 1.25 g of acetanilide in 7.5 mL of glacial acetic acid in a 50-mL Erlenmeyer flask. To this solution, slowly add with constant swirling and shaking 5 mL of brominating solution. Cool the reaction in an ice bath if the mixture becomes very much warmer than room temperature. When the addition is complete, let the mixture stand in the hood for ten minutes with occasional shaking. Pour the reaction mixture into 35 mL of cold water in a 100-mL beaker, and rinse the flask with an additional 20 mL of water, adding it to the beaker. Discharge any orange or yellow-brown color by the dropwise addition of 20% sodium bisulfite, with stirring.

See Figures 2.6A, B; 2.7 (pages 52–55).

Collect the white crystals by suction filtration in a Hirsch funnel (this may be done outside the hood); wash them thoroughly with cold water, and allow them to dry.

Recrystallize the product from methanol or ethanol. Dry and weigh the product, and determine its melting range, which should be near 165–168°C. Calculate the percent yield assuming acetanilide is the limiting reagent.

Summary

1. (Hood) Add the reactant to the acid and then add the brominating agent.

2. Control the temperature and let stand.

3. Pour into water, rinse, and decolorize if necessary.

4. Collect and wash the crude product.

5. Recrystallize from a minimum amount of hot alcohol.

6. Analyze the product.

Questions

ANSWER QUESTIONS 1–6 FROM EXPERIMENT 30A.

7. To what extent are the other two isomers of bromoacetanilide formed during the reaction? How are they removed?

8. Why is there only monosubstitution in this experiment and trisubstitutions under similar conditions in Experiment 30A?

9. What observation is or could be made to demonstrate that substitution, not addition, is taking place?

10. How could *p*-methoxybromobenzene be made using this preparation?

Disposal of Materials

Bromine (HW, I)

Glacial acetic acid (D or D, N)

Bromine solutions (D, N with bisulfite or thiosulfate)

Acetanilide (S, I)

Product (I)

Methanol or ethanol solutions (HO, I)

Additions and Alternatives

You could use thin layer chromatographic analysis to investigate the question of other products being present before or after recrystallization or of the length of time for reaction. You could test the evolution of HBr.

These conditions may be used for other aromatic compounds that have some ring activation. Try to hydrolyze the substituted amide and isolate p-bromoaniline (m.p. = 63–64°C).

See also *J. Chem. Educ.*, 1964, *41*, 341 and 1971, *48*, 405.

EXPERIMENT 31B

Preparation of *p*-Bromoacetanilide (M)
References: Sections 1.3E, 2.3, 4.2C, and Experiments 30 and 31A.

Prepare the brominating solution. If it has not been done already, prepare the solution described for Experiment 31A.

Initiate the reaction. This procedure should be carried out in the *hood*. Using a calibrated Pasteur pipet add (by drops) 0.75 mL of glacial acetic acid to 0.125 g of acetanilide in a 3-mL reaction vial fitted with a cap and a stir vane. Stir the reaction mixture to produce a solution. Using a calibrated pipet, slowly add 0.5 mL of the brominating solution to the reaction mixture, with stirring. Should the reaction mixture become much warmer than room temperature during the addition, cool the mixture in an ice bath. When addition is complete, cap the vial and stir the mixture for ten minutes. BE CAREFUL WHEN UNCAPPING THE VIAL. HBr GAS MAY BE RELEASED. Pipet the reaction mixture into 4 mL of cold water in a 10-mL or 25-mL beaker. Rinse the reaction vial with an additional 1 mL of water, and add the rinse water to the beaker. Discharge any orange or yellow-brown color in the reaction mixture by the dropwise addition of 20% sodium bisulfite, with stirring.

CAUTION

Collect the crystals on a Hirsch funnel using suction filtration (this may be done outside the hood); wash them with 5–10 drops of cold water, and allow the crystals to dry.

See Figures 2.8A, B, C (pages 56–58).

Recrystallize most of the product from methanol or ethanol in a Craig tube (or test tube). Dry and weigh the product, and determine its melting range, which should be near 165–168°C. Calculate the percent yield assuming acetanilide is the limiting reagent.

Summary

1. (Hood) Add the acid to the reactant and then add the brominating agent.
2. Control the temperature and stir.
3. Pipet into water, rinse, and decolorize if necessary.
4. Collect and wash the crude product.
5. Recrystallize from a minimum of hot alcohol.
6. Analyze the product.

Questions   ANSWER THE QUESTIONS AS DIRECTED FOR EXPERIMENT 31A.

Disposal of Materials

Bromine (HW, I)

Glacial acetic acid (D or D, N)

Bromine solutions (D, N with
 bisulfite or thiosulfate)

Acetanilide (S, I)

Product (I)

Methanol or ethanol solutions
 (HO, I)

Additions and Alternatives

See Experiment 31A.

If reaction vials and Craig tubes are not available, you can carry the reaction out in test tubes. The reaction mixture can stand in the hood with occasional shaking or mechanical stirring. Both collections of crystals can be done with the Hirsch funnel.

E X P E R I M E N T   3 2 A

Preparation of 2-Bromo-4-Nitrotoluene (SM)
References: Sections 1.3F–H and 2.3.

Assemble the apparatus and initiate the reaction. This procedure should be carried out under the *hood*. Place 3.4 g (0.025 mole) of *p*-nitrotoluene, a magnetic stir bar, and 0.1 g of iron powder or steel wool in a 50-mL round-bottom flask fitted with a Claisen adaptor having both a reflux condenser and a separatory funnel; see (Figure 8.1). Use a pressure-equalizing funnel if one is available. Attach a tube to the reflux condenser, leading to a trap with the tube just above the liquid level of a 1M sodium hydroxide solution (or use an inverted funnel; see (Figure 8.2). Heat the mixture to 70–80°C in a water bath, and begin vigorous magnetic stirring. Add 1.5 mL (4.6 g, 0.03 mole) of bromine from the funnel over a period of 12 to 15 minutes. After the bromine addition is complete, maintain the reaction mixture at 75–80°C with stirring for 40 to 50 minutes.

Isolate and purify the product. Quickly pour the molten reaction mixture with vigorous stirring into 40 mL of ice-cold 10% sodium hydroxide

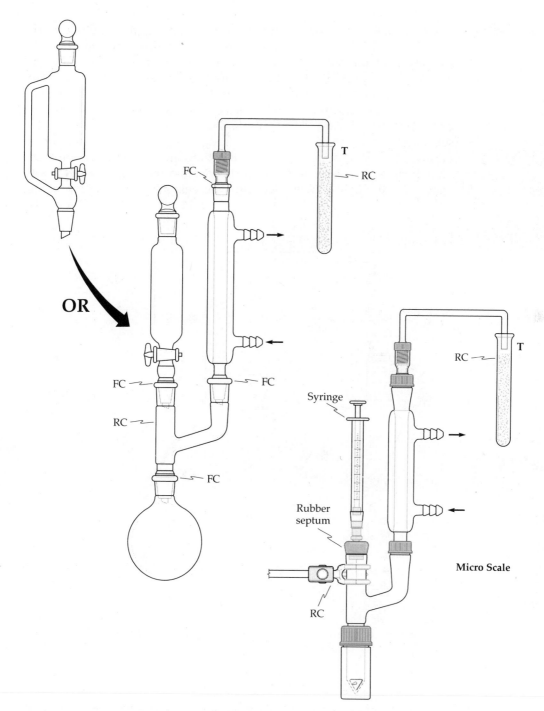

Figure 8.1 A Claisen tube equipped with a reflux condenser and addition funnel. The glassware is standard taper ground jointed.

solution, leaving the iron behind; allow the solid to settle, and decant the liquid. Add 15 mL of glacial acetic acid to the solid residue, and heat the mixture until the solid is completely melted. Thoroughly mix the two liquid phases by stirring; cool the mixture to 5°C in an ice bath, and decant the liquid. Add 25 mL of 10% acetic acid to the solid. Heat this mixture, stirring thoroughly, until the solid melts; cool it to room temperature, and decant the liquid. Add 25 mL of 1% sodium hydroxide solution to the solid, and repeat the cycle. (Stir vigorously while cooling to avoid obtaining the product as a solid cake.) Collect the solid 2-bromo-4-nitrotoluene on a Büchner or Hirsch funnel with suction, and thoroughly wash it with water. Dry the light brown solid until the next lab period; weigh it, and

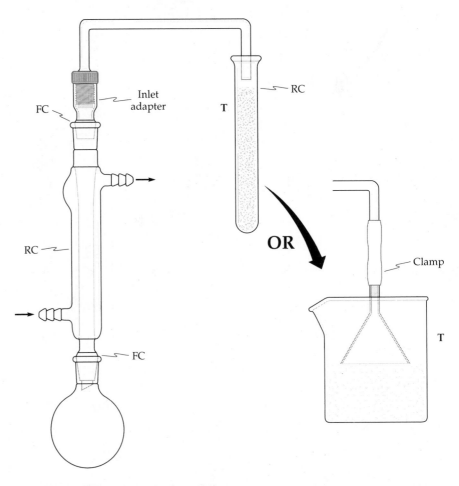

Figure 8.2 Simple reflux with fume trap.

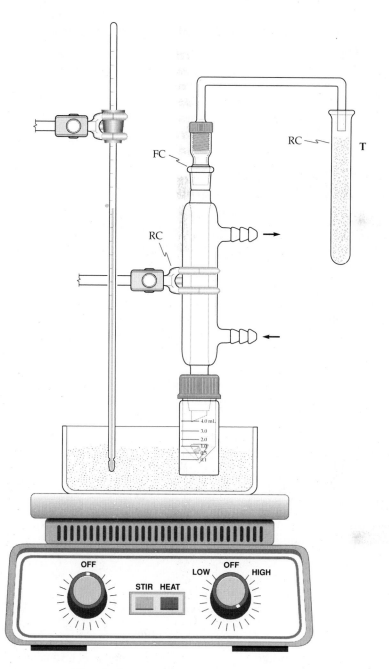

Figure 8.2 (*continued*)

calculate the percent yield. Determine the melting range, which should be between 70° and 80°C.

Summary

1. Assemble the apparatus with one reactant and the catalyst.

2. Heat and add the bromine gradually.

3. Heat to complete the reaction.

4. Pour or pipet the liquid reaction mixture into ice-cold base.

5. Solidify and separate.

See Figures 2.6A, B; 2.7 (pages 52–55).

6. Wash the product twice with acid and once with base in the same manner (melt and resolidify).

7. Collect the solid product with suction.

8. Analyze the product.

Questions

1. Why is the reflux condenser connected to a trap containing sodium hydroxide solution?

2. What is the purpose, in the work-up procedure, of each of the following: (a) the 10% sodium hydroxide, (b) the glacial acetic acid, (c) the 10% acetic acid, and (d) the 1% sodium hydroxide solution?

3. Why must the solid product be melted and stirred well with each of the following: the glacial acetic acid, the 10% acetic acid, and the 1% sodium hydroxide solution?

4. How likely are the other possible substitution products? How would they be removed?

5. How do these reaction conditions compare with those in Experiments 30 and 31? Why should they be different?

Disposal of Materials

Bromine (HW, I)

Iron (steel wool) (S)

p-Nitrotoluene (Do not handle, OW, I)

Iron or steel wool from the reaction mixture (wash with bisulfite or thiosulfate solution before S)

aq. NaOH (D or N)

Product (I)

Glacial acetic acid and acetic acid solutions (D or D, N)

Additions and Alternatives

If the iron and stir bar are poured into the sodium hydroxide, you can remove them with a stir bar retriever (or tweezers) when everything is melted.

You could use bromobenzene, or other aromatic compounds with little or no activation of the ring, in place of *p*-nitrotoluene.

See also *J. Chem. Educ.*, 1971, *48*, 405 and 1964, *41*, 566.

E X P E R I M E N T   3 2 B

Preparation of 2-Bromo-4-Nitrotoluene (M)
References: Sections 1.3F–H, 2.3, and Experiments 30, 31, 32A.

Assemble the apparatus and initiate the reaction. Place 700 mg (0.005 mole) of *p*-nitrotoluene, 0.02 g of iron powder or steel wool, and a spin vane in a 5-mL conical vial that has been fitted with a reflux condenser, Claisen tube, and septum. Attach a rubber tube to a glass tube in the inlet adaptor on the reflux condenser, leading to a small inverted funnel placed just above the liquid level of a 1M sodium hydroxide solution or below cotton moistened with this solution (Figures 8.1 and 6.1B). Heat the mixture to 75–80°C by means of a water bath, sand bath, or aluminum block, and begin vigorous magnetic stirring. Draw up 300 μL (0.3 mL) of bromine into a syringe, and introduce the needle through the septum. Add 1 drop of bromine from the syringe. After two minutes, add a second drop of bromine. Continue in this fashion until all of the 0.3 mL (0.92 g, 0.006 mole) of bromine has been added. After the bromine addition is complete, maintain the reaction mixture at 75–80°C with vigorous stirring for 30 minutes.

Isolate and purify the product. Quickly pour the molten reaction mixture into a 50-mL beaker containing 1.5 mL of ice-cold 10% sodium hydroxide solution, with vigorous stirring. Rinse the vial with 1 mL of boiling water and add the rinse to the basic solution. If the product is highly colored, discharge the color with bisulfite solution as in the previous experiments. Cool thoroughly, allow the solid to settle (centrifuge if necessary), and decant or pipet off the supernatant liquid. Add 2.5 mL of glacial acetic acid to the residue, and heat the mixture until the solid is completely melted. Thoroughly mix the two liquid phases by stirring. Cool the mixture to 5°C in an ice bath and decant or pipet off the supernatant liquid through a Hirsch funnel. Add 5 mL of 10% acetic acid. Heat this mixture until molten while stirring thoroughly. Cool it to below room temperature, and decant or pipet off the supernatant liquid through the same funnel. Add 5 mL of 1% sodium hydroxide solution, and repeat the cycle. Stir vigorously while cooling to avoid obtaining the product as a solid cake. Collect the solid

See Figure 1.8 (page 26).

2-bromo-4-nitrotoluene on the Hirsch funnel with suction, and wash it with a few drops of cold water. Dry the light brown solid until the next lab period; weigh it, and calculate the percent yield. Determine the melting range. The expected melting range is below 80°C.

Summary

See Experiment 32A, steps 1–3 and 6–8.

4. Pipet the reaction mixture into ice-cold base; leave the steel wool behind.

5. Solidify, centrifuge, and separate.

ANSWER THOSE GIVEN IN EXPERIMENT 32A.

Disposal of Materials

Bromine (HW, I)	aq. NaOH (D or N)
Iron (steel wool) (S)	Product (I)
p-Nitrotoluene (OW, I)	Glacial acetic acid and acetic acid solutions (D or D, N)
Iron or steel wool from the reaction mixture (wash with bisulfite or thiosulfate solution before S)	

Additions and Alternatives

You could use bromobenzene and other aromatic compounds in place of *p*-nitrotoluene.

If a syringe is not available, replace the septum and syringe with an inlet adapter holding a Pasteur pipet positioned to drop the bromine directly into the flask. The pipet should be closed with a well-rolled number 00 cork stopper when additions are not being made. Keep the measured bromine in a stoppered test tube, and pipet a drop or two through the pipet for each addition. The spin vane and excess steel wool can be recovered when everything has melted. Try recrystallization with aqueous ethanol or methanol.

8.3   Nitration

Nitric and sulfuric acids are believed to react together to produce the nitronium ion, NO_2^+, by the following reaction:

$$HNO_3 + H_2SO_4 \longrightarrow H_3O^+ + 2HSO_4^- + NO_2^+$$

The nitronium ion, NO_2^+, serves as an **electrophile** to the electron-rich benzene ring, with substitution of a proton to form nitrobenzene.

The nitrobenzene molecule, now having an electron-deactivating group on the ring, is less susceptible to further substitution in the ring unless the reaction conditions are made substantially more vigorous than those for the initial nitration. In the following experiment, the nitration of nitrobenzene, vigorous conditions are necessary to effect the reaction, i.e., concentrated fuming acid and a high temperature (100°C). By contrast, the nitration of benzene, a molecule without a deactivating group, may be achieved under milder conditions, i.e., concentrated but not fuming acid and a lower temperature (60–70°C).

In the following three reactions, nitration of the following types of compounds occurs: a strongly deactivated ring that demands vigorous conditions (nitrobenzene, Experiment 33), a weakly activating ring that must be monitored in order to minimize polynitration products (naphthalene, Experiment 34), and a weakly deactivating ring that can be nitrated under mild conditions to produce the mononitro product (bromobenzene, Experiment 35). To consider nitration of an activated ring, see *J. Chem. Educ.*, 1971 *48*, 635 (thin layer chromatography and column chromatography are used for analysis).

EXPERIMENT 33A

Preparation of *m*-Dinitrobenzene (SM)
References: Sections 1.3E, 2.3, 4.2C, 5.1, and 8.3.

Initiate the reaction. In the *hood*, to an ice bath–cooled 125-mL Erlenmeyer flask containing 5 mL (7.5 g) of fuming (95%) nitric acid, add (cautiously and in small portions) 7 mL (12.5 g) of concentrated (96%) sulfuric acid. Thoroughly mix and cool the acids to 10°C. Place the flask and contents in a 400-mL beaker containing about 50 mL of water, and then bring to a boil. To the acid mixture in the Erlenmeyer flask, add, stirring, eight 0.5-mL portions of nitrobenzene (d. = 1.203 g/mL) stepwise

CAUTION

with a calibrated pipet. NITROBENZENE IS TOXIC AND CAN BE ABSORBED THROUGH THE SKIN. Continue heating for ten minutes after the last addition.

Isolate and purify the product. Allow the flask and contents to cool slightly, and then cool under running water. Pour the mixture into approximately 100 mL of cold water in a 250-mL beaker, and collect the precipitated dinitrobenzene by suction filtration. Place the crude product in a small Erlenmeyer flask; cover the solid with water, and bring the water to a boil to *melt* the solid. Then remove the heat, and allow the product to resolidify. Separate the crude product and recrystallize it from 70% ethyl alcohol (ordinary laboratory ethyl alcohol is 95%); dry it by suction, wash with a few drops of cold 70% ethyl alcohol, redry by suction, and air-dry it until the next laboratory period. Determine the yield and the melting range of the dry material, which should be near 89–90°C.

See Figures 2.6A, B; 2.7 (pages 000–000).

Summary

1. Mix the two acids, with cooling.

2. Place the flask in a water bath, which is brought to a boil.

3. Add nitrobenzene in eight portions over a ten-minute period.

4. Heat to complete reaction.

5. Cool and pour into cold water.

6. Collect the solid and melt it under fresh water.

7. Solidify and collect.

8. Recrystallize from a minimum of hot 70% aqueous ethanol.

9. Analyze the product as directed.

Questions

1. What is the function of the sulfuric acid?

2. Why heat the reaction mixture after the nitrobenzene has been added?

3. Why remelt the product under water?

4. What do the reaction temperatures for the three nitration experiments suggest about the relative ease of substitution?

5. Calculate the mole fraction of the starting material and acids in each nitration experiment. Do not forget that there are four components of the reaction mixture, including water. What do your calculations suggest about the relative ease of substitution for the three substrates?

Disposal of Materials

Nitric acid (*Caution*, D, N)   Sulfuric acid (*Caution*, D, N)

Nitrating acid mixture (*Caution*, D, N)   Wash water solutions (N, I)

Recrystallization solvent (OW, I)   Product (I)

Additions and Alternatives

You may record the infrared spectrum of the product as a solidified melt, as a mull, or in a KBr pellet. You could also analyze the product mixture by thin layer chromatography.

A slightly larger scale should be used if the product is to be reduced (Experiment 29).

You could use the product of Experiment 35 for this experiment. The expected product would then be 2,4-dinitrobromobenzene, which can undergo interesting nucleophilic displacements. See *J. Chem. Educ.*, 1965, 42, 267.

E X P E R I M E N T   3 3 B

Preparation of *m*-Dinitrobenzene (M)
References: Sections 1.3E, 2.3, 4.2C, 5.1, 8.3, and Experiment 33A.

Prepare the nitrating acid mixture. Check to see if it is already available. If not, add (in small portions and cautiously) 14 mL (25 g) of concentrated (98%) sulfuric acid to an ice-cooled Erlenmeyer flask containing 10 mL (15 g) of fuming (95%) nitric acid. Thoroughly mix and cool the acid mixture to 10°C, and store it in the hood. This mixture will be sufficient for 18–20 students.

Initiate the reaction. In the hood, transfer 0.6 mL (calibrated Pasteur pipet) of the nitrating acid mixture to a 5-mL reaction vial containing a stir vane. Equip with a reflux condenser and place the vessel in a sand bath or aluminum block equipped with a thermometer. Heat the setup until the thermometer reads about 125°C (for sand or 95–110°C for aluminum). To the stirred acids, add 0.2 L of nitrobenzene (calibrated pipet) in two equal portions. Continue heating for ten minutes after the last addition.

Separate and purify the product. Remove the sand bath or block, and allow the vessel and contents to cool until the vessel can be handled. Pipet or pour the reaction mixture into a 25-mL Erlenmeyer flask containing 10 mL of cold water. Collect the precipitated product by suction filtration on a Hirsch funnel. Place the crude product in a small Erlenmeyer flask; cover it with water, and bring the water to a boil to melt the solid. Then remove the heat, and allow the product to resolidify. Separate the crude product by decantation or by suction filtration with a Hirsch funnel.

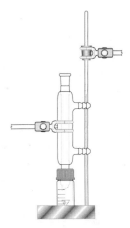

Figure 1.8   (page 26).

See Figures 2.8A, B, C
(pages 56–58).

Recrystallize as much of the crude dinitrobenzene as will dissolve in 1.8 mL of *hot* 70% ethyl alcohol (ordinary laboratory ethyl alcohol is 95%) in a Craig tube, and air-dry it on a watch glass until the next laboratory period. If the Craig tube is not available, use a test tube or 25-mL Erlenmeyer flask, a Hirsch funnel, and a minimum of hot aqueous ethanol. Wash the product on the funnel with a drop or two of 70% ethyl alcohol. Determine the yield and the melting range of the dry material, which should be near 85–90°C.

Summary

1. Place the nitrating mixture in the reaction vessel and heat to 80–88°C.

2. Carefully add the nitrobenzene directly to the hot acids in two portions, with a bit of time between additions.

3. Continue heating and then cool.

4. Pipet the reaction mixture into cold water.

5. See Experiment 33A, 6–9.

Questions   SEE EXPERIMENT 33A.

Disposal of Materials

Nitric acid (*Caution*, D, N)   Sulfuric acid (*Caution*, D, N)

Nitrating acid mixture (*Caution*, D, N)   Wash water solutions (N, I)

Recrystallization solvent (OW, I)   Product (I)

Additions and Alternatives

See Experiment 33A.

You could run the reaction in a test tube or conical tube and stir with a stirring rod or spin vane. A thermometer may be needed to monitor the reaction temperature.

If you want to double the quantities, the reaction mixture can be stirred magnetically or with a stirring rod in a 25-mL Erlenmeyer flask.

An alternative way to use the Craig tube would involve recrystallizing all of the product in a test tube and transferring 1.8–2.0 mL of the *hot* solution with a pipet preheated in 70% ethanol heated to the same temperature. It is also possible to allow the solution in the test tube to cool and remove the solvent from the crystals with a pipet or filter through a Hirsch funnel with suction.

Preparation of α-Nitronaphthalene (SM)
References: Sections 1.3E–H; 2.3; 4.2C; 5.1, 2; and Experiment 33.

In overly simplified terms, naphthalene may be regarded as an ortho-dialkyl-substituted benzene. One of the rings may be considered to be an aliphatic group attached by both ends to positions ortho to each other in the benzene ring. Such groups are ortho-para directing and activating. Consequently, most reactions of naphthalene must be carried out at relatively low temperatures to effect monosubstitution. Low temperatures favor ortho substitution, the α-substituted naphthalene product (the kinetic product), while high temperatures can effect para substitution, the β-substituted naphthalene product (the equilibrium product).

Since activating groups promote polysubstitution, there is a strong tendency for dinitration to occur in naphthalene. The ring that is nitrated first becomes more resistant to further nitration, but the second ring is deactivated very slightly if at all. A second nitro group can enter the molecule at either of the α-positions of the other ring. In the experiment that follows, the extent of dinitration is minimized by the use of a dilute nitrating mixture and by careful control of temperature.

$$\text{naphthalene} + HNO_3 \xrightarrow{H_2SO_4} \text{α-nitronaphthalene (NO}_2) + H_2O$$

Prepare nitrating acid mixture. Check to see if this 1:1 mixture has been prepared. If not, add cautiously and in small portions 54 mL (31.32 g) of concentrated (95%) sulfuric acid to an ice-cooled Erlenmeyer flask containing 54 mL (25.2 g) of concentrated (69%) nitric acid. Thoroughly mix and cool the acid mixture to 25°C in the hood. This mixture will be sufficient for ten students.

Introduce some reagents and initiate the reaction. Place 10 mL of 6N sulfuric acid in a clamped vessel (50-mL Erlenmeyer flask, 22 × 200-mm test tube, or boiling tube), and stir either *magnetically* (strongly recommended) or by hand while adding 3.3 g of naphthalene and then 3 mL of the 1:1 nitrating acid mixture. Stir throughout addition and heating. Maintain the reaction temperature at 25–30°C by monitoring the reaction with a thermometer (*Don't let the stir bar hit it!*) and immersing the vessel in an ice water bath as necessary. When removal of the reaction mixture from the ice bath no longer results in a temperature rise, add another 3-mL portion of the acid mixture and repeat the process. Add a third 3-mL portion of the acid mixture to the stirred reaction mixture at 25–30°C. The formation of polynitration products is kept to a minimum by the low temperature.

Heat the reaction mixture slowly and maintain it at 55–60°C for one-half hour with continuous stirring.

Question

6. Why heat it at all?

Isolate the product. Discontinue stirring, and cool the reaction mixture in an ice bath to solidify the product. Decant the liquid acid mixture into a beaker of cold water, and dispose of it as directed. Rinse the crude solid product (which was left behind) first with cold water, and then rinse and stir it with a mixture of 1 mL of 6N sodium hydroxide and 10 mL of water to neutralize any residual acid. Allow any oily product to settle, and cool to solidify the product. Decant the basic liquid, and wash and stir the solid product with water to remove the base. Decant the water; remove the solid cake, and dry it with filter paper or filter it by gravity.

Recrystallize the product. Place the solid in a 50-mL or 100-mL round-bottom flask equipped with a reflux condenser, and add 20 mL of low-boiling hydrocarbon solvent (hexane, petroleum ether, ligroin). Warm the mixture to boiling for a few minutes to dissolve the solid. Filter the solution hot through a small amount of dry absorbent cotton in a dry funnel. If some product remains as an undissolved oil beneath the solution, remove only the solution. Reflux the lower oily layer with another 15-mL portion of solvent. Filter the resulting solution as before, and discard any residual oil (probably polynitrated naphthalene) in an appropriate container, unless it is to be analyzed. Cool the filtrates, and collect the precipitate by gravity filtration. Keep the funnel covered with a watch glass during the filtration. Wash the crystals with a few drops of cold solvent, and allow the product to dry until the next laboratory period. Determine the yield and the melting range, which should be near 55°C.

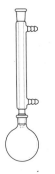

Figure 1.7, Unit I
(page 24).

Summary

1. Stir the naphthalene into sulfuric acid.

2. Gradually add the nitrating mixture with cooling as needed.

3. Warm a while while stirring.

4. Cool and solidify the product.

5. Separate the solid and wash it with base and water; then collect the product.

6. Recrystallize from warm low-boiling petroleum solvent.

7. Extract the product from any oil, if necessary.

8. Cool and collect the product without suction.

9. Analyze the product as directed.

ANSWER QUESTIONS 1, 4, AND 5 FROM EXPERIMENT 33A AND QUESTION 6 ABOVE.

7. Why is the thermometer kept in the reaction mixture?

8. Why is stirring so important? Why is the acid mixture added in three portions?

9. Why is the product sometimes an oil and sometimes a solid?

10. Write the structures for probable dinitro and trinitro products. Discuss the directive influences that led to your choices.

11. What impurities are removed in the course of this experiment? How are they removed?

12. Why is gravity filtration rather than suction filtration used to collect the crystals formed from the low-boiling petroleum solvent? Why is a watch glass used during this gravity filtration?

Disposal of Materials

conc. Sulfuric acid (D, N)	Mixed acids for nitrating (D, N)
Naphthalene (S, I)	Used acids for nitrating (D, N)
aq. NaOH (N)	Wash water (N)
Hydrocarbon solvent (OW, OS)	Hydrocarbon solutions (evaporate in hood, S)
Polynitrated residue (S, I)	
Product (I)	

Additions and Alternatives

Obtain the infrared spectrum as a solidified melt, as a mull, or in a KBr pellet. Analyze the product and product residue by thin layer chromatography. Such analysis could also determine the length of reaction time. If a nuclear magnetic resonance spectrometer is available, compare the unrecrystallized oil with the product.

Acetanilide is a bit more reactive but could be used with these conditions, with care. You can remove the protecting group if that product is hydrolyzed. Use thin layer chromatography here also. Phenol is even more reactive and can be nitrated by concentrated nitric acid itself.

Compare this experiment with those described in *J. Chem. Educ.*, 1971, *48*, 635 and 1990, *67*, 69.

You can purify naphthalene by sublimation before use.

E X P E R I M E N T   3 4 B

Preparation of α-Nitronaphthalene (M)
References: Sections 1.3E–H; 2.3; 4.2C; 5.1, 2;
and Experiments 33 and 34A.

Prepare the nitrating acid mixture. Check to see if this mixture has been prepared. If not, the directions given in Experiment 34A will provide enough for 100 or more students.

Introduce some reagents and initiate the reaction. To 1.0 mL of 6N sulfuric acid placed in a 5-mL reaction vial equipped with a small stir vane, add 0.33 g of naphthalene with stirring and then 0.3 mL of the nitrating acid mixture (use a calibrated pipet) with stirring. Maintain the reaction temperature at 25–30°C by monitoring the reaction with a thermometer (position it so it will not be hit by the stir vane, or stop stirring to measure the temperature) and by immersing the vial in an ice water bath as necessary. When removal of the ice bath from the vial no longer results in a temperature rise, add another 0.3 mL of the acid mixture, and repeat the process. Add a third 0.3-mL portion of the acid mixture to the stirred reaction mixture at 25–30°C. (The formation of polynitration products is kept to a minimum by the low temperature.) Slowly heat and maintain the reaction mixture at 55–60°C for 15 minutes with continuous stirring.

Isolate the product. Discontinue stirring, and cool the reaction mixture in an ice bath to solidify the product. Pipet off the liquid acid mixture into a beaker of cold water, and dispose of as directed. Rinse the crude solid product remaining by adding 5 drops of cold water. Pipet off the liquid. Add 0.1 mL of 6N sodium hydroxide and 25 drops of cold water to neutralize any residual acid. Allow the oily product to settle, and cool the reaction mixture to solidify the product. Pipet off the basic solution, and add with stirring 0.5 mL of water to the solid to remove the base. Filter the solid by suction on a Hirsch funnel.

Recrystallize the product. Place the solid in a 5-mL vial with a water condenser, and add 2 mL of *low*-boiling hydrocarbon solvent (hexane, petroleum ether, ligroin). Warm the reaction mixture in the sand bath or aluminum block for a few minutes to dissolve the solid. Transfer the hot *solution* with a pipet to the larger Craig tube or a small test tube. If some product remains as an undissolved oil beneath the solution, remove only the solution. Reflux the lower oily layer with another 1.0-mL portion of the solvent. Transfer the resulting solution as before to the other Craig tube or the same small test tube, and discard any remaining residual oil (unless it is to be analyzed) in an appropriate container; it is probably polynitrated naphthalene. Cool the solution, and collect the precipitate by centrifugation or by gravity filtration, using a little cotton or glass wool in a funnel. Keep the funnel covered with a watch glass during the filtration.

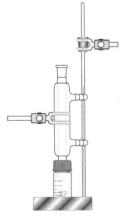

Figure 1.8 (page 26).

A Hirsch funnel and suction may also be used. Wash the crystals with 3 drops of cold solvent, and allow them to dry until the next period. Determine the yield and the melting range, which should be near 55°C.

Summary

See Experiment 34A.

SEE EXPERIMENT 34A.

Questions

Disposal of Materials

conc. Sulfuric acid (D, N)

Naphthalene (S, I)

aq. NaOH (N)

Hydrocarbon solvent (OW, OS)

Polynitrated residue (S, I)
Product (I)

Mixed acids for nitrating (D, N)

Used acids for nitrating (D, N)

Wash water (N)

Hydrocarbon solutions
   (evaporate in hood, S)

Additions and Alternatives

If reaction vials are not available, you could run the reaction in a 5-mL or 10-mL flask. Magnetic stirring is recommended because mechanical stirring is hard to maintain. Use a water bath at room temperature to surround the reaction vessel at all times, with ice added as needed. The recrystallization could be done in a boiling tube with an air condenser.
   See Additions and Alternatives in Experiment 34A.

EXPERIMENT 35A

Preparation of *p*-Bromonitrobenzene (SM)
References: Sections 1.3E, 2.3, 4.2C, 5.1, and Experiments 33 and 34.

Bromobenzene, with a slightly deactivated benzene ring, can be mononitrated between 25° and 45°C to produce both the ortho- and para-substituted nitrobromobenzene. The para-substituted isomer is formed in the larger amount and is separated from the ortho-substituted isomer by recrystallization. The more symmetrical para isomer has stronger intermolecular attractive forces and is less soluble than the ortho isomer.

Introduce the reagents and initiate the reaction. In a cooled, large Pyrex test tube or boiling tube (22 or 25 × 150 or 200 mm) cautiously mix 4 mL (5.7 g) of concentrated nitric acid and 4 mL (7.4 g) of concentrated sulfuric acid, and cool the mixture to below room temperature in an ice bath. Remove from the bath and add 2.1 mL of bromobenzene drop by drop to the mixed acids with thorough stirring and cooling as necessary to keep the temperature from rising above 50–60°C. When the addition is complete and the reaction temperature no longer rises due to the heat of the reaction, heat the tube in a boiling water bath for 30 minutes.

Isolate the product. Allow the tube and reaction mixture to cool to room temperature, and pour the reaction mixture with stirring into 40 mL of cold water in a 100-mL beaker. Filter the resulting precipitate with suction (small Büchner or large Hirsch funnel); wash the crystals with cold water, and continue to dry the crystals by suction for a few minutes.

Recrystallize the product from a minimum volume of *hot* 95% ethyl alcohol (about 18–25 mL). Determine the yield and the melting range of the crystals, which should be about 125°C. The filtrate contains the *o*-bromonitrobenzene, a more soluble isomer than *p*-bromonitrobenzene.

See Figures 2.6A, B; 2.7
(pages 52–55).

Summary

1. Carefully mix the two acids and cool below room temperature.

2. Add bromobenzene dropwise with stirring, keeping the temperature below 50°C.

3. Heat in boiling water for 30 minutes.

4. Cool to room temperature and pour into water.

5. Collect, wash, and dry the crystals.

6. Recrystallize from a minimum of hot 95% aqueous ethanol.

7. Analyze the product as directed.

ANSWER QUESTIONS 1, 4, AND 5 FROM EXPERIMENT 33.

6. Why are stirring and cooling so important?

7. What is the structure of the major dinitro product?

8. How are the impurities removed?

Disposal of Materials

conc. Sulfuric acid (D, N)   conc. Nitric acid (D, N)

Mixed acids for nitrating (D, N)   Used acids for nitrating (D, N)

Bromobenzene (HW, OS, I)   Water solution of acids (D, N)

aq. Ethyl alcohol (OS, I)   Product (I)

Ethyl alcohol used for recrystallization
    (HW or evaporate with care, I, S, HW)

Additions and Alternatives

Take the infrared spectrum as a mull or in a KBr pellet. Use thin layer chromatographic analysis to search for by-products.

Evaporate the alcohol from the recrystallization filtrate and compare the solid with the original product by infrared or thin layer chromatographic analysis. Alternatively, swamp the filtrate with water to try to force a solid out.

With minor modifications, you could nitrate chloro- and iodobenzene in this manner. Mildly deactivating esters and acids could also be nitrated in this fashion. Even with substrates such as methyl benzoate, it is very important to keep the temperature below room temperature, with stirring, to minimize unwanted additional nitration.a

A second nitro group could be added by the method used in Experiment 33A. The product is subject to interesting nucleophilic displacement. For example, see Ault, *J. Chem. Educ.*, 1965, 42, 267.

E X P E R I M E N T   3 5 B

Preparation of *p*-Bromonitrobenzene (M)
References: Sections 1.3E, 2.3, 4.2C, 5.1,
            and Experiments 33, 34, and 35A.

Preparing the nitrating acid mixture. Check to see if this mixture has been prepared. If not, add cautiously and in small portions 8 mL (14.8 g) of concentrated sulfuric acid to an ice-cooled Erlenmeyer flask containing 8 mL (11.4 g) of concentrated nitric acid. Thoroughly mix and cool the acid mixture to 25°C in the hood. This mixture should be sufficient for 18–20 students.

Figure 1.8 (pages 26–29).

CAUTION

See Figures 2.8A, B, C
(pages 56–58).

Introduce the reagents and initiate the reaction. In a 5-mL or 10-mL vial or flask with a spin vane or bar and a thermometer, place 0.8 mL of the nitrating acid mixture. *Do not let the stirrer hit the thermometer.* While stirring and cooling this mixture in an ice bath, add 0.25 mL of bromobenzene dropwise; do not allow the temperature to rise above 50–60°C. When the addition is complete and the reaction temperature no longer rises due to the heat of the reaction, equip the vial with an air condenser and heat the reaction vessel (dry the outside) in a sand or water bath or aluminum block to 100–125°C, and maintain the temperature for 15 minutes.

Isolate the product. Remove the heat source IT IS HOT, and allow the reaction mixture to cool to room temperature. Transfer the reaction mixture (pipet) with stirring to 4 mL of cold water in a small Erlenmeyer flask. Filter the resulting precipitate by suction on a Hirsch funnel; wash the crystals with 3 drops of cold water, and continue to dry the crystals by suction for a few minutes.

Recrystallize the product from approximately 2.0 mL of hot 95% ethyl alcohol in a Craig tube, or use a test tube and the Hirsch funnel. When dry, determine the percent yield and the melting range of the crystals, which should be about 125°C. The filtrate contains *o*-bromonitrobenzene, a more soluble isomer than *p*-bromonitrobenzene.

Summary

1. Add the mixed acids to a small vessel equipped for stirring and with a thermometer.

2. Add bromobenzene dropwise with stirring, keeping the temperature below 50°C.

3. Heat between 95° and 100°C for 15–20 minutes.

4. Cool to room temperature and add to cold water.

5. Collect, wash, and dry the crystals.

6. Recrystallize from a minimum of hot 95% aqueous ethanol.

7. Analyze the product as directed.

Questions

ANSWER THOSE FROM EXPERIMENT 35A.

Disposal of Materials

conc. Sulfuric acid (D, N)   conc. Nitric acid (D, N)

Mixed acids for nitrating (D, N)   Used acids for nitrating (D, N)

Bromobenzene (HW, OS, I)   Water solution of acids (D, N)

aq. Ethyl alcohol (OS, I)   Product (I)

Ethyl alcohol used for recrystallization
   (HW or evaporate with care, I, S, HW)

Additions and Alternatives

See Experiment 35A.

8.4   Sulfonation

Sulfonation of naphthalene can be effected at either of two positions, depending upon the conditions used. At a lower temperature, 80°C, the reaction of naphthalene and sulfuric acid produces α-naphthalene sulfonic acid,

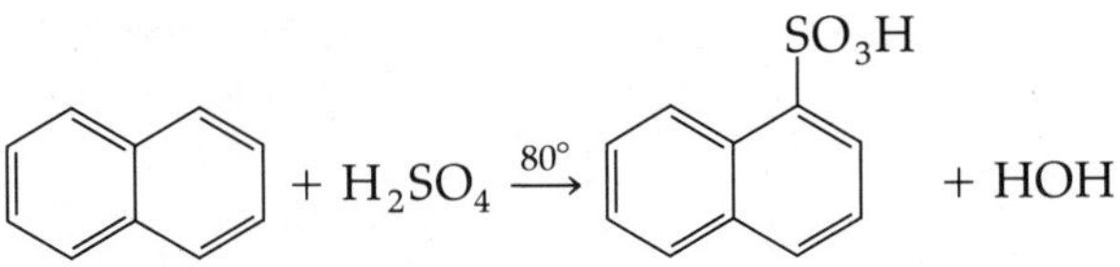

while more rigorous conditions, 160°C, result in β-naphthalene sulfonic acid.

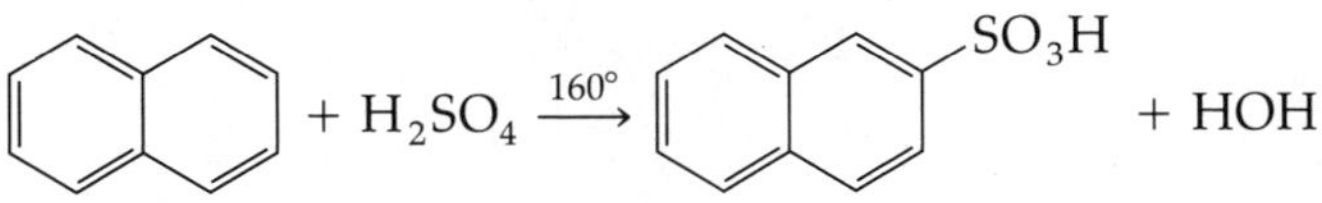

It is possible to convert α-naphthalene sulfonic acid into β-naphthalene sulfonic acid by heating the former at 160°C.

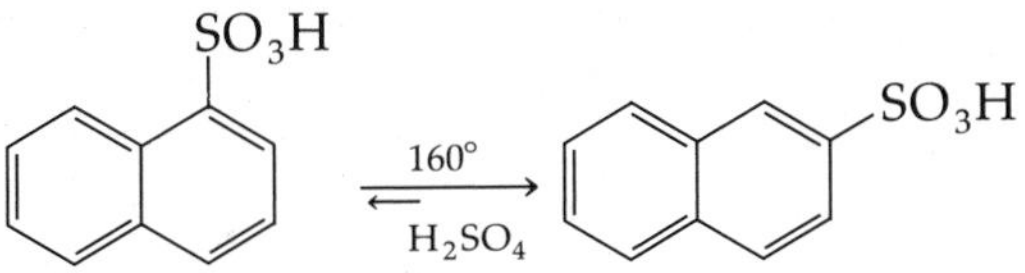

This implies that sulfonation is a reversible reaction:

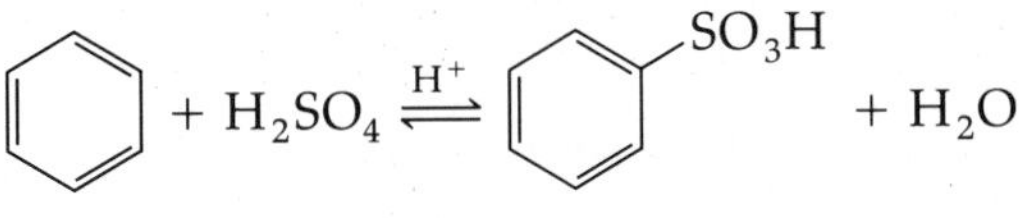

If benzene is reacted with an excess of concentrated sulfuric acid, benzene sulfonic acid and water are produced. If the water is removed as soon as it is formed, it is possible to produce a good yield of benzene sulfonic acid. If, on the other hand, dilute acid and steam distillation (i.e., a large excess of water) are used during the reaction, the benzene sulfonic acid is desulfonated to produce predominantly benzene and sulfuric acid.

The reversibility of the sulfonation reaction can be used effectively in the preparation of o-bromophenol (Experiment 37). If phenol is brominated at a low temperature, a mixture of *p*-bromophenol and *o*-bromophenol will result, with the para isomer being the predominant product. Since the more symmetric para isomer is less soluble than the less symmetric ortho isomer, it is possible to obtain *p*-bromophenol in pure form by crystallization from an appropriate solvent. It is not possible, however, to obtain *o*-bromophenol in pure form from such a solution. In order to obtain pure *o*-bromophenol, phenol is first disulfonated, placing a sulfonate group in the para and in one of the ortho positions. The sulfonated phenol is then brominated, with bromine entering the one available ortho position to yield an ortho-monobrominated, ortho-, para-disulfonated phenol. Finally, the sulfonate groups are hydrolyzed by steam distillation and pure *o*-bromophenol is obtained.

Experiment 38 illustrates sulfonation by rearrangement, which is essentially the method used to make sulfanilic acid (the precursor of sulfanilamides) from aniline. In this case, however, the starting material contains a nitro group that is reduced to an aromatic amine, while the sulfonate group is produced by oxidation.

E X P E R I M E N T   3 6 A

Preparation of Sodium β-Naphthalenesulfonate (SM)
References: Sections 2.3, 4.2C, 5.1, and 8.4.

Summary

1. Place 5 g of naphthalene in a large Pyrex test tube (25 × 200 mm).

2. Suspend the tube within a 250–300-mL Erlenmeyer with no glass-to-glass contact (air bath).

3. Heat (externally) so that half or more of the solid melts.

4. Add 6 mL of conc. H_2SO_4.

5. Heat to 160°C, measured internally, and stir carefully for five minutes.

6. Cool (< 50°C) and pour into 40 mL of cold water in a 250-mL beaker. (Decolorize, if necessary, with pelletized, activated carbon.)

7. Add 16 g of NaCl; *boil* for five minutes; maintain the water level by the addition of water.

8. Decant and cool the solution.

9. Collect the product by suction filtration; wash and dry.

10. Analyze the product as directed.

1. Maintaining the temperature at 160°C has two functions; what are they? **Questions**

2. Why is a melting point determination *not* made for this product?

3. Explain how the sodium chloride functions.

4. How are small amounts of sodium chloride removed from the product?

5. Is β-naphthalenesulfonic acid a strong or weak acid? What in this experiment supports your answer?

Questions 6 and 7 below.

Disposal of Materials

Naphthalene (S, I)   conc. Sulfuric acid (D, N)

aq. Acidic solutions (N)   Sodium chloride (S)

Product (I)

Additions and Alternatives

See Section 8.4.

You or part of the class could try this experiment with half the amount of sulfuric acid, keeping the temperature at 60°C. Stirring time should be increased to 15–30 minutes.

6. What should be the product of this experiment? **Question**

You could try adding sodium bicarbonate (with caution) instead of sodium chloride. Do not add an excess or boil.

7. What should be the reaction when this basic salt is added? **Question**

Use thin layer chromatography to reveal the presence of other isomers in the solution or filtrate before the addition of sodium chloride. If enough of the alternative solution or filtrate (before NaCl) is available, divide the solution in half, add the rest of the sulfuric acid to one half, heat that half at 160°C for five minutes, and then isolate the sodium salt from both halves. The infrared spectrum (KBr pellet) should show the difference in products. See Schaffrath, *J. Chem. Educ.*, 1970, *47*, 224 for an experiment involving sulfonation, nitration, and desulfonation. Compare this experiment (36) with the sulfonation of *p*-xylene described by Weber, *J. Chem. Educ.*, 1950, *27*, 384.

E X P E R I M E N T   3 6 B

Preparation of Sodium β-Naphthalenesulfonate (M)
References: Sections 2.3, 4.2C, 5.1, 8.4, and Experiment 36A.

Summary

1. Place 0.5 g of naphthalene in a 5-mL vial with a spin vane and a thermometer.

2. Mount the thermometer so it will not be hit by the spin vane.

3. Heat to melt half or more of the solid.

4. Add 0.6 mL of conc. H_2SO_4; heat at 160°C (internal) with stirring (five minutes).

5. Cool and add 3 mL of cold water. (Decolorize, if necessary, with activated carbon.)

6. Transfer the liquid to a larger vessel; rinse the vial with 0.5 mL of water, and transfer the rinse water.

7. Add 1.6 g NaCl, and *boil* for five minutes; maintain the volume by adding water if needed.

8. Transfer the liquid away from the excess salt with a pipet, and cool thoroughly.

9. Collect the crystals with suction; wash and dry them.

10. Analyze the product as directed.

Questions   ANSWER THE QUESTIONS FROM EXPERIMENT 36A.

Disposal of Materials

Naphthalene (S, I)   conc. Sulfuric acid (D, N)

aq. Acidic solutions (N)   Sodium chloride (S)

Product (I)

Additions and Alternatives

See Experiment 36A.

E X P E R I M E N T   3 7

Preparation of *o*-Bromophenol (SM)
References: Sections 1.3F–H; 2.6; 3.5; 4.1, 2C; and 8.4.

Summary

PHENOL AND BROMINE CAUSE VERY DANGEROUS BURNS.

CAUTION

1. Equip a 250-mL three-neck flask with a condenser, funnel (center), and stopper.

2. Place 9.4 g of phenol and 19 mL of conc. H_2SO_4 in the reaction flask.

3. Heat and stir in a boiling water bath for two to two and one-half hours.

4. Cool, then place the flask in an ice water bath.

5. With stirring, slowly add a solution of 28 g NaOH in 70 mL of water to make the solution basic.

6. Replace the stopper with a thermometer arranged to monitor the temperature of the reaction mixture without being hit by the stir bar.

7. Add 16 g of Br_2 over 20–30 minutes, keeping the temperature below 50°C.

8. Stir for another 30 minutes or overnight.

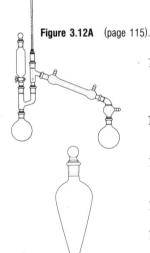

Figure 3.12A (page 115).

9. Replace the funnel with a Pasteur pipet that reaches as far as possible into the solution. Remove the condenser.

10. (Hood) Without stirring, pass clean air through the solution as it is being heated at 150–155°C until most of the water is removed (30–45 minutes).

11. Cool; replace the pipet with a clean funnel, stir, and gradually add 80 mL of conc. H_2SO_4 (under the hood, HBr is evolved).

12. Add 100 mL of water, and steam distill the mixture. (Modify Figure 3.8 or 3.12A; replace the thermometer with a stopper.)

13. Collect 80 mL or more of distillate.

14. Let the distillate stand until clear (about one hour).

15. Extract the distillate twice with an equal volume of ether.

16. Evaporate the ether extract.

Figure 4.1 (page 129).

See Figure 3.11A
(pages 104–106).

17. Distill at reduced pressure. Distillation at atmospheric pressure would occur at about 200°C with decomposition.

Q u e s t i o n s

1. Why is the product of the disulfonation of phenol converted to the sodium salt before bromine is added to the reaction mixture?

2. Why is it necessary to concentrate the reaction mixture containing the brominated sulfonated phenol before adding sulfuric acid to it?

3. How is it possible to keep the desulfonization reaction from reversing itself?

Disposal of Materials

Phenol (HW, I)   conc. H_2SO_4 (D, N)

Bromine (HW, I)   Sulfuric acid residue (N, I)

Steam distillate (N, I)   Product (I)

Additions and Alternatives

You could use thin layer chromatography to follow the reaction or to analyze the product.

E X P E R I M E N T   3 8

Preparation of Naphthionic Acid (Piria Reaction) (SM)
References: Sections 1.3E–H and 8.4.

Many aromatic nitro compounds react under reflux with sodium bisulfite or ammonium bisulfite to yield the analogous 2- or 4-aminosulfonic acids. Some of these reactions occur rather rapidly and are essentially complete in 30–40 minutes. Others give poor yields after several hours. The materials used in this exercise give average yields and require an average amount of time. Hunter and Sprung (*J. Am. Chem. Soc.*, 1931, *58*, 1432) have studied this reaction in detail. The directions given below are based in part on their results and in part on information gained from a laboratory study made by Clifford German (Univ. of Conn., 1944). The time allowed for the reaction is known to be too short. In most cases sufficient yield will be obtained to permit use of the product in the preparation of other compounds. When time, space, and equipment permit, it is recommended that the heating period be extended to approximately four hours. This can be done by starting the preparation at the beginning of one laboratory period and finishing it the next. Preparations such as *m*-nitroaniline or experiments from Chapter 9 can occupy most of the time of these two periods. Some aminonaphthalenes are listed as carcinogens, and there is a possibility of forming hazardous materials in this experiment.

Procedure

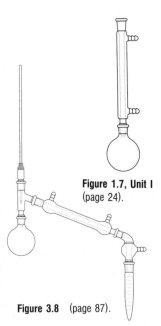

Figure 1.7, Unit I
(page 24).

Figure 3.8 (page 87).

CAUTION

Question

Prepare 2.6M sodium bisulfite solution. Check to see if this solution has been made. If not, dissolve 67.75 g of NaHSO₃ in water to make 250 mL of solution. The solution should be made fresh, at least each week. This solution should supply 40 students.

Introduce the reagents and reflux them. Dissolve 0.5 g of α-nitronaphthalene in 13 mL of 95% ethanol in a 50-mL flask. Add 6 mL of 2.6M aqueous sodium bisulfite solution, 13 mL of water, and 0.5 mL of 6M aqueous sodium hydroxide (or 3 mL of 1M NaOH). Equip the flask with a water condenser (Unit I, Figure 1.7); run cold water through the condenser, and initiate gentle reflux. The mixture should be refluxed for *at least* one hour (see above).

Remove the solvent by distillation. When the refluxing has ended, cool the apparatus. When the liquid has drained back into the flask and the glassware can be handled comfortably, rearrange the apparatus for simple distillation (Figure 3.8) with a 25-mL flask or two graduated receivers to collect the distillate. Distill off the solvent until bumping begins and/or until the 13 mL of ethanol or more than two-thirds of the solution has been distilled.

Isolate and wash the product. Turn off the heat, and add 1 mL of concentrated hydrochloric acid (carefully!) to the hot mixture. The product should precipitate. AVOID CONTACT WITH THIS MATERIAL; DO NOT GET ANY ON YOUR SKIN. Crush and mix the solid with the water, and let it settle. Decant the liquid through a funnel with filter paper. Add 5–6 mL of warm water to the solid mixture, mix, and decant through the filter. Repeat this procedure two more times with 5–6 mL of warm water each time.

1. What impurities are being removed?

The last time, transfer the solid product to the same filter paper. Dry the product, and determine the yield. Turn in the product with a proper label to have it checked by the instructor.

Summary

1. Dissolve the substrate in ethanol and add base, water, and reducing agent.

2. Reflux for an hour or more.

3. Cool and remove almost all of the solvent by simple distillation.

4. Add conc. HCl to the hot mixture (residue).

5. Crush the solid; mix with water and decant the liquid.

6. Wash the solid three times with warm water.

7. Collect the crystals and dry them.

8. Analyze and/or use the product.

QUESTION 1 ABOVE.

2. What property of bisulfite allows it to convert $-NO_2$ to $-NH_2$? What happens to the bisulfite?

3. What other organic product(s) could be formed?

4. Why is no melting point taken?

5. Would the product be expected to be somewhat water soluble? Why?

6. Would the addition of more hydrochloric acid increase the amount of product? Explain.

Disposal of Materials

Sodium bisulfite (S)   aq. Sodium bisulfite (save for use in disposal)

1-Nitronaphthalene (S, I)   aq. Ethanol (OS, I, D)

aq. NaOH (N)   Distillate (I, D)

conc. HCl (D, N)   Water washes (acidify, evaporate, I, S)

Product (I)

Additions and Alternatives

Take an infrared spectrum (KBr pellet) of the product. Use thin layer chromatographic analysis to determine the length of reflux and the presence of by-products.

The product may be coupled with a diazonium slat to produce a dye/indicator, or it can be diazotized and coupled with something else.

You could try the reaction with nitrobenzene, dinitrobenzene, or β-nitronaphthalene, or ammonium bisulfite could be used. You could try this reaction with different concentrations of alcohol (70–100%) or with other alcohols (e.g., 2-propanol).

Compare this reaction with the standard preparation of *p*-aminobenzenesulfonic acid from aniline and sulfuric acid.

8.5   Friedel–Crafts Reaction

The Friedel–Crafts reaction provides the means of introducing either an alkyl or an acyl group onto the aromatic ring, as shown in the reactions below.

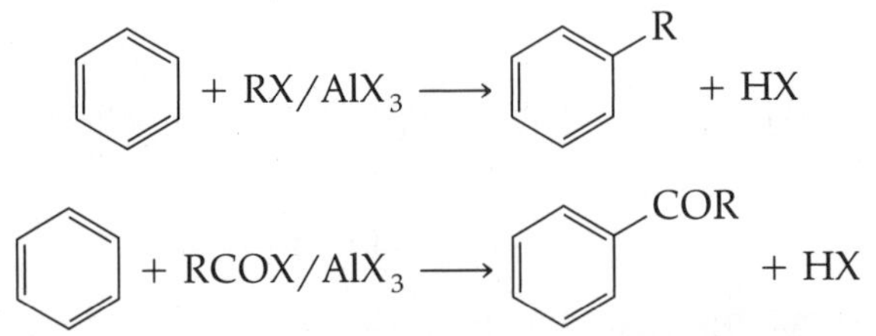

The electrophile is usually a carbocation or a polarized species coupled with the Lewis acid catalyst. Aromatic rings more deactivated than halobenzenes are not expected to undergo the Friedel–Crafts reaction.

E X P E R I M E N T   3 9 A

Preparation of *t*-Amylbenzene (M)
References: Sections 1.3F–H; 3.4; 4.1, 2D, 4G; 5.1; 7.1; and 8.5.

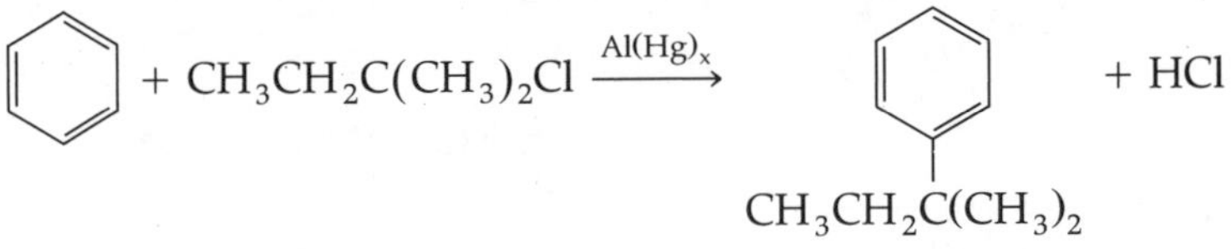

Discussion

The catalyst in this reaction is amalgamated aluminum rather than aluminum chloride. The advantage of the amalgamated aluminum is that the catalyst is prepared as needed, whereas the aluminum chloride, which is purchased ready to use, must be kept anhydrous in order to be effective. The disadvantage of the amalgamated catalyst lies principally in its relatively slow action when used with primary and secondary alkyl halides. This catalyst also creates different waste disposal problems.

Aluminum chloride seems to be formed during the early stages of the reaction. The mechanism whereby this is brought about has not been studied, but the evidence for it is as follows:

a.   The reaction is autocatalytic and increases in rate until depletion of the alkyl halide causes the reaction to slow down.

b. A sluggish catalyst is rendered more active by the addition of a small piece of active catalyst from another preparation.

c. A sluggish catalyst is activated by a very small quantity of anhydrous aluminum chloride.

d. The appearance of the reaction mixture is similar to that obtained from the conventional Friedel–Crafts reaction.

e. The same characteristic rearrangements occur in the alkyl groups being added as occur when aluminum chloride is used as a catalyst. Thus, if *n*-butyl chloride is used with either catalyst, the reaction products consist of mixtures of *n*-butyl and *sec*-butyl substitution products of benzene.

With toluene, both the ortho- and para-substituted tertiary-amyl-substituted benzenes are theoretically possible in this reaction. The para-substituted product is highly favored because of the large size of the entering tertiary-amyl substituent.

Preparation of the Catalyst

Immerse a 0.5-inch length of aluminum wire or screen, held at one end by tongs (which should not touch the solution), into a 6N sodium hydroxide solution until a vigorous evolution of hydrogen occurs. Rinse the wire briefly under running water, and quickly suspend it in a 0.2M mercuric chloride solution. After a few seconds, a mercury deposit will appear at the lower end of the wire and proceed rapidly upward. When the deposit has reached the surface of the solution, remove the wire, rinse it quickly, shake it just to remove adhering water, and transfer it to the reaction mixture. One gram of the metal has been shown to be sufficient for at least 10 g of tertiary-amyl chloride. Do not handle any metal that has been amalgamated with mercury, which is toxic.

Procedure

BENZENE IS A KNOWN, CUMULATIVE CARCINOGEN. DO *NOT* DO THIS EXPERIMENT WITHOUT SUPERIOR VENTILATION IN AN APPROPRIATE HOOD. THE EXPERIMENT AND MATERIAL HANDLING MUST BE DONE IN A HOOD. IF SUCH EQUIPMENT IS NOT AVAILABLE, SEE THE ALTERNATIVE.

CAUTION

Assemble the apparatus and initiate the reaction. Place 1.8 mL of benzene and 0.6 mL of tertiary-amyl chloride in a 5-mL conical vial equipped with an air condenser and tubing connected to a trap filled with 5–10% sodium hydroxide solution, by an inverted funnel or test tube containing cotton soaked in sodium hydroxide solution (Figure 8.2). Drop in pieces of

Figure 1.6A and Figure 4.5
(pages 19 and 148).

catalyst as they are prepared. (Two pieces should result in about the right rate.) Brown spots on the metal appear shortly and spread over the surface. The mixture turns yellow, then brown while the evolution of hydrogen chloride first becomes rapid, then slackens. Heat the bath or block as needed to reflux the mixture gently for ten minutes.

Isolate and purify the product. Remove the bath or block, and allow the reaction mixture to cool. Using a Pasteur pipet, transfer the liquid to a clean test tube with a spin vane. Add 0.5 mL of concentrated hydrochloric acid, and stir well. Allow the layers to separate. Use a Pasteur pipet to pick up both layers of liquid, being careful not to mix the layers; return the lower layer to the original vessel. Place the upper organic layer in another clean test tube. Add 1 mL of cold water and stir. Allow the layers to separate. Use a Pasteur pipet to pick up both layers of liquid carefully so as not to mix the layers, and return the lower layer to the original vessel. Pipet the upper organic layer through a drying pipet containing a small amount of anhydrous magnesium sulfate or sodium sulfate. Collect the liquid in a clean, dry 3-mL reaction vial, which is then connected to a Hickman still or short path head for distillation.

Isolate the product by distillation. Benzene, probably cloudy because of a trace of moisture, distills first. Remove the benzene from the receiver with a Pasteur pipet before the temperature reaches 180°C. If you are using the short path head, simply change receivers. Collect the distillate in the range of 180–195°C; measure the volume or weigh the distillate in a tared receiver, and determine the percent yield. The density of *t*-amyl benzene is 0.874 g/mL.

Any residue remaining probably consists of polyalkylated benzenes, which have high boiling points and can be distilled only under vacuum.

Figure 3.9 (pages 91–95).

Summary

1. Assemble the apparatus, adding the organic reactants.

2. Prepare catalyst pieces and add them to the reactants.

3. When the reaction slows way down, reflux for ten minutes.

4. Cool, transfer the liquid, and add HCl to it.

5. Separate the upper organic layer and wash it with water.

6. Separate the upper organic layer and dry it.

7. Distill the mixture—first the solvent and then the product.

8. Analyze the product as directed.

Questions

1. Write the equation for the reaction of aluminum with sodium hydroxide and with mercuric chloride.

2. What is the purpose of washing the aluminum in sodium hydroxide?

3. What is amalgamated aluminum?

4. Why is the concentrated hydrochloric acid added to the mixture after it has been refluxed?

5. Why is the organic layer (after separation from the HCl layer) washed with water?

6. The product is dried in two ways; what are they?

7. How would distillation at reduced pressure help?

8. Most Friedel–Crafts reactions are inhibited by water. How serious a problem is that here?

9. What other monosubstituted benzenes could be formed? Why are they unlikely?

Disposal of Materials

Aluminum (S)

aq. NaOH (N)

Wash water (N)

Benzene and distillate
(HW, special container, I)
aq. Mercuric chloride (and used)
(HW, special containers, I)
Amalgamated aluminum
(and used catalyst) (HW, special
containers, I) Distillation residue (OW, I)

t-Amyl chloride (HO, I)

conc. HCl (D, N)

Drying agent (evaporate, care, S)

Product (I)

Additions and Alternatives

You can run the initial reaction in a side-arm test tube (which fits in the block) closed with a cork stopper and attached to a trap by a side arm. This will accommodate the catalyst.

Record the infrared spectrum, and analyze the product by gas liquid chromatography (20% carbowax, ca. 160°C). The nuclear magnetic resonance suggests some rearrangement.

You can run the reaction of toluene, *t*-amyl chloride, and this catalyst at the micro scale with the following changes:

a. Use 1.6 mL of toluene.

b. Reflux the reaction at 100–135°C.

c. Either distill the product over 200°C or cool the apparatus after all the toluene has distilled off and distill the product at reduced pressure.

d. Increase the column temperature for gas liquid chromatography.

You could do similar preparations using *t*-butyl chloride as the halide. Other halides to be tried include ethyl iodide, 1- or 2-bromo- or iodopropane (Experiment 22), and 2-bromobutane (Experiment 20C).

You may wish to do this reaction with just aluminum metal, although the yields may not be as good. When doing this, carefully clean the metal surface with sand paper or steel wool. Heat the reaction to gentle reflux to start the reaction. Discontinue heating when the reaction is well started until it is time for reflux.

Consider some of the experiments described in *J. Chem. Educ.*, 1961, *38*, 306; 1988, *65*, 367; or 1989, *66*, 176. See also *J. Am. Chem. Society*, 1941, *63*, 3527.

EXPERIMENT 39B

Preparation of *p-t*-Amyltoluene (SM)
References: Sections 1.3F–H; 3.4; 4.1, 2D, 4G; 5.1; 7.1; 8.5; and Experiment 39A.

Although both ortho- and para-substituted tertiary-amyl-substituted benzenes are theoretically possible in this reaction, only the para-substituted product is practically possible because of the large size of the entering tertiary-amyl substituent. The meta substituent can predominate with thermodynamic equilibration.

Preparation of the Catalyst

Follow the directions given in Experiment 39A, but use three-inch pieces of aluminum wire or screen (partially bent over).

Procedure

Assemble the apparatus and initiate the reaction. In a 2.5-mL round-bottom flask equipped with a reflux condenser and a gas trap filled with sodium hydroxide solution (Figure 8.2), place 8 mL of toluene and 2.5 mL of tertiary-amyl chloride. Drop in pieces of catalyst as they are prepared. (Three pieces will usually result in about the right rate.) Brown spots on the metal appear shortly and spread over the surface. The mixture turns yellow, then brown while the evolution of hydrogen chloride becomes rapid and then slackens. Reflux the mixture gently for ten minutes to complete the reaction.

Isolate and purify the product. Cool and then transfer only the liquid to a small Erlenmeyer flask; add 5 mL of concentrated hydrochloric acid (Hood), and stir the mixture thoroughly. Transfer the reaction mixture to a separatory funnel. No metal should get into the separatory funnel. Separate the layers. Wash the organic layer once with cold water, and dry it over anhydrous magnesium sulfate. Decant the dried liquid into a clean, dry 25-mL round-bottom flask, which is then connected to a clean, dry simple distillation setup (Figure 3.8).

Toluene, probably cloudy because of a trace of moisture, distills first. Change the receiver (tared) before 180°C, and collect the distillate, whose range will usually be over 200° or 210°C. Distill at reduced pressure if possible. Determine the yield, and calculate the percent yield.

Any residue remaining probably consists of polyalkylated compounds that have high boiling points and can be distilled only with greatly reduced pressure.

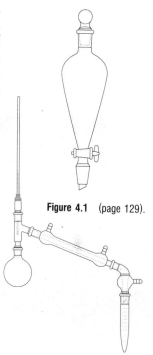

Figure 4.1 (page 129).

Figure 3.8 (page 87).

Summary

See Experiment 39A.

ANSWER QUESTIONS 1–8 IN EXPERIMENT 39A.

Disposal of Materials

Q u e s t i o n s

Aluminum (S)	*t*-Amyl chloride (HO, I)
aq. NaOH (N)	conc. HCl (D, N)
Wash water (N)	Toluene (OS, OW, I)
Drying agent (evaporate, S)	Product (I)

aq. Mercuric chloride (and used) (HW, special containers, I)

Amalgamated aluminum (and used catalyst) (HW, special containers, I)

Distillation residue (OW, I)

Additions and Alternatives

If possible, stop the distillation after the toluene has been removed, and distill the product at reduced pressure.

If *t*-butyl chloride has been made by the method used in Experiment 21, you could use it here. Other modestly active aromatic compounds could be substituted for toluene.

Try a gas liquid chromatographic analysis with collection.

See the Additions and Alternatives given for Experiment 39A.

EXPERIMENT 40A

Preparation of *p-t*-Butylphenol (SM)
References: Sections 2.3, 4.2C, 5.1, and 8.5.

MAJOR

Introduce the reagents and initiate the reaction. The first part of the experiment must be done *under the hood*. In a 50-mL Erlenmeyer flask, place 3.3 mL of *t*-butyl chloride and 2.4 g of phenol. PHENOL CAN CAUSE VERY BAD "BURNS." SEE CHAPTER 1. Stir this mixture until the phenol is almost dissolved. With the flask still in the hood, slowly add 0.25 g of *anhydrous* aluminum chloride from a fresh bottle. A lot of hydrogen chloride gas evolves when the reaction starts. DON'T BREATHE IT! If the flask becomes warm, cool it in an ice bath. Stir the reaction mixture with a stirring rod occasionally to expose fresh catalyst surface and to obtain a uniform slurry. The mixture should solidify in one-half hour or less if the reagents were carefully measured. If the mixture does not solidify, separate any aqueous layer (dispose of properly) from the oily layer, and try cooling and scratching. If this still does not work, let the mixture stand until the next laboratory period.

Isolate and purify the product. When the mixture is solid, add 12 mL of water, and break the solid up into finely divided particles with a spatula. (From this point on, the procedure may be done out of the hood.) Use a stirring rod to test a drop of the slurry on litmus paper. If the liquid is not acidic to litmus paper, add up to 1 mL of concentrated hydrochloric acid,

CAUTION

CAUTION

dropwise, with testing, until the solution is clearly acidic. Break up all of the lumps, and filter the uniform slurry with suction, drawing air through the material for several minutes. Press the solid dry with a spatula or a clean cork, and air-dry it for five minutes.

Recrystallize the solid from a slight excess of hot solvent (petroleum ether or ligroin). White crystals of *p-t-*butylphenol should appear upon cooling. Dry and weigh the crystals. Determine the melting range, which should be just under 100°C, when dry. See Hart, *J. Chem. Educ.*, 1950, *27*, 398.

See Figures 2.6A, B; 2.7 (pages 52–55).

Summary

1. Mix the organic reactants and cautiously add AlCl$_3$. (HOOD)

2. Cool as needed and stir occasionally.

3. Allow the product to solidify.

4. Add water to the solid, break it up, and acidify if necessary.

5. Collect the product by filtration and dry.

6. Recrystallize from hot petroleum ether.

7. Analyze the product as directed.

Questions

1. If this reaction is not done in an efficient hood, how should the glassware available to you be set up to carry out the reaction and trap the evolved hydrogen chloride gas?

2. Why does the aluminum chloride fume when the bottle is opened?

3. Why should the mixture (after the addition of water) be acidic? Why does it need to be acidic?

4. What use is made of 5% aqueous phenol? What does this suggest about the action of phenol itself?

Disposal of Materials

t-Butyl chloride (OS, OW, I)   conc. HCl (D, N)

Aqueous solutions (N)   Product (I)

AlCl$_3$ (Hood, *Caution*: reacts violently with water, D, N)

Hydrocarbon solvent and solutions (OS, OW, I)

Phenol (*Caution*, HW, special container, I)

Additions and Alternatives

Record the infrared spectrum and analyze the product by thin layer chromatography.

You could use *t*-butyl chloride made by the method used in Experiment 21A; the *t*-amyl chloride from Experiment 21A can be used here with appropriate modification.

A neat microscale alkylation that does not use aluminum chloride is suggested in *J. Chem. Educ.*, 1987, *64*, 440.

E X P E R I M E N T   4 0 B

Preparation of *p-t*-Butylphenol (M)
References: Sections 1.3E, 2.3, 4.2C, 5.1, 8.5,
and Experiments 3B and 40A.

Introduce the reagents and initiate the reaction. In a 5-mL conical vial equipped with a spin vane, place 0.33 mL of *t*-butyl chloride and 0.24 g of phenol (See the warnings in Experiment 40A and Chapter 1.). Stir the mixture until the phenol is almost dissolved, and then, with the flask *in the hood*, slowly add 0.025 g of anhydrous aluminum chloride. Evolution of hydrogen chloride gas begins with this addition. Cool the vial, if it becomes warm, in an ice water bath. Stirring the reaction mixture will expose fresh catalyst surface and produce a uniform slurry. The mixture should solidify in about one-half hour or less if the reagents were carefully measured.

Isolate and purify the product. When the mixture has solidified, add 1.6 mL of water, and break the solid up into finely divided particles. Test a drop of the solution on litmus paper, and, if the solution is not acidic to the litmus paper, add 3 drops (about 0.1 mL) of concentrated hydrochloric acid dropwise, repeating this test until the solution is clearly acidic. Break up all of the lumps, and filter the solid by suction on a Hirsch funnel, drawing air through the material for several minutes. Press the solid dry with a spatula or clean cork, and air-dry it for five minutes.

See Figures 2.8A, B, C (pages 56–58).

Recrystallize as much of the solid as possible in a Craig tube from a slight excess of hot solvent (petroleum ether, ligroin, or cyclohexane). White crystals of *p-t*-butylphenol should be obtained. A test tube and Hirsch funnel may be used instead of the Craig tube. Dry and weigh these crystals, and determine the percent yield. When dry, determine their melting range, which should be about 99°C.

Summary

See Experiment 40A.

Questions

SEE THOSE FOR EXPERIMENT 40A.

Disposal of Materials

t-Butyl chloride (OS, OW, I)

Aqueous solutions (N)

$AlCl_3$ (Hood, *Caution*: reacts
violently with water, D, N)

Hydrocarbon solvent and
solutions (OS, OW, I)

Phenol (*Caution*, HW, special
container, I)

conc. HCl (D, N)

Product (I)

Additions and Alternatives

You may substitute a 10-mL test tube for the 5-mL conical vial and use a
small stirring rod for mixing. If no hood is available, invert a funnel over
the vial or tube and connect the funnel to a water aspirator to suction off
the gas.

For recrystallization, dissolve the solid in hot solvent in a test tube,
adding several drops of excess solvent. With a preheated pipet, transfer
about 2 ml of the *hot* solution to the Craig tube to cool. See Experiment
33B.

See the Additions and Alternatives for Experiment 40A.

E X P E R I M E N T   4 1 A

Preparation of *o*-(4-Methylbenzoyl) Benzoic Acid (SM)
References: Sections 1.3F–H, 2.3, 3.6, and 8.5.

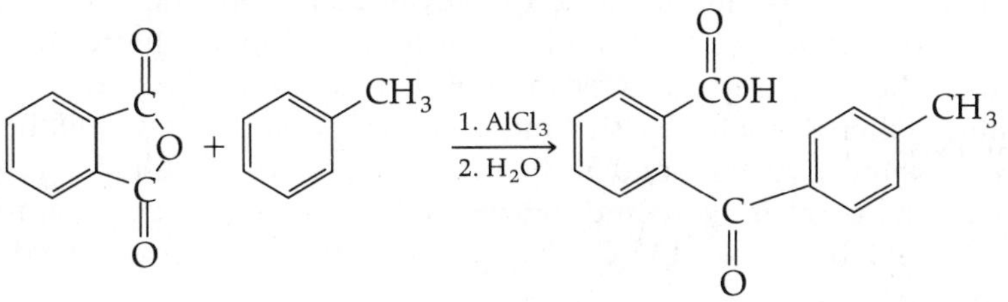

Assemble the apparatus and introduce the organic reagents. Place 4 g (0.027
mole) of pure phthalic anhydride and 17 g (0.017 mole) (19 mL) of dry
toluene (put 15 mL in now, and use the other 4 mL to rinse the catalyst
into the flask later) in a 100-mL round-bottom flask containing a magnetic
stirrer and equipped with a reflux condenser connected to a trap containing 1M sodium hydroxide (Figure 8.2).

CAUTION

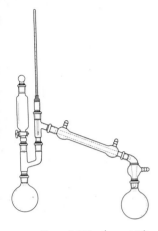

Figure 3.12A (page 115).

See Figures 2.6A, B; 2.7
(pages 52–55).

Initiate the reaction. Add 8.5 g of powdered anhydrous aluminum chloride (FORMS HCl WITH MOIST AIR) from a stoppered test tube, in four portions. Use a funnel to add the catalyst through the condenser. Immediately replace the connection to the trap. Don't get any catalyst on the ground joint or on your skin! If the reaction does not begin after adding the first 2.1 g of AlCl$_3$, warm the reaction mixture slightly. Have an ice water bath ready in case the reaction should become too vigorous. After all of the aluminum chloride has been added, rinse any catalyst from the funnel and condenser into the flask with the reserved 4 mL of toluene. When the evolution of hydrogen chloride gas decreases, warm the solution gently, and then reflux until the evolution of hydrogen chloride almost ceases (about two hours). Check at the gas trap exit with moist blue litmus paper.

Isolate and purify the product. Cool the flask to room temperature and then cool it in an ice bath. Add ice to the reaction mixture until the dark mass of material decomposes and the flask is about one-half filled with the mixture. Keep reconnecting the gas trap. The experiment can be left at this point or after the addition of acid until the next period.

Add 6–7 mL of concentrated hydrochloric acid until the mixture coagulates and then clears. Steam distill (Figure 3.12A) the mixture to remove toluene. Cool the residue in the flask, and most of the product will solidify. Decant the aqueous solution through a Büchner funnel with suction. Wash the residue with a little cold water; decant the water again through the Büchner funnel, and return any solid on the filter paper to the flask. Add a preheated solution of 4 g of sodium carbonate in 38 mL of water to the reaction mixture, and pass steam into the reaction mixture and/or heat gently until all of the solid, except aluminum hydroxide and some tar, dissolves. Cool the solution slightly. Add a few pellets of decolorizing charcoal. Reheat the solution, and filter by suction. A filter aid may help. Place the filtrate in a 100-mL beaker. Cool the solution, and slowly and cautiously add 2–3 mL of concentrated hydrochloric acid with stirring. The acid separates as an oil but upon cooling and stirring solidifies. Filter the solid by suction, and wash it with a few milliliters of ice-cold water. Dry the solid until the next period. Weigh the product, calculate the percent yield, and determine the melting range. The melting point should be below 140°C. The product may be recrystallized from toluene.

Summary

1. Assemble the apparatus with the organic reagents.

2. Add the catalyst in four batches; be sure the reaction goes in a controlled manner each time.

3. Reflux until no more HCl comes off.

4. Cool completely and decompose the addition complex with ice.

5. Add conc. HCl to complete the process.

6. Steam distill the mixture to remove toluene.

7. Filter the remaining *aqueous solution* with suction.

8. Wash the solid with a bit of water; suction filter the liquid and return any solid to the reaction flask.

9. Add very warm Na_2CO_3 solution to the reaction mixture.

10. Pass steam into the mixture until the product dissolves, but not the impurities.

11. Cool, decolorize, reheat, and suction filter hot.

12. Cool and *cautiously* add conc. HCl; cool to solidify the oil.

13. Collect the solid with suction; wash and dry.

14. Analyze as directed.

Questions

1. If the phthalic anhydride is not pure, what could very likely be mixed with it?

2. Why is it desirable to keep aluminum chloride in a stoppered test tube until it is all added to the reaction mixture?

3. Why are ice and then concentrated hydrochloric acid added to the reaction mixture?

4. After the reaction mixture is steam distilled to remove the toluene, the acid product is reacted with sodium carbonate to form the sodium salt and then converted back to the acid again by adding concentrated hydrochloric acid. Why is the acid converted to the salt and then back to the acid again before its isolation as final product?

Disposal of Materials

Phthalic anhydride (S, I)

aq. NaOH (N)

Aqueous solutions (N)

Decolorizing carbon (S)

Toluene (OS, OW. I)

Aluminum chloride (Hood, *Caution*: reacts violently with water, D, N)

Toluene (OS, OW, I)

conc. or dil. HCl (D, N)

Sodium carbonate (S, I)

aq. Na_2CO_3 (N)

Toluene solutions (OW, I)

Additions and Alternatives

You could use the succinic anhydride made in Experiment 54 in place of the phthalic anhydride.

5. What would the product be?

See Campbell and Kline, *Semimicro Experiments in Organic Chemistry*, D. C. Heath, Boston, 1964, p. 109. Compare with *J. Chem. Educ.*, 1989, *66*, 1056.

You could cyclize this product to an anthroquinone with polyphosphoric acid or possibly concentrated sulfuric acid.

E X P E R I M E N T   4 1 B

Preparation of *o*-(4-Methylbenzoyl) Benzoic Acid (M)
References: Sections 1.3F–H, 2.3, 3.6, 8.5, and Experiment 41A.

Assemble the apparatus and introduce the organic reagents. Place 400 mg (0.003 mole) of pure phthalic anhydride and 1.65 g (0.018 mole or 2 mL) of dry thiophene-free toluene in a 10-mL pear-shaped flask with a spin vane and a reflux condenser connected to a trap containing 1M sodium hydroxide (Figure 8.2).

Initiate the reaction. Add 0.85 g of powdered anhydrous aluminum chloride (see CAUTION in Experiment 41A) from a stoppered test tube, in four portions. Don't get any on the ground joint! Use a funnel. Reconnect the trap immediately after each addition. If the reaction does not begin after adding the first 0.21 g, warm the reaction mixture slightly in a water bath. Have an ice water bath ready in case the reaction should become too vigorous. After all of the aluminum chloride has been added, rinse the funnel into the vial with 0.25 mL of toluene. When the evolution of hydrogen chloride gas decreases, dry the outside of the flask, heat the flask and solution gently, and then reflux until the evolution of hydrogen chloride almost ceases. Check with moist blue litmus.

Remove the heater; cool the flask to room temperature and then cool it in an ice bath. Add a little ice to the reaction mixture until the dark mass of material decomposes and the flask is about two-thirds filled with the mixture. The reaction can be stopped here if necessary.

Add 0.6–0.7 mL of concentrated hydrochloric acid until the mixture coagulates and then clears. Place a distilling adapter or short path head on the flask, and heat the reaction mixture with an electric heater in order to remove the toluene by steam distillation.

Isolate and purify the product. Cool the residue in the flask, and most of the product will solidify. With a Pasteur pipet, transfer the aqueous solution through a Hirsch funnel with suction. Wash the residue in the flask with a small amount of cold water. Pipet the water again through the Hirsch funnel. Return any solid that may have gotten on the filter paper to the flask. Add a preheated solution of 0.5 g of sodium carbonate in 3.7 mL of water to the reaction mixture, and heat the mixture gently until all of the solid, except aluminum hydroxide and some tar, dissolves. Transfer the liquid to a 12-mL glass centrifuge tube with a (plugged) pipet, and slowly and cautiously add 0.3 to 0.4 mL of concentrated hydrochloric acid with stirring. The acid separates as an oil but upon cooling and stirring solidifies. Filter the solid through a Hirsch funnel, and wash it with a few drops of ice-cold water. Dry the solid until the next period to obtain anhydrous *o*-(4-methylbenzoyl) benzoic acid. Weigh the product. Calculate the percent yield, and determine the melting point. The literature suggests a melting range of 138–139°C and a yield of 0.65 g (95%). The product may be recrystallized from toluene.

See Figures 2.8A, B, C (pages 52–55).

Summary

See Experiment 41A for 1–9.

10. Heat gently and stir to dissolve as much as possible.

11. Transfer the hot liquid to a clean tube and cool.

12–14. in Experiment 41A.

SEE THOSE FOR EXPERIMENT 41A.

Questions

Disposal of Materials

Phthalic anhydride (S, I)

aq. NaOH (N)

Aqueous solutions (N)

Decolorizing carbon (S)

Toluene (OS, OW. I)

Aluminum chloride (Hood, *Caution*: reacts violently with water, D, N)

Toluene (OS, OW, I)

conc. or dil. HCl (D, N)

Sodium carbonate (S, I)

aq. Na_2CO_3 (N)

Toluene solutions (OW, I)

Additions and Alternatives

See those for Experiment 41A.

CHAPTER 9

Diazonium Ions and Their Reactions

9.1   Diazonium Ions

Primary aromatic amines react with nitrous acid, HNO_2, at low temperatures to yield stable aryldiazonium ions, $Ar\!-\!N^+\!\!\equiv\!N$, which are among the most versatile intermediates in organic synthesis. The diazo group can be easily replaced with nucleophiles, including halogens and hydroxyl, cyano, and azido groups (Sandmeyer reactions, Figure 9.1). Replacing this group with an iodide ion from KI is perhaps the simplest reaction because no special catalyst is required (Experiment 42A). In addition to their reactivity in Sandmeyer substitution reactions, diazonium ions undergo a coupling reaction with activated aromatic rings to yield azo compounds. Some of them are the brightest and most useful compounds in the synthetic dye industry (equation 1).

methyl orange
deprotonated form

$$(1)$$

Alkylamines also react with nitrous acid, but the products of these reactions, alkyldiazonium ions, are too reactive to isolate since they lose nitrogen instantly. Alkyldiazonium ions generated from processes involving ingested nitrates and nitrites have been found to modify DNA or RNA and hence are carcinogenic.

Replacing the diazo group with an azido group (Experiment 42B) to form light-sensitive aryl azides is widely used in preparing photoaffinity labels for biological studies. The reaction mechanism for this replacement is quite different from other replacements.

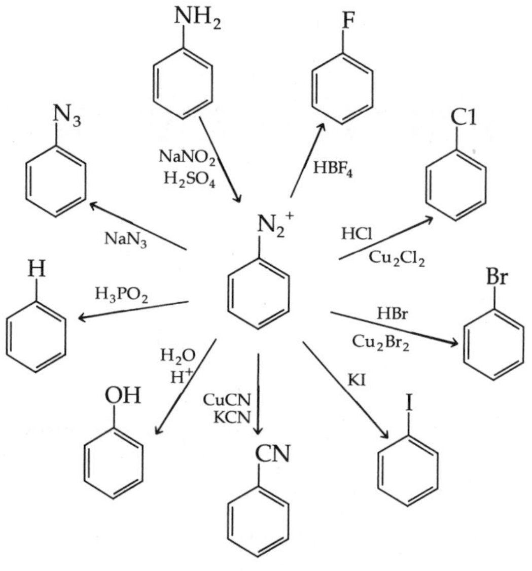

Figure 9.1 Sandmeyer reactions.

Photolysis of aryl azides produces a set of fascinating reactive intermediates, including triplet nitrenes. The nitrene may dimerize to form diazobenzene, react with oxygen to give nitrobenzene, or label proteins or nucleic acids. Experiment 42B, in which photolysis of *p*-nitrophenyl azides gives 4,4'-dinitroazobenzene, offers an example of preparing azo compounds in which both aryl groups are deactivated. The authors are indebted to Dr. Yuzhuo Li and his students (Clarkson University, 1991–1992) for their work in the development of this experiment.

EXPERIMENT 42A

Part One: Preparation of *p*-Iodotoluene (SM)
References: Sections 2.3, 3.6, and 4.3

$$CH_3C_6H_4NH_3^+ + HNO_2 \longrightarrow CH_3C_6H_4N_2^+ + 2H_2O$$
$$CH_3C_6H_4N_2^+ + KI \longrightarrow CH_3C_6H_4I + K^+ + N_2$$

Prepare solutions and diazotize the amine. Prepare two solutions, one consisting of 8 g of KI in 15 mL of water and the other consisting of 3 g of NaNO$_2$ in 10 mL of water (or use stock solutions as directed). Place 15 mL of water in a 250-mL beaker, and carefully add 6 mL of concentrated H$_2$SO$_4$. The reaction will give off heat. While the mixture is still hot, add 4.25 g of *p*-toluidine. Add 30 mL of finely crushed ice (measured in a graduated cylinder) with stirring, and continue stirring until the temperature drops to 0°C. Although the finely divided amine sulfate precipitates as the solution cools, it will react more readily with nitrous acid in this form than in its original, coarsely crystalline state.

Add the NaNO₂ solution dropwise to this suspension; stir the mixture constantly, and add enough additional ice to maintain the temperature below 8°C. Carry out this diazotization step as rapidly as possible without exceeding this temperature. Higher temperatures or allowing the suspension to stand for a great length of time in water increases the possibility of the reaction:

$$CH_3C_6H_4N_2^+ + H_2O \longrightarrow CH_3C_6H_4OH + N_2 + H^+$$

As the diazotization progresses, the precipitated amine sulfate should dissolve completely. After all of the NaNO₂ solution has been added, add 0.5 g of solid urea to the solution, with stirring, to destroy the excess nitrite ion. The reaction is:

$$2\,HNO_2 + H_2NCONH_2 \longrightarrow CO_2 + 3H_2O + 2N_2$$

Withdraw about 1 mL of the *p*-toluenediazonium acid sulfate solution *for the coupling reaction below.* This portion must be kept cold until used.

Initiate the replacement. Pour the remaining solution, with stirring, into a 400-mL beaker containing the KI solution, and allow the mixture to stand for half an hour. (During this time, work on the coupling reactions.) Then heat the mixture to 60°C in a water bath, with stirring. Decant the acidic, aqueous solution from the oily product (d., 1.68 g/mL at 40°C). Wash the product successively with warm aqueous 5% Na₂CO₃, 5% NaHSO₃, and several portions of water in the same manner.

Questions

1. Why must the product be kept warm during the washing process?

2. What is the purpose of each of these washes?

Add 20 mL of crushed ice and water to the oil, and stir. The product should solidify but may require standing. The product may be purified by steam distillation, recrystallization, or sublimation, but the melting point is usually satisfactory without purification, even though the product is tan. Omit the solidification if the product is to be purified by steam distillation. The melting point range should be about 34–36°C.

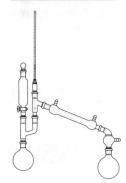

Summary

1. Prepare two stock solutions and one solution containing the amine.

2. Cool the amine solution to 0°C by adding ice and stirring.

Figure 3.12A (page 115).

3. Add the nitrite solution, dropwise, with temperature control, without delay.

4. Destroy the excess nitrite.

5. Save a portion and pour the rest into the iodide solution.

6. Allow to stand for 30 minutes and then warm.

7. Decant off the aqueous solution.

8. Wash the oil with warm rinses of base, bisulfite, and water.

9. Each time, separate the aqueous layer by decantation and/or pipet.

10. Solidify the product by adding crushed ice.

10a. Transfer the oil to a round-bottom flask and rinse it in with hot water.

11. Collect the solid and purify by recrystallization or sublimation, as directed.

11a. Steam distill the oil, separate, and solidify as in (10) and (11).

12. Dry and take the melting point, or analyze as directed.

Part Two: Coupling of a Diazonium Salt (M)

A small ice-cold test tube or a large dent in the top of an ice cube can serve as the vessel for this microscale coupling experiment. Test one compound from each of the following groups by combining a few drops of its solution with a few drops of the diazonium salt solution.

a. Test a 5% solution of *N,N*-dimethylaniline in alcohol. If this mixture gives a pink color, add 5% Na_2CO_3 dropwise until the color changes.

Question 3. What causes this change?

b. Test a 5–10% solution of sulfanilic acid or naphthionic acid in 5% Na_2CO_3. After reaction, make the mixture acidic with conc. HCl and then basic again with solid sodium carbonate.

Question 4. What would be expected?

c. Test a 5–10% solution of β-naphthol or a nitro- or alkoxyphenol in 5% NaOH (avoid skin contact with phenols).

Questions QUESTIONS 1–4 ABOVE.

5. Why is the excess HNO_2 destroyed?

6. Summarize the results of the coupling tests that you performed.

Disposal of Materials

aq. KI (D or evaporate, S)

aq. Sodium bisulfite solution (save for disposal of other materials) conc. H_2SO_4 (N)

p-Toluidine (HW, special container, I)

N,N-Dimethylaniline solution (evaporate, OW, I)

aq. $NaNO_2$ (evaporate or add diluted acid) and urea, N)

Basic sulfanilic acid or napthionic acid (N, evaporate, HW, I)

p-Toluenediazonium solution (heat, N)

aq. Acid and base solutions (N)

Product (I)

Basic *β*-naphthol solution (evaporate, OW, HW, I)

Additions and Alternatives

See Part Two of Experiment 43. Try using the coupling reactions above for dyeing.

Other aromatic amines such as *p*-nitroaniline or *m*-nitroaniline (Experiment 29, 5.5 g) could be used in Part One.

You could use the *p*-iodotoluene in the Ullman reaction to produce *p,p'*-dimethylbiphenyl or for the formation of a Grignard reagent.

E X P E R I M E N T   4 2 B

Part One: Preparation of *p*-Nitrophenyl Azide (SM)
References: Section 2.3 and Chapter 5

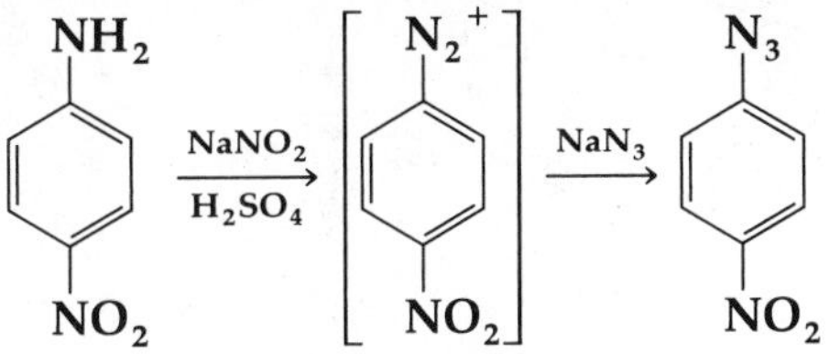

$$\text{(2)}$$

CAUTION

MOST ORGANIC AZIDES ARE EXPLOSIVE. THIS AZIDE IS VERY STABLE, BUT THE HAZARD OF AN EXPLOSION FROM A SPARK, IMPACT, GRINDING, OR REACTION WITH STRONG ACIDS AND OXIDANTS SHOULD BE REMEMBERED.

Assemble the apparatus and introduce the reagents. Make a slurry with 740 mg of *p*-nitroaniline (5.36 mmol) in 10 mL of acetic acid and 5 mL of

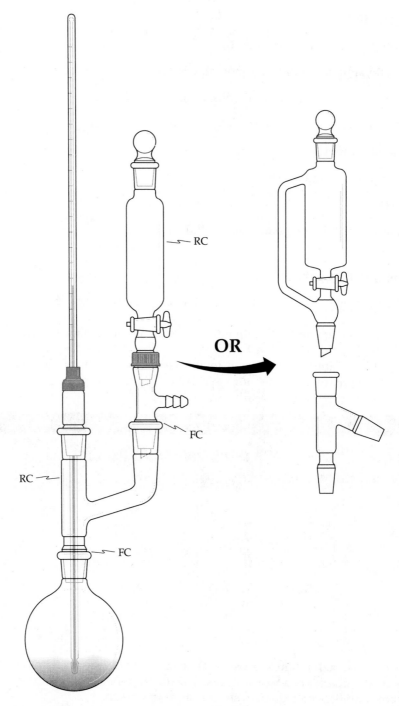

Figure 9.2 Addition with temperature control.

concentrated sulfuric acid in a 50-mL round-bottom flask, equipped with a Claisen adaptor, a separatory funnel, a thermometer, and a magnetic stir bar (Figure 9.2). When spinning, the magnetic stirrer must not strike the thermometer.

Initiate the diazotization. Cool the slurry well *below* 10°C in an ice bath. Prepare a solution with 500 mg of sodium nitrite (7.25 mmol) and ice-cold water (5 mL). Put this solution into the separatory funnel. While stirring the slurry, add the sodium nitrite solution dropwise over a 40–45-minute period. Be careful to keep both solutions as cold as possible during this addition. Allow the solution to sit (cold) for an additional 40–45 minutes, stirring. When the waiting period is almost over, make a slurry with 1.1 g of sodium azide (17 mmol), sodium acetate (10 g), and ice (10 g) in water (10 mL). Add solid urea to the *cold* reaction mixture, with swirling, until no more bubbles of gas form. Suction filter this solution and keep the filtrate as cold as possible. Add the filtrate slowly to the newly prepared azide slurry; nitrogen will evolve vigorously and the nitrophenyl azide will precipitate. You can stop at this point if necessary and store this mixture in the refrigerator until the next laboratory period.

Isolate the product. Filter the precipitate with suction and wash it with cold water three times. The crude product should give a reasonable melting range around 70°C, an infrared spectrum, a proton nuclear magnetic resonance spectrum, and a UV-Vis spectrum. The infrared spectrum can be obtained in chloroform or as a KBr pellet, and the nuclear magnetic resonance spectrum can be obtained in chloroform-d. The product can be recrystallized from a minimum of hot 95% ethyl alcohol.

See Figures 2.6–2.8 (pages 52–58).

Summary

1. Assemble the apparatus and make the amine slurry in the flask.

2. Cool thoroughly and prepare the cold nitrite solution.

3. Add the cold solution to the cold slurry, dropwise.

4. Allow to stand (cold) and make a fresh, cold azide slurry.

5. Destroy any excess nitrite with urea.

6. Filter the reaction mixture and add the cold filtrate to the cool azide solution.

7. Collect the product with suction filtration and wash it three times with ice water.

8. Analyze the product as directed.

Part Two: Preparation of 4,4′-Dinitroazobenzene by Photolysis
References: Sections 1.3l and 5.1

$$(3)$$

Prepare the solution and photolyse it. Prepare 10 mL of a 0.01M solution of *p*-nitrophenyl azide in toluene. Purge the solution with nitrogen for five minutes in a 25-mL pyrex photochemical cell equipped with a septum. Photolyze the solution in the cell with 350-nm light for one hour.

See Figure 1.9
(page 30)

Isolate and analyze the product. Transfer the solution into a small Erlenmeyer flask and evaporate the solvent (see Figure 1.9). The crude product can be analyzed by infrared spectroscopy, proton nuclear magnetic resonance, and melting range determination (just above 80°C). Further purification can be accomplished by recrystallization from 95% ethanol.

See Figures 2.6A, B; 2.7
(pages 52–55).

Summary

1. Prepare the azide solution.
2. Photolyse for at least one hour.
3. Evaporate the solvent.
4. Analyze the product.

Questions

1. Why does the reaction mixture have to be chilled before the sodium nitrite solution is added?

2. Why is nitrous acid used in this reaction formed in situ rather than prepared before and then added?

3. How and why is the wavelength selected for photolysis? (Hint: Consider the UV-Vis spectrum.)

4. How is the glassware selected for photolysis? (*Hint*: Different glassware absorbs different wavelengths of light.)

5. Rank the stability of the following diazonium ions:

6. Why must the reaction mixture be purged with nitrogen before photolysis?

Disposal of Materials

p-Nitroaniline (HW, S, I)

conc. H_2SO_4 (D, N)

Sodium azide (S, I)

Sodium acetate (S)

Urea (S)

Chloroform (HW, OS, I)

Toluene (OS, OW, I)

Ethanol (OS, D, I)

aq. Sodium nitrite (add diluted acid and evaporate, S, I)

p-Nitrophenyl azide (S, HW, special container, I)

Filtrate and wash solutions (evaporate, S, HW, I)

Toluene solution (evaporate, HW, I)

Ethanol solutions (evaporate, HW, I)

Additions and Alternatives

You can monitor the photolysis by infrared or UV-Vis spectra to optimize the photolysis time. Azides absorb strongly at 2100 cm^{-1}, while diazo compounds absorb at 1900 cm^{-1}.

The diazotization in Part One can be done in a 50-mL or 125-mL Erlenmeyer flask.

E X P E R I M E N T   4 3

Part One: Production of Methyl Red (SM)
References: Sections 2.3, 5.1, and Experiment 42A

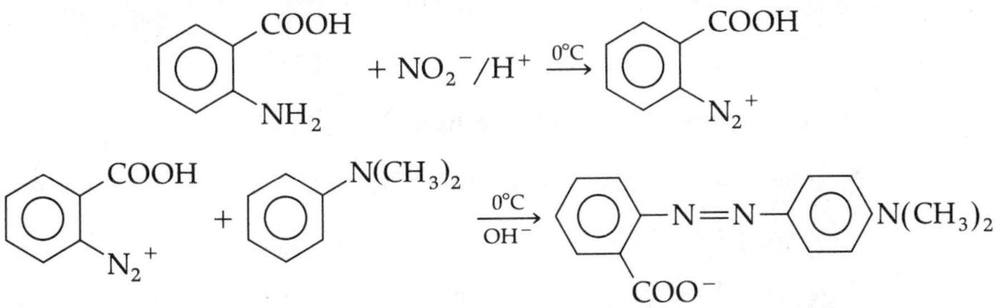

Prepare the acid solution of the reagent. Place 0.41 g of anthranilic acid in a clamped 50-mL Erlenmeyer flask with a magnetic spin bar or vane. Cautiously add 12 mL of water to which 0.3 mL of concentrated sulfuric acid has been added. Stir and warm with a water bath as necessary to dissolve the starting material. Cool to room temperature and add a few chips of ice. Stir and cool in an ice bath until the temperature has been below 5°C for five minutes. This may take 20 minutes. The formation of crystals should not be a problem.

Diazotize the acid and couple it with the amine. To the ice-cold solution, add 1.6 mL of cold 2M sodium nitrite and stir the mixture for five minutes more in the ice bath. A yellowish solution should form. Then add a few milligrams of urea with gentle stirring and continued cooling. If a gas is generated, add a bit more urea.

Meanwhile, prepare a cold solution of 1 mL of water, 0.35 mL of concentrated hydrochloric acid, and 0.35 mL of dimethyl aniline. Add this cold solution dropwise to the cold diazotization mixture, maintaining the temperature below 5°C. Stir at ice temperature for a few minutes.

Isolate the indicator/dye. Neutralize the cooled reaction mixture by carefully adding 2.5M sodium hydroxide, dropwise. It should require 4–6 mL of base or less. Do not add too much base. As the addition progresses, observe that where a drop of base is added, more color appears. Stir very gently. When a drop of base no longer causes the appearance of more color and much or most of the base has been added, assume the end point has been reached and stop the addition.

Add 8 mL of saturated sodium chloride solution and heat to a brief boil. Cool, and then ice the mixture. Collect the colored product by suction filtration on a Hirsch funnel. Wash the product twice with 2-mL portions of fresh saturated salt solution. Allow the product to dry. Determine the yield and test for indicator properties. Record the infrared spectrum if directed to do so.

See Figures 2.6A, B; 2.7 (pages 52–55).

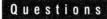

QUESTION 1 ABOVE.

2. Why must the mixtures be kept cold until the product is isolated?
3. What is the purpose of the urea? What is the gas?
4. Why is saturated sodium chloride added? Why is it used for washing?
5. Why should anthranilic acid dissolve in acid?

Summary

1. Dissolve the anthranilic acid in acid and cool.

2. Add $NaNO_2$ solution and keep the mixture very cold.

3. Test for excess reagent with urea.

4. Add the cold base dissolved in acid, with the mixture still in ice.

5. When coupling is complete, neutralize the cold mixture with base.

6. Add saturated NaCl, boil, and cool thoroughly.

7. Collect by suction and water with saturated NaCl.

8. Dry, and analyze as directed.

Disposal of Materials

Anthranilic acid (I, S)	aq. H_2SO_4 (D, N)
aq. $NaNO_2$ (I, D, or oxidize with bleach cautiously)	aq. HCl (D, N)
	Dimethylaniline (OW, I)
sat. NaCl (D)	Filtrate (N, aq. OW, I)
Product (I)	Urea (S)

Additions and Alternatives

Try substituting sulfanilic acid for anthranilic acid to make methyl orange. Anthranilic acid from Experiment 44 could be used.

Analyze the product with thin layer chromatography and record the UV spectra of the product in both the acidic and basic forms.

Part Two: Dyeing

Organic dyes are principally attached to cloth in one of two ways. Direct dyes attach themselves to the cloth itself without another compound to bind the two together. Mordant dyes require the presence of a third substance that can combine with the textile and the dye simultaneously, completing the dyeing process. This third substance is called a mordant. Congo red, like many dyes made from diazotized benzidine, is a direct dye toward cotton.

Direct dyeing: Try dyeing wool (and/or silk) and cotton (or cheesecloth) with triphenylmethane dyes. Solutions (0.05%) of two such dyes (e.g., malachite green and methyl violet) will be made available. Several students can use the same dye "bath." For direct dyeing, place the cloth samples in the beaker of dye solution, which has been brought to a boil, for about two minutes. Remove the samples, and wash them thoroughly with water. Return the cooled dye solutions to the stock bottle.

For *mordant dyeing*, place a sample of cotton or cheesecloth in a 0.2% tannic acid solution (shared with other students and then returned to the stock bottle), and boil it for one minute. Remove the cloth, and, after you press out the excess solution with a spatula, boil the sample for two minutes in a malachite green solution (or in a methyl violet solution). This solution should be shared but not returned to the stock bottle.

6. Why should it not be returned?

Wash the dyed sample thoroughly as before.

Ingrain or development dyeing may be illustrated by the formation of para red within fibers of cheesecloth during the dyeing reaction. Grind about 0.1 g each of *p*-nitroaniline and sodium nitrite with 5 mL of water to make a paste. Then add 6 mL of dilute hydrochloric acid to about 10 g of crushed ice, and stir this mixture with the stirring rod used to make the paste. When most of the ice has melted, add the paste to the ice-cold acid, and stir the mixture well. Immerse the cheesecloth, and stir it in this cold solution. Remove the cloth, and free it from excess solution by pressing it with a spatula. Immerse the cloth in a solution formed by adding about 500 mg of β-naphthol to 20 mL of 5% sodium carbonate. When the color has fully developed, remove the cloth, and wash it thoroughly.

QUESTION 6 ABOVE.

7. Compare the "fastness" to washing and the intensity of color for the samples dyed. What difference is found in the color?

8. Explain how aluminum hydroxide and other oxides or hydroxides would function as mordants.

Disposal of Materials

Basic solutions (NaOH, Na$_2$CO$_3$) (N)

Triphenylmethane dye solutions (I)

Tannic acid solution (N, I)

p-Nitroaniline (S, HW, I)

dil. HCl (N)

β-Naphthol solution (evaporate, S, OW, I, avoid skin contact)

Additions and Alternatives

Consider other interesting experiments with vat dyes that are given by Shapiro et al., *J. Chem. Educ.*, 1960, *37*, 526. For two other uses of coupling, see *J. Chem. Educ.*, 1971, *48*, 413 and 1975, *52*, 195.

Prepare the directions to use methyl red or methyl orange as a development dye. Carry out the process.

CHAPTER 10

Rearrangement

10.1 Introduction

Most organic reactions are characterized by the complete retention of a fundamental carbon skeleton and by the complete retention of the location of functional groups of the principal molecule. It is then of considerable interest to study those reactions that result in relocations and altered structures. The five rearrangements dealt with in this text should be carefully compared to determine, at least, in what sense a rearrangement has occurred and how it has occurred. Experiments 44 and 45 involve a 1,2 shift of the "Whitmore" type. That is, one group or part of the molecule, complete with a pair of bonding electrons, migrates from one atom to an adjacent, electron-deficient atom. Experiment 45 involves the rearrangement of a reduced compound, whereas Experiment 46 involves the rearrangement of an oxidized compound.

The Piria rearrangement (Experiment 38) is not included in this section, but rather it is grouped with other sulfonations of the aromatic ring. The Wittig rearrangement (Experiment 64) is included with other named reactions.

E X P E R I M E N T   4 4

Preparation of Anthranilic Acid (Hofmann Rearrangement) (SM)
References: Sections 2.2 and 2.3

The amides of most of the simpler carboxylic acids can be converted into amines with the loss of carbon dioxide. The acid is converted into the amide; this is treated with bromine or chlorine in a strongly alkaline solution. The reaction may be more rapid and may give better yields if chlorine is used in the form of sodium hypochlorite solution. This can be prepared by dissolving chlorine in aqueous sodium hydroxide. Commercially available hypochlorite (laundry) bleaches can also be used. If bromine is used, the mixture of bromine and sodium hydroxide (sodium hypo-bromite) must be prepared immediately before use because the solution loses much of its activity upon standing.

In this exercise, phthalimide reacts with sodium hydroxide to yield the sodium salt of phthalamidic acid:

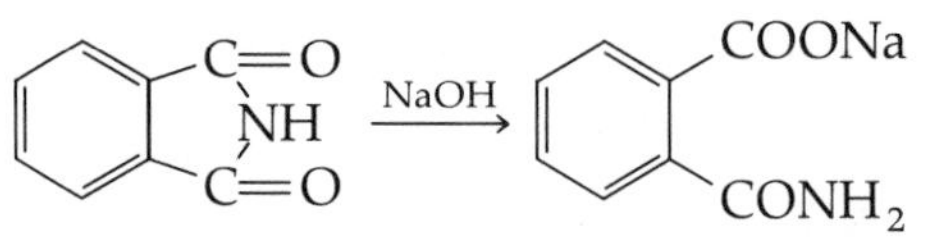

The amide portion of the molecule then reacts with the hypochlorite to produce the N-chloroamide of the sodium salt:

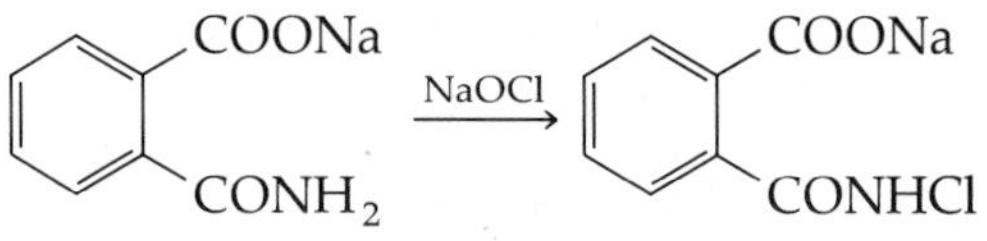

The elements of HCl are then lost in the presence of the alkali. The unstable intermediate rearranges immediately to form an isocyanate:

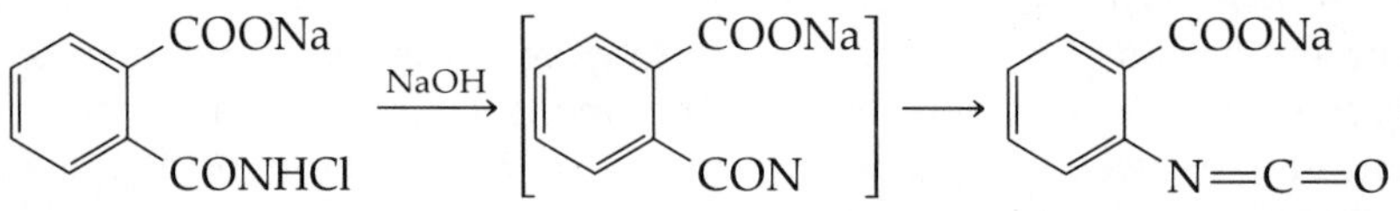

The isocyanate hydrolyzes to yield the sodium salt of *o*-aminobenzoic acid (sodium anthranilate):

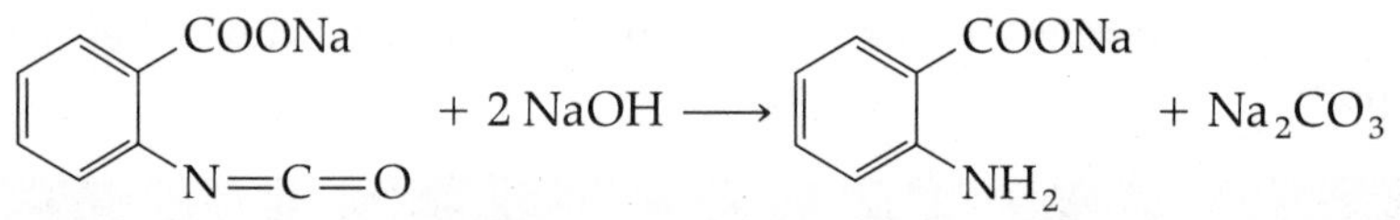

From this salt, anthranilic acid is precipitated by the use of a slight excess of dilute acid. Phosphoric acid is more suitable than hydrochloric or sulfuric acid. The product is amphoteric, and the weaker phosphoric acid is less likely than either of the stronger acids to form a salt with the amino group.

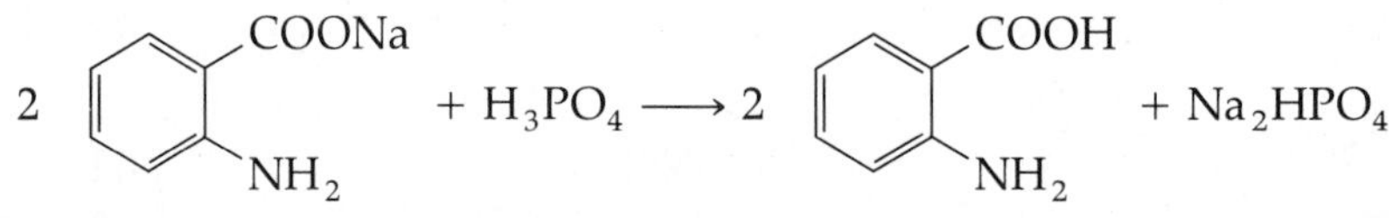

Procedure

Initiate the reaction by mixing the reagents. Place 2.5 g of a good grade of phthalimide in a 125-mL Erlenmeyer flask, and add 6 mL of fresh 6N sodium hydroxide solution. This will dissolve nearly all of the imide. The undissolved residue will go into solution as the reaction progresses. Add 30 mL of fresh 5% sodium hypochlorite solution. Any of the hypochlorite laundry bleaching solutions can be used for this purpose if they have this concentration. If they are more dilute, as indicated on the label, make a correction in volume to ensure that a small excess of the hypochlorite is used. Place the flask in a beaker of warm water. Heat the water to its boiling point, and maintain it at this temperature for 15 minutes. Swirl or stir occasionally. During this process, a precipitate of silky needles may form and redissolve. Filter the solution hot, and then cool it.

Neutralize the solution by the cautious addition of dilute phosphoric acid (one volume of concentrated phosphoric acid to three volumes of water). During this operation, a precipitate forms and increases in amount as the acid is added drop by drop. The effect produced by each drop is quite noticeable if the flask is shaken between additions. Discontinue the additions when two or three successive drops produce no apparent increase in the amount of precipitate. If an excess of acid is added, the precipitate will start to dissolve.

Question

1. Why might excess acid cause the product to dissolve?

With phosphoric acid, which is a rather weak acid, there is a leeway of several drops before this becomes serious. If sulfuric or hydrochloric acid is used to liberate the anthranilic acid, the dissolution process starts immediately when the mineral acid is present in excess.

Collect the precipitated anthranilic acid by suction filtration. After draining it, dry-pack it down with a spatula blade, and discontinue the suction. Moisten it with about 1 mL of water and drain it again.

The solubility of the product is so low in water that it is usually impractical to try to recrystallize it from this solvent. It can be dissolved in sodium hydroxide solution (about 1N) and reprecipitated. Do this only if the product is quite noticeably discolored. From time to time, brown, acidic tars become troublesome. If this occurs, give more attention to all of the reagents; it may be necessary to add less hypochlorite and/or to dilute it. In such a case, it may also be worthwhile to recrystallize the product from hot water (or base and acid) with decolorizing carbon.

See Figures 2.6A, B; 2.7 (pages 52–55).

Analyze the product. Allow the product to dry until the next laboratory period. It should then be dry enough to weigh. Determine the melting range (below 148°C) and yield.

Summary

1. Dissolve, as far as possible, the amide in basic solution.

2. Add bleach, warm, and heat near 100°C for 15 minutes.

3. Filter hot, and cool.

4. Cautiously neutralize with dilute phosphoric acid.

5. Collect the product by suction, wash it, and air-dry it.

6. Recrystallize and decolorize as needed.

7. Analyze the product.

QUESTION 1 ABOVE.

2. Suggest a relationship between the hypochlorite solution and tar formation.

3. How could the starting material be prepared from naphthalene? From phthalic acid or anhydride?

4. What advantage does this rearrangement have over the Hofmann reaction $(RX + NH_3)$ for the preparation of amines?

5. What would the product be if succinimide were used instead of phthalimide?

Disposal of Materials

Phthalimide (S, I)

aq. NaOH (D, N)

Sodium hypochlorite or bleach (save for other disposal)

Phosphoric acid and solutions (D, N)

Product (I)

Additions and Alternatives

Obtain the infrared spectrum and thin layer chromatographic analysis of the product. Carry out this preparation at the micro scale using 0.25 g of phthalimide, 0.6 mL of NaOH, and 3 mL (or another volume, depending on strength) of bleach in a 5-mL vial or flask or in a 10-mL or 25-mL Erlenmeyer flask. When purifying by dissolving in base, use the concentration and volumes of reagents actually used to calculate the stoichiometric volume of dilute phosphoric acid.

The product of this experiment may be used in Experiment 43.

Design and carry out this reaction using succinimide in place of phthalimide.

EXPERIMENT 45

Pinacol can be prepared by a bimolecular reduction of acetone with magnesium metal. This reduction proceeds through the hydrolysis of the intermediate

$$O-Mg-O$$
$$(CH_3)_2C\text{————}C(CH_3)_2$$

to give the glycol. (This method of reduction is not particularly satisfactory with other aliphatic ketones. However, similar procedures give good yields with alkylaryl or diaryl ketones. See Additions and Alternatives, Experiment 45.) Pinacol hydrate (Experiment 45) is rearranged to the ketone pinacolone, also by means of an acid catalyst. This rearrangement involves a 1,2 shift of the "Whitmore" type. It is suggested that the attachment of a proton to an oxygen, followed by the loss of water, yields a carbocation within which an alkyl group, complete with a pair of bonding electrons, may migrate to the adjacent, electron-deficient carbon. The resulting carbocation may be stabilized by the loss of the alcoholic hydrogen as a proton.

Figure 1.7, Unit I (page 24).

Preparation of Pinacolone (SM)
References: Sections 1.3H; 3.6; 4.2C, D; and 5.1

$$(CH_3)_2COHCOH(CH_3)_2 \longrightarrow (CH_3)_3CCOCH_3 + H_2O$$

Initiate the reaction by mixing the reagents. Place in a 25-mL round-bottom flask 8 mL of water, and cautiously add 2 mL of concentrated sulfuric acid (these amounts may be adjusted for the amount of pinacol hydrate available if you prepared your own). Dissolve in this warm mixture 2 g of pinacol hydrate. Add a reflux condenser (Unit I, Figure 1.7), and reflux for 15 minutes. Allow the flask to cool and rearrange for distillation (Unit III, Figure 1.7, or Figure 3.8); a stopper can replace the thermometer.

Steam distill the mixture using a separatory funnel as a receiver.

Separate the product. Separate and save the nonaqueous layer, and extract the aqueous layer with 4–5 mL of methylene chloride. Dry the combined nonaqueous layers with anhydrous magnesium sulfate.

Collect the product by distillation. Assemble an apparatus for distillation (Figure 3.8) or an apparatus for simultaneous addition and distillation (Figure 7.1) with a 10-mL flask and a boiling stone. The latter is used in case the volume to be placed in the flask exceeds 7 mL. Decant (or decant with filtration) the dried nonaqueous solution into the 10-mL flask, and rinse the residue with 0.5 mL of methylene chloride. If the volume will exceed 7 mL, place 5 mL in the flask and the rest in the addition funnel, to be added during distillation. Carefully distill the methylene chloride into a cooled receiver, and then collect a product fraction boiling near 102–107°C, using a clean, dry receiver. Determine the yields, infrared spectrum, and other properties as directed (d., 0.8 g/mL).

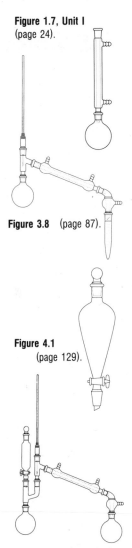

Figure 3.8 (page 87).

Figure 4.1 (page 129).

Figure 7.1 (page 257).

Summary

1. Dissolve the pinacol in warm diluted H_2SO_4.

2. Reflux for 15 minutes.

3. Steam distill the mixture.

4. Separate, extract, combine, and dry the organic material.

5. Recover the product by fractional distillation.

6. Analyze the product.

Questions

1. How is simultaneous distillation and addition carried out (see Experiment 27)?

2. Name pinacol and pinacolone by the IUPAC system.

3. Suggest a different synthetic route to pinacolone using a Grignard reagent and oxidation.

4. Write equations to show how trimethylacetic acid could be made from pinacolone.

5. Outline the suggested mechanism for this rearrangement.

6. Analyze the infrared spectrum of the product. Contrast the features of this spectrum with those expected for the starting material.

Disposal of Materials

conc. H_2SO_4 (D, N)	dil. H_2SO_4 (N)
Pinacol hydrate (S, I)	Methylene chloride (OS, HO, I)
$MgSO_4$ (evaporate, care, S)	Product (I)

Additions and Alternatives

An important alternative would be for you to produce the pinacol hydrate that will be used for the rearrangement. Consider the directions given by Vogel (*Elementary Practical Organic Chemistry*, Longman, New York, 1959, p. 182) or Campbell and Kline (*Semimicro Experiments in Organic Chemistry*, D. C. Heath, Boston, 1964, p. 88).

Ethyl ether could be used in place of methylene chloride; note the difference in density.

You will find the analysis of pinacolone with thin layer chromatography and gas liquid chromatography quite interesting. Pinacol and pinacolone could be tried with the iodoform test (Exercise 10, Chapter 17). You can prepare trimethylacetic acid by the iodoform method or by oxidizing the product of this reaction with bleach.

One rewarding alternative is the photochemical reduction of benzophenone to benzopinacol, followed by rearrangement to benzopinacolone. Very high yields have been obtained by following the method of Bachmann (*Organic Synthesis*, Coll. Vol. II, Wiley, New York, 1943, pp. 71 and 73) but using one-twentieth the amount called for and allowing the reduction mixture to stand in a window with a southern exposure for two to three weeks in the winter, less in other seasons. The initial procedure takes only a few minutes, and the work-up after standing takes much less than one period. Variations of this reaction, using other alcohols and another light source, have been studied by students (*J. Chem. Educ.*, 1959, *36*, 615). This reaction can also be used to demonstrate the effectiveness of suntan lotions. Design such a test. The benzopinacolone may be used to demonstrate the formation of a relatively stable, colored carbanion (*J. Chem. Educ.*, 1961, *38*, 307).

Another important alternative would be for you to develop the directions and carry out this reaction on the micro scale.

For a theoretical study, you could consider a version of this experiment that emphasizes migratory aptitudes (*J. Chem. Educ.*, 1971, *48*, 257).

EXPERIMENT 46A

Preparation of Benzilic Acid (from Benzoin (*M*)
References: Sections 1.3H; 2.2, 3; and 5.1

$$2C_6H_5COCH(OH)C_6H_5 + BrO_3^- + 2OH^- \longrightarrow$$

$$OBr^- + 2(C_6H_5)_2C(OH)COO^- + 2H_2O$$

Make a solution of the reagents. In a 5-mL vial, place a spin vane, 0.25 g of sodium hydroxide (or 0.1 g more if potassium hydroxide is used), 0.06 g of sodium bromate (or 0.005 g more if you use the potassium salt) and 0.6 mL of water. Stir and warm in a hot water bath to bring the solids into solution. To this solution, add 0.21 g of benzoin in small portions. The vial should be supported with a clamp. Attach an air condenser loosely between additions. The slurry should be stirred and heated in a hot water bath near its boiling range (88–93°C). Bring the benzoin into solution, using a stirring rod if necessary. Add water as needed to maintain a fluid slurry. This may take 0.3–0.5 mL. Continue stirring, adding water, and heating for one to three hours. When the solution becomes yellow, test a drop of slurry with 5–10 drops of water to see if a solution is formed. If all or almost all material goes into solution, discontinue heating after 5–15 minutes more.

Isolate the acid. Transfer the slurry to a small Erlenmeyer flask containing 2 mL of water. Rinse the vial and vane with 0.5 mL of warm water. Add

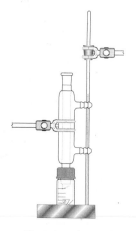

Figure 1.8 (page 26).

the rinsings to the flask. Separate any solid or oil that does not dissolve. Acidify the solution with 3M sulfuric acid to a pH of 3 or below. Between 6 and 7 mL will be required. Add the last portions of acid dropwise with stirring. If the end point is exceeded, bromine may appear. If it does, add a drop or two of 6M base. Collect the acid on a Hirsch funnel with suction and wash thoroughly with ice water.

Analyze the product. When dry, determine the melting range and the yield. The melting range should be just below 150°C. The product may need to be recrystallized from a minimum of hot water. Record the infrared spectrum if directed to do so.

See Figures 2.8A, B, C
(pages 56–58).

Q u e s t i o n s

1. What is the function of the bromate ion? (*Hint*: This function is the first step of a two-step reaction.)

2. Why should the slurry be kept fluid?

3. What is the function of the sulfuric acid?

4. Bromine and hypobromite are in equilibrium. Why is the base added if bromine appears during the addition of acid?

5. How does this experiment differ from Experiment 46B?

Summary

1. Assemble the apparatus.

2. Make an aqueous solution of base and bromate.

3. Add benzoin in small portions, with heat and stirring.

4. Maintain the slurry with water and controlled heating.

5. Stir and heat to complete the reaction.

6. Test for completeness of reaction.

7. Dissolve the salt in water and separate insoluble material.

8. Acidify the solution.

9. Collect the acid by suction filtration and wash.

10. Dry and analyze the product.

Additions and Alternatives

You could use the crude benzoin from Experiment 55.

Test for completeness of the reaction. Develop a thin layer chromatographic analysis. If the reaction is heated too strongly, benzhydrol (diphenylcarbinol) forms and is largely insoluble in the water.

6. How could benzhydrol (diphenyl carbinol) be formed?

Q u e s t i o n

This by-product could be detected by the TLC analysis.

E X P E R I M E N T   4 6 B

Preparation of Benzilic Acid (from Benzil) (M)
References: Sections 1.3H; 2.2, 3; 5.1; and Experiments 3A and B

Assemble the apparatus for reflux with a 5-mL vial and spin vane.

Figure 1.8   (page 26).

Introduce the reagents and reflux. Place in the vial, 0.21 g of benzil, 0.625 mL of 8M aqueous potassium hydroxide and 1 mL of 95% aqueous ethanol. Reflux for 12–15 minutes.

Transfer the warm, dark solution with a Pasteur pipet to a small beaker or evaporating dish.

1. Why should the solution be kept warm during transfer?

Q u e s t i o n

Cool to room temperature and then to ice temperature. The potassium salt of the product should crystallize almost completely. In some cases, it may be necessary to wait until the next period. Break up the solid, collect the solid by suction filtration, and wash twice with 0.5–1.0 mL of 95% aqueous ethanol as cold as possible.

Dissolve the salt in 5–6 mL of hot water contained in a centrifuge tube within a boiling water bath. If any solid remains, separate the liquid from it with a pipet after centrifugation or with filtration, using a Hirsch funnel. Cool the solution to room temperature and acidify by adding 0.2 mL of 6M HCl. The pH should be clearly below 3, as shown with a drop on external indicator paper. If not, add more acid dropwise. When the solid product has formed, cool the mixture in ice.

Isolate and purify the product. Collect the solid by suction filtration with a Hirsch funnel. Wash thoroughly with ice water to remove all the potassium chloride.

2. Where did the potassium chloride come from?

Q u e s t i o n

More product may be obtained by concentrating the filtrate. The product may have a light color and should have a melting range close to 150°C. The product may be recrystallized from a minimum of hot water. Pelletized activated carbon should be used if the product is dark or has a tendency to oil.

See Figures 2.8A, B, C
(pages 56–58).

Analyze the product. Allow the product to dry and determine the melting range and the yield. Record and analyze the infrared spectrum (KBr) if directed to do so.

QUESTIONS 1 AND 2 ABOVE.

Q u e s t i o n s

3. Compare Experiments 46A and B.

4. Propose a possible mechanism for this rearrangement.

Summary

1. Set up the apparatus for reflux.

2. Mix benzil, aqueous base, and ethanol.

3. Reflux for 15 minutes.

4. Crystallize solid potassium benzilate.

5. Dissolve this salt in hot water and separate any insoluble solid.

6. Cool, and acidify with diluted HCl.

7. Collect the product and wash with cold water.

8. If needed, recrystallize with hot water and activated carbon.

9. Analyze as directed.

Disposal of Materials

Benzil (I, S)   aq. KOH (D, N)

aq. Ethanol (D)   aq. HCl (D, N)

aq. Filtrate (N, I)   Pelletized carbon (I, S)

Product (I)

Additions and Alternatives

Develop a thin layer chromatographic analysis.

You could make the benzil needed by using hot concentrated nitric acid as the oxidizing agent as in Experiment 26. This procedure would illustrate oxidation and rearrangement as two separate stages.

CHAPTER 11

Acids

11.1   Types of Reactions

The experiments in this section illustrate not only some of the preparations and reactions of esters, amides, anhydrides, and nitriles but also certain interesting principles and types of reactions. The first two preparations (Experiments 47A and 48) involve the displacement of an equilibrium to promote product formation. The similar reactions of esters, acyl halides, and acid anhydrides with water (hydrolysis), with alcohols or phenols (alcoholysis), and with ammonia or amines (ammonolysis) are also illustrated in some of the experiments in this chapter and elsewhere. In Experiment 49, an ester is treated with ammonia, whereas Experiment 9B (Chapter 4) and Experiment 70 (Chapter 16) involve the hydrolysis of esters. In the latter cases, essentially the same equilibrium is involved as in the esterification reaction of Experiment 47, but the equilibrium is displaced in the reverse direction by the addition of a base (saponification). In Experiment 50, an ester is formed by the reaction of an anhydride with a phenol. The ammonolysis and alcoholysis of acetyl chloride are illustrated in Chapter 17, Exercises 5 and 11, and the ammonolysis of acetic anhydride is exemplified in Experiment 3C. The dehydration described in Experiments 51 and 52 involves both intra- and intermolecular dehydration and may be compared with Experiments 15, 16, and 17. The mechanism of the intramolecular dehydration in these last three dehydrations may be compared with the mechanisms of sulfuric acid–catalyzed esterification (Experiment 47) and with the intermolecular dehydration of *n*-butyl alcohol to *n*-butyl ether. You may wish to refer to *J. Chem. Educ.*, 1962, *39*, 212 for the details of the latter mechanism. Incidentally, this experiment has been performed by students using the same apparatus as used in Experiment 47A (Figure 11.1A). Experiment 53 illustrates bimolecular decarboxylation of an acid, whereas Experiment 61 illustrates monomolecular decarboxylation.

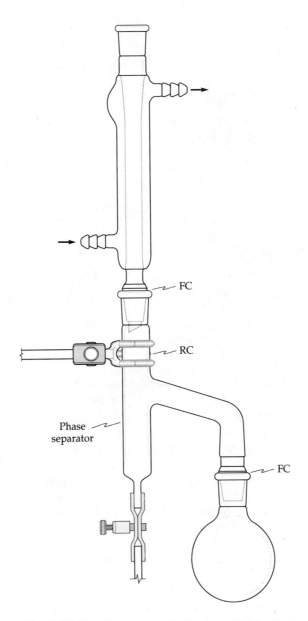

FC

RC

Phase
separator

FC

Figure 11.1A Phase separator for use with standard
taper ground jointed glassware. (SM)

11.2   Displacement of Equilibrium

11.2A   Concentration Effects

Relatively few reactions used for the preparation of organic compounds proceed to completion. In many cases, equilibrium is attained while considerable portions of the reactants are still unconsumed.

In the study of general chemistry, you learned that in the reaction

$$nAB + pCD \rightleftharpoons rAD + sCB$$

the relationship between the equilibrium concentrations (AB), (CD), (AD), and (CB) is given by the expression,

$$\frac{(AD)^r(CB)^s}{(AB)^n(CD)^p} = K$$

where K is the **Equilibrium Constant**. You also learned that the concentration of a product, say AD, in the reaction mixture at equilibrium can be increased by several methods, among which are: (1) The increase in the concentration of one of the reactants, AB or CD. The reactant chosen for this purpose is usually the one that is cheaper or more readily available. (2) The removal of a reaction product from the reaction mixture by volatilization, precipitation, or the formation of a nonreactive product that, in the case of ionic reactions, may be a nonionized product.

Method (1) is not particularly effective in reactions in which no solvent is used. An increase in the concentration of one reactant reduces by dilution the concentration of the other reactant. Method (2), however, is capable of producing astonishing results if proper conditions can be provided. A reduction in the concentration of a product results in an increase, rather than a decrease, in the concentrations of the remaining components of the reaction system; the only dilution effect is due to the accumulation of the other product. In this way, reactions can be brought about by seemingly most unlikely conditions. For one such case, refer to the article by Othmer and Kleinhans on the nitration of toluene (*Ind. Eng. Chem.*, 1944, *36*, 447).

11.2B   Removal of a Product

In Experiment 47A, benzene or toluene would form constant-boiling mixtures with water and are almost complete nonsolvents for water. The addition of either to the reaction mixture and maintenance of a slow rate of distillation results in the removal of water as rapidly as it is formed. Any butyl alcohol that distills with the hydrocarbon and water dissolves in the nonaqueous layer upon separation. If the hydrocarbon and water layers can be allowed to separate and if the nonaqueous layer (probably containing a small amount of butyl alcohol) is returned to the reaction vessel, the use of method (2) causes this type of reaction to proceed to completion. At the microscale level in esterification, advantage is taken of the fact that the ester forms an azeotrope with water (Naff and Naff, *J. Chem. Educ.*, 1967, *44*, 680).

The phase separator is not needed for Experiment 48. It is generally accepted that carboxylic acids react with ammonia or amines to form addition compounds of a transient nature, which break down or are decomposed by heat to give amides and water. Slow distillation accomplishes this decomposition and, by removing the volatile product, shifts the equilibrium toward the organic products. The reverse of this process would be hydrolysis.

E X P E R I M E N T   4 7 A

Preparation of Butyl Propanoate (SM)
References: 1.3F1–H, 3.4, 4.1, 5.1, and 11.2B

$$n\text{-}C_4H_9OH + C_2H_5COOH \longrightarrow n\text{-}C_4H_9OOCC_2H_5 + H_2O$$

Assemble the apparatus shown in Figure 11.1A or 11.1B. For the assembly in Figure 11.1B, clamp a distilling head to a ringstand. Tilt it so that it is approximately at a 45° angle. Place 10 mL of a saturated salt solution in a 10-mL round-bottom flask. Attach this flask to the side arm of the distilling head using a horseshoe or similar clamp, being careful not to get the grinding of the joint wet. Carefully turn the adaptor so that any liquid near the top of the adaptor would drain into the side arm on its way down. Using a bent tip Pasteur pipet, add more saturated salt solution (about 6 mL) to the 10-mL flask. The liquid should be discharged slowly and carefully from the pipet so that it enters the side arm. The volume of liquid should extend beyond the joint but be contained well within the side arm. Fill the phase separator in an analogous manner, leaving enough room for the water produced and some toluene.

Introduce the reagents and initiate the reaction. To a 25-mL round-bottom flask, add 6 mL of *n*-butyl alcohol (1-butanol), 5 mL of propanoic acid, 8 mL of toluene, 2–3 drops of concentrated sulfuric acid, and a boiling stone. Use a horseshoe clamp to attach this flask to the distilling head. Add a water condenser, start the water flow, and begin a slow distillation. Two layers should form in the side arm or phase separator.

Question

1. What are the two layers?

Observe the rate at which water is evolved, as well as the ability of the system to continue to produce droplets of water. Toluene should flow

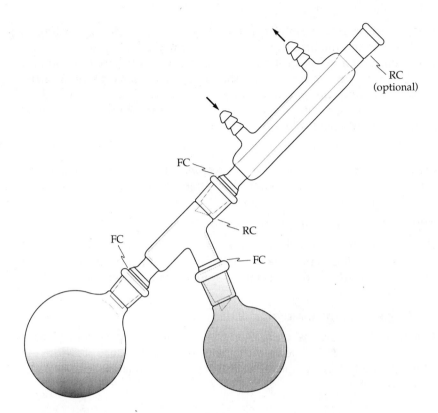

Figure 11.1B Phase separation with standard equipment. (SM)

back into the flask. During this time (at least 45 minutes), calculate the amount of water to be expected from the esterification and the height to which this quantity of water should fill the side arm or separator. Cessation of the production of water is to be expected when the interface of the two layers reaches a little above this point.

The water level (lower) will remain constant when the reaction is complete. When the water level appears to stay the same, let the distillation continue for a few more minutes and observe the water level to make sure it is not increasing. It is better to let it distill a little too long than not long enough. Stop the distillation by removing the heat source, and allow the apparatus to cool and drain.

Purify the product. Carefully straighten the apparatus so that most of the cloudy, top layer (if there is one present) in the side arm drips back into the 25-mL flask. *Don't let the aqueous solution drip back.* Pour the contents of the flask (excluding the boiling stone) into a separatory funnel containing an equal volume of saturated sodium bicarbonate solution to which a drop

Figure 4.1 (page 129).

of Congo red or bromcresol green indicator has been added. Venting as needed, shake the mixture thoroughly to neutralize the sulfuric acid and any residual propanoic acid. If one extraction is not sufficient to neutralize the acid, repeat this extraction until the indicator remains the color of the basic solution and the evolution of carbon dioxide has ceased. Allow the two layers to separate. The aqueous layer must be the same color as the original bicarbonate solution.

Question

2. What does the color staying the same demonstrate?

Collect the product by fractional distillation. Transfer the upper, non-aqueous layer to a 25-mL flask, and arrange for a fractional distillation (Unit IV, Figure 1.7, or Figure 3.10A).

Question

3. Why is it not necessary to dry the nonaqueous layer?

Add a boiling stone, and distill the mixture. Collect the fraction in the range of 136–145°C, and retain this portion as the product. A small amount of the ester (d., 0.883 g/mL) is sometimes obtained by redistilling the *forerun*.

Question

4. What else is in the forerun?

Calculate the theoretical and percent yields for the report. Also include an infrared spectrum to indicate the purity of the product. Turn in the product (properly labeled, including the observed distillation temperature) to your instructor.

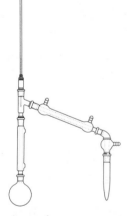

Figure 3.10A (page 96).

Summary

1. Assemble the apparatus and partially fill the phase separator.

2. Introduce the reagents and reflux for 45 minutes or more.

3. Add all of the organic materials to a bicarbonate solution.

4. Separate and ensure the neutralization of any acid.

5. Fractionally distill the organic solution.

6. Analyze the product as directed.

Questions

QUESTIONS 1–4 ABOVE.

5. What are the functions of the toluene?

6. This arrangement might well give almost quantitative results. Student results seldom approach these. Why not? What problem for all preparative experiments does this illustrate?

7. Sometimes more than the theoretical amount of water is collected. What could be the source of this extra water?

8. Compare the important features of the infrared spectrum of the ester with those expected for the acid and alcohol.

Disposal of Materials

sat. Salt solution (I, D)

Propanoic acid (OW, N, I)

conc. H_2SO_4 (D, N)

aq. Sodium bicarbonate (D)

n-Butyl alcohol (1-butanol) (OS, OW, I)

Boiling stone (evaporate, S, I)

Toluene and forerun (OS, OW, I)

Product (I)

Additions and Alternatives

You might want to join with two other students to work together and pool your experiences. One student could use normal butyl alcohol, one secondary butyl alcohol, and one tertiary butyl alcohol. (The results with the latter could be quite different. Why?) Suitable adjustments must be made, of course, in the temperature range for the collection of any products obtained. Much greater care will be needed in the fractionation of the secondary product. This collaboration makes possible a direct comparison of esterification of primary, secondary, and tertiary alcohols.

For an alternative water separator, see *J. Chem. Educ.*, 1983, *60*, 595.

Another interesting comparison would be of the three methods of preparation (suggested by Gliesher, private communications). One student could use this method (47A). One student could omit the toluene and use straight reflux (Unit I, Figure 1.7) for 40–60 minutes. The third student could change the proportions—6 mL of butanol, 10 mL of propanoic acid, and no toluene—and use straight reflux. The differences in the percent yield should be explained. The 2-butanol reaction could be run without the toluene, monitoring water collection. (See Experiment 47B.)

You could make a considerable variety of esters in a similar manner. Esters such as methyl benzoate, methyl salicilate, or diethyladipate can be made with the procedure used in this experiment (use a bit more sulfuric acid). The starting acids are solids. The distillation of the products could be used to illustrate distillation with reduced pressure. Alternatively, use the proportions in Experiment 47B, extraction with ether or methylene chloride, and solvent evaporation.

If acidic ion exchange resins are available, try one of them instead of the sulfuric acid with the three different alcohols. Decant the liquid in the reaction flask through filter paper in a funnel to transfer it to the separatory funnel. Less bicarbonate wash should be needed.

You can also make esters of phenols by this approach (*J. Chem. Educ.*, 1964, *41*, 39).

Preparation of Butyl Propanoate (M)
References: Sections 1.3F–H, 3.4, 4.1, 5.1, 11.2B, and Experiment 47A

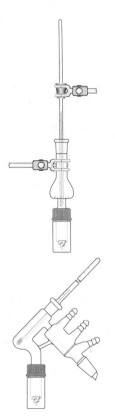

Figure 3.9   (pages 91–95).

Introduce the reagents and initiate the reaction. In a 5-mL conical vial, place 1.2 mL of *n*-butyl alcohol (1-butanol), 1.5 mL of propanoic acid, 3 drops of concentrated sulfuric acid, and a spin vane. Attach this vial to a Hickman still (Figure 3.9A), and place it in a sand bath or aluminum block. Begin a slow distillation. Monitor the temperature of the bath or block and remove most of the distillate as it is formed. Continue distillation until the production of water has ceased (20–45 minutes). Remove water as the distillation continues. Allow the apparatus to cool and drain.

Neutralize and dry the organic layer. Transfer the contents of the vial (excluding the spin vane) and any organic layer in the distillate to a small flask containing an equal volume of saturated sodium bicarbonate solution to which a drop of Congo red or bromcresol green indicator has been added, and mix using a pipet (draw the liquid in and out of the pipet). Remove the aqueous layer with a Pasteur pipet. Repeat this extraction with the saturated sodium bicarbonate solution again. If the aqueous layer is not the same color as the original base and carbon dioxide formation has not ceased, repeat the extraction until this is so. Remove the lower (aqueous) layer using a Pasteur pipet. Test this layer to be certain it is aqueous. Transfer the remaining organic layer to a vial or small test tube containing a small amount of anhydrous calcium chloride. After a few minutes, transfer the liquid from this vessel, using a pipet, into a small vessel containing a small amount of anhydrous sodium or magnesium sulfate (or use a drying pipet that has anhydrous sulfate overlayed with calcium chloride for both steps).

Collect the product by fractional distillation. After a few minutes, transfer the liquid into a 3-mL conical vial containing a spin vane, ready for distillation. Attach the vial to a Hickman still or short path head (Unit III, Figure 1.8B, or Figure 3.9), and begin a slow distillation. Have a Pasteur pipet ready to remove any liquid that might distill before 135°C. Collect a fraction that is in the range of about 136–145°C as it forms. Record the observed distillation temperature.

Determine the theoretical and percent yields, and turn the product in to your instructor. Record the infrared spectrum as directed.

Summary

1. Assemble the apparatus and introduce the reagents.

2. Remove the water by distillation.

3. Neutralize the acids and dry the organic layer.

4. Collect the product by distillation.

5. Analyze the product as directed.

ANSWER THE QUESTIONS IN EXPERIMENT 47A.

9. What is the purpose of using both the calcium chloride *and* the sodium sulfate?

10. Why is an excess of propanoic acid required in this experiment but not in the one on the semi-micro scale?

11. Why is toluene not used? What is its purpose that it can be excluded in this version of the experiment?

Disposal of Materials

n-Butyl alcohol (OS, OW, I)   Propanoic acid (OW, N, I)

conc. H$_2$SO$_4$ (D, N)   aq. Sodium bicarbonate (D)

Forerun (OW, I)   Product (I)

Additions and Alternatives

You can do the final distillation with a short path head.

You could include toluene (1 mL) in the reaction vial for this experiment. If so, the portions of the remaining constituents would be reduced slightly (*n*-butyl alcohol to 1 mL and propanoic acid to 1.3 mL). Two layers would form in the reservoir of the Hickman still during the first distillation, and a first fraction would definitely be present in the final distillation. This would make the experiment analogous to the one on the semi-micro scale.

Alternatively, you could use slow, careful distillation as in Experiment 48.

Compare to the preparation of diethyl adipate discussed in *J. Chem. Educ.*, 1990, *67*, 341.

EXPERIMENT 48A

Preparation of Dipropionylethylenediamine (SM)

References: Sections 1.3F–H, 2.3, 3.4, and 11.2B

2 CH$_3$CH$_2$COOH + NH$_2$CH$_2$CH$_2$NH$_2$ $\longrightarrow$

CH$_3$CH$_2$CONHCH$_2$CH$_2$NHCOCH$_2$CH$_3$ + 2 H$_2$O

Assemble the apparatus, introduce the reagents, and initiate distillation. Set up

Figure 3.10A (page 96). an apparatus for fractional distillation (Unit IV, Figure 1.7, or Figure 3.10A) using a 50-mL round-bottom flask. In the flask, place 4.5 mL of 98% ethylenediamine (d., 0.89 g/mL) and 2 mL of water. If commercial ethylenediamine (60–70%) is used, add 6 mL (d., 1.02 g/mL) to the flask and omit the water. Carefully add 9.2 mL of propionic acid to the flask in 1-mL increments. Add a boiling stone, and reassemble the apparatus. Distill the mixture very slowly. It should require approximately 20 minutes to collect 4 mL of distillate.

Q u e s t i o n

1. What is the distillate?

By this time the temperature may be fluctuating erratically, well above 100°C.

Solidify the reaction mixture. Remove the heat, and turn the flask on its side as far as possible without losing its contents; leave it in this position until it is cool enough to be held in your hand.

Q u e s t i o n

2. Why is it helpful if it solidifies with the maximum surface area exposed?

It may be wiped with a moist cloth to hasten cooling.

Purify the product. When the flask has cooled, add 15 mL of 95% ethanol, and attach it to a water condenser. Reflux the mixture for several minutes. Take care not to lose any solvent. Clean the joint of the flask before you decant the solution into a 125-mL Erlenmeyer flask, leaving any solid or oil behind. Clean and regrease the joint of the 50-mL flask. If any undissolved solid or oil remains in the flask, add an additional 5 mL of ethanol and reflux. Combine the solutions, and cool them to room temperature.

Collect, by suction filtration, the product that has crystallized, and wash it with three 1-mL portions of cold ethyl acetate. Drain the crystal mass as completely as possible by suction, and draw air through the crystals for several minutes. Allow the filtered cake to dry until the following laboratory period when you will determine the weight and melting range (near

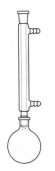

Figure 1.7, Unit I
(page 24).

190–195°C) and make the usual report. Record the infrared spectrum if directed to do so.

Summary

1. Assemble the apparatus and introduce the reactants.

2. Slowly distill the mixture.

3. Solidify the reaction mixture.

4. Recrystallize from 95% aqueous alcohol.

5. Analyze the product as directed.

QUESTIONS 1 AND 2 ABOVE.

3. What causes this reaction to go nearly to completion?

4. If the mixture of ethylenediamine and propionic acid is allowed to cool before distillation, a solid separates. Is this the same compound as the product obtained after distillation? Explain.

5. The diacetyl derivative of ethylenediamine melts at 175.6°C. How should its solubility in ethanol compare with that of the product obtained in this exercise? How could you check this?

6. What is the purpose of using ethyl acetate?

Disposal of Materials

Ethylene diamine (OW, I)

Propionic acid (propanoic acid) (OW, NO, I)

Distillate (D or N)

Ethanol (D)

Ethanol solutions (I, evaporate, S)

Ethyl acetate (OS, OW, I)

Product (I)

Additions and Alternatives

Develop a thin layer chromatographic analysis.

This preparation was adapted from Amundsen's preparation of diacetyl ethylenediamine, *J. Chem. Educ.*, 1937, *14*, 141. You could make the original compound using those directions or by adapting these. It would also be interesting to explore the reaction of oxalic acid with ethylene diamine or of urea with malonic acid, though considerable modification would be needed. You could try these experiments with toluene for azeotropic distillation (Experiment 47).

As an interesting alternative, you could make a component of insect repellent. See *J. Chem. Educ.*, 1974, *51*, 631.

E X P E R I M E N T 4 8 B

Preparation of Dipropionylethylenediamine (M)
References: Sections 1.3F–H, 2.3, 3.4, 11.2, and Experiment 48A

Assemble the apparatus and introduce the reagents. Place 0.3 mL of 98% ethylenediamine and 0.25 mL of water in a 5-mL conical vial or flask. Slowly add 0.6 mL of propionic acid in small increments. Include a spin vane or bar, and attach the vial to a Hickman still with a water condenser or to a short path head (Figure 1.8B, Unit III, or Figure 3.9).

Figure 1.8 (page 26).

Figure 3.9 (pages 91–95).

Initiate distillation. Begin a slow distillation by gradually raising the temperature of the sand bath or aluminum block. Remove the distillate as formed (one-half of a milliliter or so).

Question

7. What is it? Test it!

Maintain the temperature in this range for 15–20 minutes; then increase the distillation temperature above 100°C to complete the distillation. Remove the distillate. Tilt the vial on its side as far as possible without the loss of its contents. Allow it to cool. To hasten the cooling process, a moist piece of cloth may be wrapped around the vial. If no solid forms after cooling, distill it for another 15 minutes.

Purify the product. When the vial is cool and the solid has formed, add 0.5 mL of 95% ethanol to the vial. Reflux, using a water condenser, to dissolve the solid (this may take several minutes), and then reflux for an additional 5–10 minutes. Stop the refluxing, and immediately transfer only the alcoholic solution to a 3-mL Craig tube. (A preheated, short-stemmed funnel or Pasteur pipet could be used for this filtration or transfer.) If all of the solid or oil has not dissolved, add an additional 0.5–0.75 mL of alcohol to the solid or oil in the vial, reflux, and transfer the alcoholic solution.

Allow the solution to cool thoroughly and crystals to form. Collect the crystals by centrifuging the Craig tube. Allow the crystals to dry until the next laboratory period when the melting range and weight are to be determined (and any other analyses as directed). Make the usual report.

See Figures 2.8A, B, C
(pages 56–58).

Summary

See Experiment 48A.

Questions

Answer the questions in Experiment 48A.

Disposal of Materials

Ethylene diamine (OW, I)

Distillate (D or N)

Ethanol solutions (I, evaporate, S)

Propionic acid (propanoic acid) (OW, NO, I)

Ethanol (D)

Product (I)

Additions and Alternatives

See Experiment 48A.

If a Craig tube is not available, use a small flask or test tube and a Hirsch funnel.

An interesting alternative would be the preparation of a component of insect repellent, *J. Chem. Educ.*, 1990, *67*, A304.

Preparation of Salicylamide (Ammonolysis) (SM)
References: Sections 2.3, 4.2C, 5.1, and 11.1

$$HOC_6H_4COOCH_3 + NH_3 \longrightarrow HOC_6H_4CONH_2 + CH_3OH$$

The reaction illustrated in this preparation is a general one for the esters of aliphatic acids. Most aromatic esters require heating in a sealed tube to bring about ammonolysis. Esters of salicylic acid are noticeably soluble in ammonium hydroxide due to the presence of the hydroxyl group in the ortho position. This may account for the ease with which the methoxyl is replaced by the amide group. You must not accept this reaction, under these conditions, as indicative of the ease of ammonolysis of the esters of aromatic acids in general. In a comparative study, methyl benzoate was shaken for one week with concentrated ammonium hydroxide, and no detectable amount of benzamide was produced.

Mix the reagents. Place 1 mL of methyl salicylate (d., 1.18 g/mL) and 10 mL of concentrated ammonium hydroxide (d., 0.89 g/mL, containing about 30% NH_3) in a glass bottle with a ground glass stopper or in an equivalent vessel. Shake the mixture thoroughly, and set it aside in your locker for a week or more. The reaction is complete when no oily ester layer is visible.

Neutralize the base and crystallize the product. Dilute the solution with an equal volume of water, and neutralize it with concentrated hydrochloric acid (just a bit more than the volume of the base). Use external indicator paper. Cool the solution as thoroughly as possible; this cooling will cause the product to separate. Collect the fine crystals by suction filtration.

Wash the crystals thoroughly. Several washings are necessary to remove the ammonium chloride. The product is appreciably soluble in water, but this difficulty is overcome by keeping the volume of water used for a washing at a minimum. When the filtrate has been drained from the crystals as thoroughly as possible, discontinue the suction and moisten the crystals with water. Reapply the suction, and drain the washing from the product. Repeat this process two or three times. The effectiveness of the washing depends more upon the completeness of draining than on the amount of water used.

Allow the product to dry on a watch glass until the next laboratory period. Determine the yields and the melting range (near 140°C). Make the usual report.

Summary

1. Mix the reagents and allow them to react for a week or more.
2. Dilute with water and just neutralize with concentrated HCl.
3. Collect the crystals with suction and wash thoroughly.
4. Analyze the product as directed.

1. The starting material has a sweet smell. What is it?
2. Compare this reaction and those of water, ethyl alcohol, and ethyl amine with methyl salicylate.
3. Explain the formation of ammonium chloride.
4. Explain why the procedure used is effective in removing ammonium chloride.
5. How is the methanol removed from the product?

Disposal of Materials

Methyl salicylate (OS, OW, I)   conc. Ammonium hydroxide (D, N)

conc. HCl (D, N)   Filtrate (D)

Product (I)

Additions and Alternatives

See Figures 4.6A, B, C (pages 154–157).

You could do this preparation on one-tenth the scale. Analyze the product using thin layer chromatography (Experiment 11), and record the infrared spectrum.

EXPERIMENT 50A

Preparation of Acetylsalicylic Acid (SM)
References: Sections 2.2, 3; and 11.1

This preparation serves to illustrate the reaction of alcohols and phenols with acyl halides and acid anhydrides. This esterification can be catalyzed by hot acids and bases. In this experiment, concentrated sulfuric acid will be used. Although acetyl chloride would work, acetic anhydride is used in practice to prepare aspirin because the acetic acid formed as a by-product is easier to handle and can be recovered to make more anhydride. Because of the wide use of aspirin as an antipyretic and analgesic, it will be interesting to compare the properties of your product with those of the commercial tablets.

1. What do the terms "antipyretic" and "analgesic" mean?

Question

Part One

Initiate the reaction. Preheat 60–90 mL of water in a 400-mL beaker to boiling. Place 3.5 g of salicylic acid, 3.5 mL of acetic anhydride (d., 1.08 g/mL), and 4 or 5 drops of concentrated sulfuric acid (some heat may be generated) or 85% phosphoric acid in a 125-mL Erlenmeyer flask. Heat the flask in a beaker of boiling water for five or six minutes. Stir. During this time, the solid should dissolve completely.

    Isolate the product. Remove the flask from the bath, and add 15 mL of ice water to it. THIS DECOMPOSES THE EXCESS ANHYDRIDE.

2. What is this reaction?

CAUTION

Question

Thoroughly cool the flask to complete the crystallization. Scratch or seed as needed. Collect the crystals by suction filtration.

    Recrystallize the product. Transfer the crystals to a clean, dry 125-mL Erlenmeyer flask. Add 8 mL of ethanol, and heat in a hot water bath until the crystals go into solution. Add 20 mL of hot water to the flask, and heat it until the solution clears. Remove the flask from the bath; cover it loosely, and allow it to cool slowly. Collect the needle-like crystals by suction filtration. The crystals may have a wide melting range, from 125 to 138°C, because of decomposition. The use of a preheated melting point device will help to minimize this decomposition. Water or petroleum ether may be used as an alternative recrystallization solvent.

See Figures 2.6A, B; 2.7 (pages 52–55).

Summary

1. Heat the reactants together.

2. Add ice, cool, and collect the crystals that form.

3. Dissolve the crystals in ethanol, add hot water and recrystallize.

4. Collect with suction.

5. Analyze the product as directed.

Part Two

While the crystals are forming, the following tests may be started.

    a. Crush an aspirin tablet, and mix it with 10 mL of water. This mixture may be warmed to determine the solubility of the tablet in hot water. Divide this solution or mixture in half, and retain it for parts (e) and (f). Similarly, test 0.5 g of your product, and reserve it for part (e).

    b. Test the solubility of half of a commercial tablet and of 0.3 g of your product in 5–10 mL of toluene.

    c. Determine the solubility of 0.1 g of your product in 10% NaOH. Carefully add concentrated HCl to the latter solution until a precipitate is formed.

Question

3. Explain the results observed in part (c).

    d. Add 5 mL of water and a drop or two of concentrated HCl to the other half of the aspirin tablet (from b).

    e. Add a few drops of 2% $FeCl_3$ solution to each of the following: one half of the aqueous solution of commercial aspirin from part (a), the water solution of your product from part (a), and a dilute solution of salicylic acid.

    f. To the other half of the aqueous mixture of commercial aspirin, add a drop or two of very dilute I_2-KI solution.

Questions

QUESTIONS 1–3 ABOVE.

4. What are the components of a commercial aspirin tablet?

5. Aspirin and salicylamide (Experiment 49) are used for much the same medical purposes. A simpler compound formed from both by hydrolysis may be the active agent. What would it be?

Disposal of Materials

Salicylic acid (S)	Acetic anhydride (D, with *care* N, I)
conc. H_2SO_4 or H_3PO_4 (D, N)	Filtrate (D, I)
Ethanol or aq. ethanol (D)	Product (I)
aq. Aspirin or product (D)	aq. NaOH (N)
Toluene (OS, OW, I)	conc. HCl (D, N)
aq. $FeCl_3$ (D, I)	aq. I_2-KI solution (treat with bisulfite or thiosulfate, D, I)

Additions and Alternatives

Determine the infrared spectrum of your dry product. Analyze the product or reaction mixture by thin layer chromatography (see Experiment 11) or use a mixture of this product and that from Experiment 49.

See Figures 4.6A, B, C
(pages 154–157).

Investigate the use of acidic or basic ion exchange resins to catalyze this reaction. Try using the method from Experiment 3C. Determine the purity of your product by saponification (see Experiment 70A or *J. Chem. Educ.*, 1969, *46*, 245, but use a different solvent, such as ethylene glycol).

E X P E R I M E N T   5 0 B

Preparation of Acetylsalicylic Acid (M)
References: Sections 2.2, 3; 11.1; and Experiment 50A

Heat the reactants together and then add ice water. Place, in the bottom half of a Craig tube, 0.2 g of salicylic acid, 0.2 mL of acetic anhydride, and 1 or 2 drops of concentrated H_2SO_4 or 85% phosphoric acid. When adding the reagents, be sure to deposit them *below* the ground glass joint. No material should be on the joint. Gently heat the tube in a hot water bath for a few minutes after the solid dissolves. Remove the tube from the heat, and add 1 mL of ice water to it. Allow the tube to cool and crystals to form.

Collect the crystals. To collect the crystals, centrifuge the Craig tube. When centrifuging, loosely pack cotton below and around the top of the Craig tube to prevent its bumping within the centrifuge tube.

See Figures 2.8A, B, C
(pages 56–58).

Recrystallize the product. To the crystals in the Craig tube add 0.15 mL of 95% ethanol. Heat the tube gently in a hot water bath until the solid goes into solution. Add 0.5 mL of hot water, and gently heat the tube again. Remove the tube from the bath; cover it loosely, and allow it to cool completely. Collect the crystals by centrifuging the Craig tube, and place them on a watch glass. Allow them to dry until the next laboratory period.

Determine the melting range of the product. It may be in the range of 125–138°C. This broad range is due to the decomposition of the acetylsalicylic acid. The use of a preheated melting point device will help to minimize this decomposition. Make the usual report.

Summary

See Experiment 50A.

ANSWER THE QUESTIONS IN EXPERIMENT 50A.

Questions

Disposal of Materials

Salicylic acid (S)

Acetic anhydride (D, with *care* N, I)

conc. H_2SO_4 or H_3PO_4 (D, N)

Filtrate (D, I)

Ethanol or aq. ethanol (D)

Product (I)

Additions and Alternatives

You can carry out this reaction in a small Erlenmeyer flask or test tube and collect the crystals on a Hirsch funnel or in a test tube or centrifuge tube using a plugged pipet for the removal of liquids.

Consider an alternative preparation given by Hammond, et al., *J. Chem. Educ.*, 1987, *64*, 440.

See the suggestions for Experiment 50A. Do part or all of Experiment 50A, Part B.

E X P E R I M E N T   5 1

Preparation of Acetonitrile (Dehydration) (SM)
References: Sections 3.4; 4.1, 4G; and 11.1

$$CH_3CONH_2 \xrightarrow[P_2O_5]{-H_2O} CH_3C\equiv N$$

In writing the complete equation for this reaction, assume that the hydration product of P_2O_5 is $H_4P_2O_7$.

Assemble the apparatus and introduce the reagents. Arrange the apparatus as shown in Figure 11.2. Grind a 5-g portion of acetamide to a very fine powder (the yield depends in part upon how carefully this is done), and mix it thoroughly with 8 g of P_2O_5 (phosphorous pentoxide) on a sheet of glassine paper. Pour the powder into the tube with steady tapping to settle it. P_2O_5 IS EXTREMELY HYGROSCOPIC AND SO WILL DAMAGE TISSUE. This mixing and transferring operation must be done rapidly, and the excess powder on the paper destroyed by flooding with water. Use gloves during the transfer or wash your hands after completing the transfer. The large tube should *not* be more than half full. Orient and tap the tube to maximize the surface area of the mixture.

CAUTION

Heat the reaction mixture by playing a low flame over it near its upper surface. For a time, there will be no change; then the mixture will appear to melt and foam as the product distills. The reaction mixture may foam over unless heated carefully, but this can be controlled by heating it from the top down and by heating the top of the layer of foam as it forms. Continue heating until no more product distills and the reaction mixture has been converted to a dark brown mass.

Isolate the product. To the turbid distillate, add half of its volume of distilled water; cool the mixture in an ice bath, and add sufficient

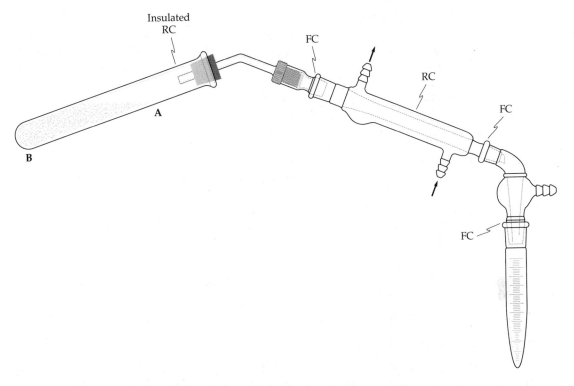

Figure 11.2 Assembly for pyrolysis using standard taper ground jointed condenser, adapter, and graduated receiver. (SM)

anhydrous K_2CO_3 (potassium carbonate) to saturate the aqueous layer. (About 1 g per mL of water is required.)

1. How can you tell when it is saturated?

Shake the cold mixture; allow the excess solid to settle, and decant the liquid into a separatory funnel. Discard the aqueous layer, and transfer the other layer into a 10-mL round-bottom flask arranged for distillation (Unit III, Figure 1.7, or Figure 3.8), containing 1 g of P_2O_5 as a drying agent.

2. The lower or upper layer? How would you test to demonstrate which is the aqueous layer?

Distill the product. Begin a slow distillation, and collect as the product the material boiling in a range near 79–83°C. Record the yield and boiling range of the product. The density of acetonitrile is 0.78 g/mL.

Clean the reaction tube. In so far as possible, transfer all cooled solids from the reaction tube to a large flask or beaker containing 150–200 mL of water. **MUCH HEAT MAY BE GENERATED. USE A SAFETY SHIELD.** Cautiously add

Question

Figure 4.1 (page 129).

Figure 3.8 (page 87).

Question

CAUTION

aqueous sodium carbonate solution to the test tube to dissolve any remaining solid. Add this solution to the beaker or flask. Rinse the tube with very warm water and combine with the previous solution. If the test tube does not come clean, dispose of it as directed by the instructor. *Cautiously*, neutralize the solution by adding solid sodium carbonate before disposal.

Summary

1. Grind the reactants together to a fine powder.

2. Assemble the apparatus, maximizing the surface area of the powder.

3. Heat the mixture to distill out the product.

4. Saturate the distillate with K_2CO_3.

5. Separate the organic layer.

6. Collect the product by distilling the organic layer from more dehydrating agent.

7. Analyze the product as directed.

Questions

QUESTIONS 1 AND 2 ABOVE.

3. What other name is used for acetonitrile?

4. The product is miscible with water. Why does the K_2CO_3 solution form a separate phase?

5. Although acetonitrile can react with water to give acetamide, this reversal of the preparation does not occur directly. What additional reagent is required to get acetamide?

6. How does the dehydrating agent used here work? Write the equation.

7. How is the $H_4P_2O_7$ removed from the product?

Disposal of Materials

Acetamide (S, I) P_2O_5 (*Cautiously* D, N, I)

K_2CO_3 (S, I)   Product (I)

E X P E R I M E N T   5 2 A

Preparation of Succinic Anhydride (SM)
References: Sections 1.3F–H; 2.3, 2; and 11.1

$$HOOCCH_2CH_2COOH + (CH_3C(O))_2O \longrightarrow 2CH_3COOH + \begin{matrix} CH_2-C{=}O \\ | \qquad \diagdown O \\ CH_2-C{=}O \end{matrix}$$

Assemble an apparatus for simple reflux (Unit I, Figure 1.7) using a 25-mL round-bottom flask.

Introduce the reagents and initiate the reaction. Place in this flask 6.5 g of succinic acid, 10 mL of acetic anhydride (d., 1.081 g/mL), and a boiling stone. Reassemble the apparatus, and heat this mixture slowly and gently until the solid has melted. Reflux the mixture for 15 minutes; remove the heat, and allow the flask to cool to about room temperature. Place the flask in an ice bath to further facilitate the formation of crystals.

Isolate and wash the product. Collect the crystals by suction filtration using a Hirsch or Büchner funnel. Wash the crystals with two 5-mL portions of ice-cold methylene chloride (or ethyl ether—*no flames!*). The anhydride may be allowed to dry on a watch glass until the next laboratory period. Determine the melting range (just below 120°C), and make the usual report. Record the infrared spectrum if directed to do so.

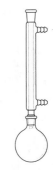

Figure 1.7, Unit I
(page 24).

Summary

1. Assemble the apparatus and introduce the reagents.

2. Melt, reflux, and cool.

3. Collect the product with suction; wash and allow to dry.

4. Analyze the product as directed.

Questions

1. In what sense is this a dehydration? How does the action of the dehydrating agent compare with that in the previous experiment? In Experiments 15, 16, and 17?

2. What is the purpose of the organic washes? Why must they be cold?

3. How is acetic anhydride prepared?

4. Which other dibasic acids should readily form a cyclic anhydride?

5. Why is the anhydride not recrystallized from water?

6. Why is it reasonable to expect this anhydride to melt at a lower temperature than the corresponding acid?

7. Why does the reaction go largely to completion in the direction written?

Disposal of Materials

Succinic acid (S, I)

Filtrate (D, N, I)

Methylene chloride (OS, HO, I)

Ethyl ether (OS, OW, I)

Boiling stone (S, I)

Acetic anhydride (D, with *care* N, I)

Methylene chloride or ether solution (evaporate, *care*, S, I)

Product (I)

E X P E R I M E N T 5 2 B

Preparation of Succinic Anhydride (M)
References: Sections 1.3F–H; 2.2, 3; 11.1; and Experiment 52A

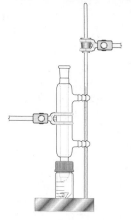

Figure 1.8 (page 26).

See Figures 2.8A, B, C (pages 56–58).

Initiate the reaction. In a 3-mL conical vial, place a spin vane, 0.5 g of succinic acid, and 0.75 mL of acetic anhydride (in that order). Attach a water condenser to the vial, and heat it gently and slowly in a sand bath or aluminum block until the solid melts. Reflux it for an additional 15 minutes after the solid melts. Use a preheated, short-stemmed funnel or pipet to transfer the hot solution to a 3-mL Craig tube. (If the Craig tube is not used, omit the transfer.) Allow the tube or vial to cool to room temperature, and place it in an ice bath to ensure the complete formation of crystals.

Isolate and wash the product. Collect the crystals by centrifuging the Craig tube, removing the solvent from the vial, with a (plugged) pipet or collecting the crystals on a Hirsch funnel. Wash the crystals once with 1 ml of ice-cold methylene chloride or ethyl ether (*no flames!*) and allow them to air-dry on a watch glass until the next laboratory period. Determine the melting range, and make the usual report. Record the infrared spectrum if directed to do so.

Summary

1. Assemble the apparatus and introduce the reagents.

2. Melt, reflux (transfer if using a Craig tube), and cool.

3. Isolate the crystals, wash them, and allow them to dry.

4. Analyze the product as directed.

Q u e s t i o n s

ANSWER THOSE IN EXPERIMENT 52A.

Disposal of Materials

Succinic acid (S, I)

Filtrate (D, N, I)

Methylene chloride (OS, HO, I)

Ethyl ether (OS, OW, I)

Acetic anhydride (D, with *care* N, I)

Methylene chloride or ether solution (evaporate, *care*, S, I)

Product (I)

Additions and Alternatives

You may use a Hirsch funnel for suction filtration and a small Erlenmeyer flask or a test tube in place of the Craig tube. A 5-mL or 10-mL flask equipped for reflux can be used for the initial reaction.

11.3   Decarboxylation

11.3A   Monomolecular Decarboxylation

Hydrocarbons can be prepared by many methods. One of the most general and certainly one of the most easily accomplished is that of removing the elements of carbon dioxide from a carboxylic acid (decarboxylation):

$$RCOOH \longrightarrow RH + CO_2$$

The ease with which this reaction takes place depends upon the structure of the acid. Sometimes it occurs upon gently heating. Alpha- and beta-substitution facilitate this process. The decarboxylation of α and β ketoacids occur in several biochemical pathways. This is illustrated by 2,4,6-trinitrobenzoic acid, which decomposes when heated to a temperature below the boiling point of water:

$$2,4,6\text{-}(NO_2)_3C_6H_2COOH \longrightarrow 1,3,5\text{-}(NO_2)_3C_6H_3 + CO_2$$

Oxalic acid yields formic acid and, finally, carbon monoxide and water by decarboxylation followed by dehydration. This reaction differs from that of the fatty acids in that it takes place in the presence of a strong, relatively nonvolatile inorganic acid such as sulfuric acid:

$$HOOCCOOH \longrightarrow HCOOH + CO_2$$

$$HCOOH \longrightarrow H_2O + CO$$

Other substituted acids, such as trichloroacetic acid, decarboxylate very easily in alkaline solution:

$$CCl_3COONa + NaOH \longrightarrow CHCl_3 + Na_2CO_3$$

This reaction is fundamental in the haloform test.

The products in the two examples given above are not hydrocarbons but represent special cases in which structure weakens the carbon–carboxyl bond. Ordinary unsubstituted carboxylic acids decarboxylate with considerable difficulty and require strong alkali to bring about the production of CO_2, even at temperatures in excess of 300°C. A mixture of sodium hydroxide and calcium oxide known as "soda lime" is used for this purpose.

Certain other compounds that are not hydrocarbons can also be prepared by decarboxylation. Furan, a heterocyclic compound, can be produced by the decarboxylation of furoic acid after the conversion of the latter to the sodium salt. Compounds like malonic and acetoacetic acids also decarboxylate readily. The reaction is, therefore, widely applicable to compounds that do not contain other groupings sensitive to heat and to acids or alkalis. These decarboxylations are also monomolecular. Another example of monomolecular decarboxylation is the preparation of methane from sodium acetate.

One famous decarboxylation was used to prepare pure benzene. When this benzene did not give a reaction thought to be characteristic of benzene, research led to the detection of thiophene, which occurs as a constant impurity in crude benzene from petroleum.

The starting material chosen for the exercise that illustrates monomolecular decarboxylation (Experiment 61, Chapter 13) is a mixture of the sodium salts of the fatty acids produced by the hydrolysis of natural fats and oils. Distillation of the resulting mixture of hydrocarbons gives a rough indication of the acids present as esters in the original fat or oil.

11.3B   Dimolecular Decarboxylations

If the calcium (or other multivalent metal) salts of carboxylic acids are subjected to similar treatment, the reaction follows a different course, and ketones (in special cases, aldehydes) are formed as the principal products:

$$RCOOCaOOCR \longrightarrow RCOR + CaCO_3$$

This reaction can be conducted as a continuous process by passing the vapors of the acid involved over a hot catalyst bed consisting of the metal oxide or carbonate. The acid reacts to form a salt of the metal, and, at the temperatures prevailing in the catalyst bed, this salt is pyrolyzed to yield the ketone, as shown above. The ketone distills out, and the carbonate that is formed is reconverted to the salt, CO_2, and water by the vapors of the acid; this salt, also, will be pyrolyzed. Water and carbon dioxide are formed along with the ketone but in different reactions, and the overall equation for the continuous process would be written:

$$2\ RCOOH \longrightarrow RCOR + CO_2 + H_2O$$

Thoria (ThO_2) and magnesia (MgO) are also used in such catalytic processes. An interesting discussion of this subject is given by Schultz and Sichels (*J. Chem. Educ.*, 1961, *38*, 300).

For the small-scale preparation of ketones, it is hardly practical to construct the equipment needed for such a continuous process. In fact, a satisfactory catalyst bed must be long, properly filled, and uniformly heated to be effective. These conditions are difficult to realize with equipment issued to undergraduate students. The exercise (Experiment 53) to be performed employs separate operations for the formation of the calcium salt of a carboxylic acid and its subsequent conversion to a ketone.

E X P E R I M E N T   5 3

Preparation of Diethyl Ketone (Dimolecular Decarboxylation) (SM)
References: Sections 3.4, 4.2D, 11.3, 17.2D, and Experiments 24 and 25

$$2\ C_2H_5COOH + CaO\ (or\ Ca(OH)_2) \longrightarrow Ca(C_2H_5COO)_2 + H_2O$$

$$Ca(C_2H_5COO)_2 \longrightarrow CaCO_3 + (C_2H_5)_2CO$$

Prepare the calcium salt of the acid. Place 1.9 g of calcium oxide (or 2.25 g of calcium hydroxide) in an evaporating dish. Under the hood, add to this mixture 3.1 mL of propionic acid (d., 0.99 g/mL). WHEN THE PROPIONIC ACID IS ADDED, DO NOT HOLD THE EVAPORATING DISH IN YOUR HAND, BECAUSE IT GETS VERY WARM. Mix the two reagents thoroughly with a stirring rod. Place the evaporating dish on a beaker of boiling water, and heat it with stirring (break up the chunks) until the calcium propionate is dry and chalky. If complete dryness is not attained at this point, difficulty with frothing will be encountered later.

C A U T I O N

Prepare the pyrolysis tube. Crush the dried material to a semi-fine powder, and transfer it to a large test tube (22 or 25 × 200 mm). With the tube nearly horizontal, spread this material to expose maximum surface and overlay the propionate with 0.5 g of calcium oxide (or 1.0 g of calcium hydroxide).

1. Why the extra CaO?

Q u e s t i o n

Attach the tube to the rest of the apparatus as shown in Figure 11.2. The pyrolysis tube should be nearly horizontal to maximize the surface area.

Initiate the reaction. Heat the tube using a Meker or Fisher burner. The heat should be intense enough to cause luminosity of the flame where it touches the tube but not intense enough to cause the glass to soften. Start heating at the top of the tube, and proceed until the material in the tube (the "charge") has turned dark gray at all surfaces and until there is no further evolution of ketone.

Separate and identify the ketone. The crude ketone will frequently separate into two layers.

2. Why should they separate? What are the layers?

Q u e s t i o n

Record the volume of the upper layer as the yield of the crude ketone (d., 0.82 g/mL). The presence and identity of the ketone are shown by converting part of it to the 2,4-dinitrophenylhydrazone derivative. Place 5 drops of the distillate in a 50-mL Erlenmeyer flask, and add 12 mL of 95% ethyl alcohol and 0.1–0.2 g of 2,4-dinitrophenylhydrazine TOXIC MATERIAL. Heat the mixture in a water bath at 75°C for two minutes. Add 0.5 mL of concentrated hydrochloric acid to it, and heat it for five minutes more. Cool the solution to room temperature. Collect the product, which

C A U T I O N

separates by suction filtration using a Hirsch funnel, and allow it to dry. Determine its melting range in the next laboratory period. Calculate the percent yield assuming that the crude product is pure. Turn in any unused crude ketone and the hydrazone derivative to your instructor for credit.

A little dilute hydrochloric acid is useful to finish cleaning the pyrolysis tube after the usual methods have been used.

Summary

1. Heat the solid acid and base together to make the solid salt.
2. Powder the salt, transfer it to the pyrolysis tube, and maximize the surface of the powder plus base.
3. Pyrolyse the solid mixture; collect the liquid that distills.
4. Separate and measure (or weigh) the ketone layer.
5. Prepare a derivative of the product.
6. Analyze as directed.

Questions

QUESTIONS 1 AND 2 ABOVE.

3. What impurities are probably present in the ketone?
4. To what type of ketones is this method of preparation limited? By what analogous method(s) might aldehydes and other ketones be prepared?
5. How could cyclopentanone be prepared in this manner? (Write the equation.)

Disposal of Materials

CaO, Ca(OH)2 (S, I)	Propionic acid (D, N, I)
Calcium propionate (S, I)	Post pyrolysis material (carefully, D, N, I)
Crude ketone (I)	2,4-Dinitrophenylhydrazine (HW, I)
Ethyl alcohol (D)	Ethyl alcohol solutions (evaporate, S, HW)
Derivative (I)	

Additions and Alternatives

You could develop a microscale experiment on one-fifth to one-tenth of this scale. Try another method for preparing a derivative. See Experiments 24 and 25 as well as Section 17.2D. Other acids—acetic, adipic, and butyric—could be tried. A method could be developed to isolate the ketone, purify it on the micro scale, and employ other methods of identification. Try analyzing and/or collecting the ketone by gas liquid chromatography.

CHAPTER 12

Reactions of Benzaldehyde

12.1   Introduction

There are many characteristic reactions of aldehydes and ketones and many compounds that could be used to illustrate them. The selection of benzaldehyde for use in the experiments given in this chapter has been made arbitrarily. The intention is to provide four more or less-typical reactions of aldehydes. The reactions of benzaldehyde are chosen not only to suggest the reactions of aliphatic aldehydes but also to illustrate the peculiarities of the reactions of aromatic aldehydes that are associated with the absence of α-hydrogens and the presence of the aromatic ring on the carbonyl carbon. Although a simple aldol condensation is not described (see Experiment 66), the two condensations and the disproportionation that are included may give you some insight into the nature of that condensation. The Cannizzaro reaction might be considered to be a competitor of the aldol condensation, whereas the benzoin condensation results in quite a different condensation product. These reactions are all conducted under basic conditions. The Perkin reaction is a condensation of much the same type as the aldol but uses an anhydride and results in the immediate loss of water to give the unsaturated acid, which is then isolated. The fourth experiment, the Grignard preparation, is generally applicable to aldehydes and ketones. Other reactions of aldehydes and ketones are illustrated in the chapters on natural products, oxidation and reduction, and qualitative organic analysis.

12.2   Disproportionation: The Cannizzaro Reaction

The Cannizzaro reaction is a disproportionation of an aldehyde in which half of the compound is reduced to an alcohol and the rest is oxidized to the corresponding acid, which is converted into the sodium salt in the presence of an excess of sodium hydroxide. In most cases, this reaction is relatively slow and remains unnoticed in the presence of the aldol condensation, which occurs under much the same conditions but much more

377

rapidly.

$$2\,RCHO + NaOH \longrightarrow RCH_2OH + RCOONa$$

$$2\,RCH_2CHO \xrightarrow{OH^-} RCH_2-\overset{\overset{\displaystyle OH}{|}}{\underset{\underset{\displaystyle H}{|}}{C}}-\overset{\overset{\displaystyle H}{|}}{\underset{\underset{\displaystyle R}{|}}{C}}-CHO$$

Only certain aldehydes, those that lack hydrogen on the alpha carbon atom or those that lack the alpha carbon atom, undergo the Cannizzaro reaction as a principal reaction. This limits its application to formaldehyde and such compounds as benzaldehyde, furfuraldehyde, and trimethylacetaldehyde.

The Cannizzaro reaction can be used straight or "crossed." The crossed reaction is brought about by using a large excess of an aldehyde much more easily oxidized or reduced than the one that it is desired to reduce or oxidize. Practically, it is employed to reduce aldehydes. For this purpose, a large excess of formaldehyde, one of the most easily oxidized of the aldehydes, is used as the reducing agent.

$$RCHO + HCHO + NaOH \longrightarrow RCH_2OH + HCOONa$$

Satisfactory yields of the alcohol are reported for this reaction.

There are many limitations to the Cannizzaro reaction. Trihalogenoacetaldehydes undergo the final stage of the haloform reaction. Triphenylacetaldehyde undergoes a similar cleavage. Aldehydes with one or more α-hydrogens undergo aldol condensations first, after which a crossed Cannizzaro-type reaction may occur.

As an interesting sidelight, certain enzymes, such as those found in the liver and in yeasts, are able to bring about a disproportionation similar to the Cannizzaro reaction when allowed to act upon aldehydes having as many as two α-hydrogens. Active platinum catalysts are reported to possess the same ability. The mechanism of the Cannizzaro reaction has been discussed by Jacobus, *J. Chem. Educ.*, 1972, *49*, 349.

EXPERIMENT 54A

Preparation of Benzoic Acid and Benzyl Alcohol (SM)
References: Sections 1.3E–I; 2.3; 3.4, 5; 4.1, 2C, 2D; and 5.1

$$2\,C_6H_5CHO + NaOH \longrightarrow C_6H_5COONa + C_6H_5CH_2OH$$

This reaction does not go to completion in the three hours available for the laboratory period. Consequently, the reaction mixture is maintained under the best conditions possible for 30–90 minutes, and the products are separated from the unchanged reactants. Examine the steps involved in this exercise and determine where and how the separation is effected.

Introduce the reagents and initiate the reaction. To a 125-mL Erlenmeyer flask containing a stirring bar, add 6 mL of benzaldehyde (d., 1.05 g/mL) and 16–20 mL of a 20% aqueous sodium hydroxide solution. Place the flask in a beaker of water on a stirrer/hot plate, and stir to get an emulsion. Continue stirring, and heat the water bath to 85–90°C. If a solid forms, add a little hot water to maintain an emulsion. Place a funnel in the flask (see Figure 2.6A), and heat and stir the mixture for 30–90 minutes, keeping the bath at this temperature. The time must be at the shorter end of this range if all of the experiment must be completed in one laboratory period.

Separate the products. Cool the flask to about 50°C, and add warm water from the beaker to just dissolve any solid. *Do not add any more water than absolutely needed*, because that may reduce the yield of benzoic acid. Make sure the solution is basic. Extract it three times (in a separatory funnel) with 8-mL portions of methylene chloride.

Isolate the acid. Transfer the aqueous layer to a 125-mL Erlenmeyer flask, and heat it to boiling until the organic layer has been removed.

1. What is the purpose of this procedure?

Acidify this solution, *when cooled*, with dilute sulfuric acid until the solution is acidic to an external indicator paper (pH of about 3). Cool the acidified solution thoroughly, and collect the benzoic acid crystals by suction. Recrystallize the benzoic acid from a minimum amount of hot water. Allow the crystals to dry before determining amount, purity, and identity.

Isolate the alcohol. Wash the nonaqueous layer twice in the separatory funnel with 5-mL portions of saturated sodium bisulfite solution.

2. What does this extraction do?

If a solid product is noticed during this operation, repeat the washing with two more portions of bisulfite solution. Dry the nonaqueous layer, and transfer it to an apparatus for simultaneous addition and distillation (Figure 7.1). Place 6–7 mL of the solution in a 10-mL flask and the rest in the addition funnel. Distill off all of the methylene chloride with addition of solution to maintain the volume in the flask at about 6–7 mL. The receiver must be cooled. When everything boiling at or below 100°C has been removed, cool and reassemble as in Unit III, Figure 1.7 or Figure 3.8; add a fresh boiling stone and one or two grams of biphenyl (as a chaser). Distill this mixture without water in the condenser. (Wrap aluminum foil around the top of the flask and the lower head.) Collect a fraction that has a boiling range near 180–205°C (d., 1.06 g/mL). Determine the percent yield, the infrared spectrum, and other properties as directed, for both products.

Figure 4.1 (page 129).

Question

See Figures 2.6A, B; 2.7 (pages 52–55).

Question

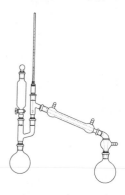

Figure 7.1 (page 257).

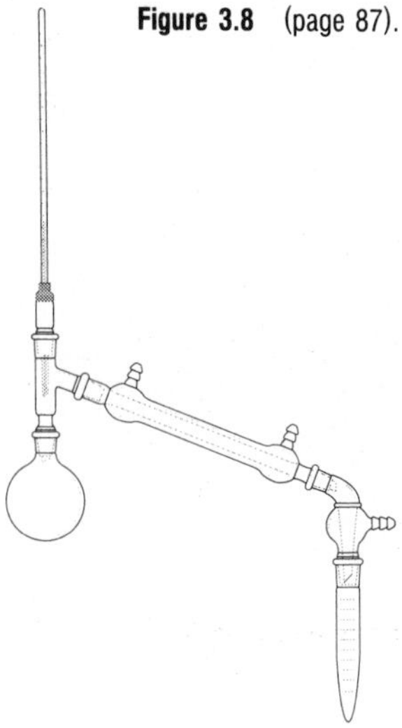

Figure 3.8 (page 87).

Summary

1. Heat the aldehyde in aqueous base, with stirring.

2. Cool the solution, but make sure no solid comes out.

3. Extract the *basic* aqueous solution three times with methylene chloride (or ether).

4. Steam distill any volatile material from the aqueous solution.

5. Cool the aqueous solution, make it acidic, and cool thoroughly.

6. Collect the crystals and recrystallize from hot water.

7. Extract the organic extracts with aqueous sodium bisulfite.

8. Dry the organic solution and distill off all of the methylene chloride (or ether).

9. Distill the product with an air condenser and a chaser or at reduced pressure.

10. Analyze the products as directed.

Questions

QUESTIONS 1 AND 2 ABOVE.

3. Suggest reasons for the lack of agreement between the percent yields for the two products.

4. Write the equations for the Cannizzaro reaction with three other compounds mentioned above as giving this reaction.

Disposal of Materials

Benzaldehyde (OS, OW, I)	aq. NaOH (D, N)
Methylene chloride (OS, HO, I)	Methylene chloride solutions (HO, I)
Ether (OS, HO, I)	Ether solutions (OW, I)
Aqueous filtrate (N, I)	aq. NaHSO$_3$ (save for other disposal)
Sodium bisulfite extractions (evaporate, S, I)	Distillation residue (OW, I)
Biphenyl (OW, I)	Products (I)

Additions and Alternatives

You could follow the progress of the reaction with thin layer chromatography to determine the required length of heating. Analyze the alcohol product by gas liquid chromatography. The alcohol can be *distilled at reduced pressure*. Try additional heating, the addition of methanol to give a homogeneous solution, or just using methanolic KOH to help the reaction. Enough water would need to be added in order to give two layers for extraction, and the water might have to be evaporated to maximize the yield of benzoic acid. Other possibilities are suggested in the introduction to this experiment.

You might want to use ether in place of methylene chloride. If so, do not use flames. Use a boiling water or steam bath when isolating the acid. Remember that ether is volatile and flammable.

E X P E R I M E N T   5 4 B

Cannizzaro Reaction (M)
References: Experiment 54A, its references, and 4.4G

Introduce the reagents and initiate the reaction. In a 5-mL vial or 10-mL flask with a spin vane or bar, place 0.5 mL of benzaldehyde and 1.25 mL of 20% aqueous sodium hydroxide solution. Equip the vessel with an air or water condenser (Unit IA, Figure 1.8A). Put the vessel in a heated sand bath or aluminum block, and begin stirring to get an emulsion. Place a small test tube of water in the sand bath or aluminum block to warm it. Continue stirring, and heat the sand bath or block to keep the temperature in the reaction about 85–90°C for 30–90 minutes, depending upon the length of the period and the amount of work needed to be done in one period. Use a few drops of hot water to maintain the emulsion.

See Figure 1.8 (pages 26–27).

Separate the products. Cool the vial to 50°C, and add warm water to dissolve any solid that may form. Make sure the solution is basic. Extract the cool, aqueous solution (with a Pasteur or plugged pipet) three times with 0.6–0.8-mL portions of methylene chloride. Save both the aqueous and nonaqueous layers.

See Figures 4.2A, B, C (pages 131–133).

Isolate the acid. Place the vessel with the aqueous layer in a sand bath or aluminum block at about 60°C (in the hood), and evaporate any volatile organic material with a stream of nitrogen or gentle suction (Section 1.3I) for three to ten minutes (compare with Experiment 54A). Transfer the solution to a small Erlenmeyer flask and cool. Acidify with dilute sulfuric acid until the solution is acidic to an external paper indicator (ca. pH = 3). Cool the acidified solution *thoroughly*, and collect the crystals on a Hirsch funnel with suction. Recrystallize the benzoic acid with a minimum amount of hot water. Allow the crystals to dry before determining amount, purity, and identity.

See Figures 2.8A, B, C (pages 56–58).

Isolate the alcohol. Extract (wash) the nonaqueous layer twice (with a Pasteur pipet) with 0.5-mL portions of saturated sodium bisulfite solution. If a solid is noticed during this operation, repeat the washing with one or two more portions of bisulfite solution. Dry (with a drying pipet and anhydrous sodium or magnesium sulfate) the nonaqueous solution, and transfer it to a 5-mL vial or flask. Place the vessel in a sand bath or aluminum block (in the hood) set at 50–80°C, and *carefully* evaporate the solvent with a stream of nitrogen or gentle suction (Section 1.3I). When no further evaporation takes place (3–10 minutes), remove and assemble for

See Figures 1.6 and 4.5 (pages 19 and 148).

See Figures 1.9 (page 30).

See Unit III, Figure 1.8B
(pages 28–29).

distillation. Use an aluminum foil wrap around the vessel and the heat source. Add about 0.3–0.5 g of biphenyl as a chaser. Collect a fraction with a distilling range in the vicinity of 185–210°C. (Remove all previously distilled material as it collects.) Determine the percent yield, the infrared spectrum, and other physical properties as directed, for both products.

Summary

See Experiment 54A, but note that steps 4 and 8 involve assisted evaporation instead of distillation.

Questions

ANSWER THOSE GIVEN FOR EXPERIMENT 54A.

5. Explain how the evaporation of the solvent is promoted by the procedure described in the last paragraph.

Disposal of Materials

Benzaldehyde (OS, OW, I)	aq. NaOH (D, N)
Methylene chloride (OS, HO, I)	Methylene chloride solutions (HO, I)
Ether (OS, HO, I)	Ether solutions (OW, I)
Aqueous filtrate (N, I)	aq. NaHSO$_3$ (save for other disposal)
Sodium bisulfite extractions (evaporate, S, I)	Distillation residue (OW, I)
Biphenyl (OW, I)	Product (I)

Additions and Alternatives

See Experiment 54A. If ether is used, it (or methylene chloride) may be removed by distillation with a short path head before the final distillation with a chaser.

EXPERIMENT 55A

Preparation of Benzoin (SM)
References: Sections 1.3E–H, 2.3, 4.2C, 5.1, and Experiment 3

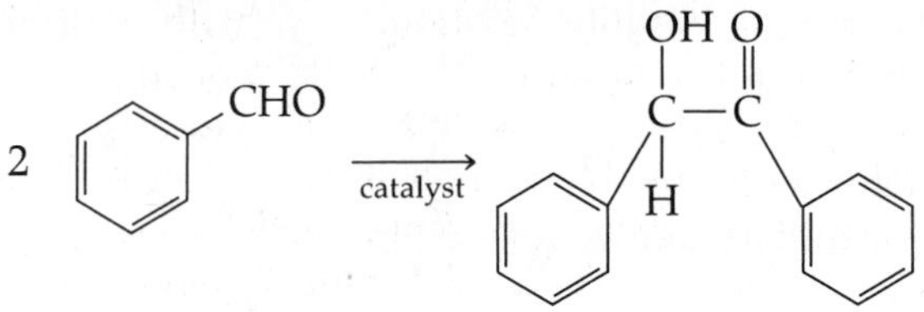

This cyanide-catalyzed reaction is common only to aromatic aldehydes and to the closely related heterocyclic aldehydes. It occurs with such compounds as furfuraldehyde as well as with benzaldehyde. The traditional catalyst is the cyanide ion. Sodium and potassium cyanides function equally well. Both are effective poisons!

One peculiarity of this reaction is its reported semireversability. Benzoin does not seem to be convertible to benzaldehyde easily, but if mixed with a substituted benzaldehyde in the presence of an alkali cyanide, it gives a mixed benzoin. Thus, benzoin and anisaldehyde (*p*-methoxybenzaldehyde) yield a mixture containing benzoyl *p*-anisyl carbinol. A rupture of the bond between the carbinol and carbonyl carbons atoms is induced under these conditions.

The acyloin reaction is not unknown in the aliphatic series, although alkali cyanides are generally not suitable catalysts for this purpose. These salts are so alkaline that they induce either aldol condensations or Cannizzaro reactions instead. Acetoin, $CH_3CHOHCOCH_3$, is apparently produced from acetaldehyde by the action of enzymes in yeast.

Breslow (*J. Am. Chem. Soc.*, 1958, *80*, 3719) has shown that thiamine (vitamin B_1) can act as a catalyst for this reaction. Thiamine is an enzyme cofactor known to react with carbonyl groups (via an enolate-like zwitterion) to form a new carbon-to-carbon bond, as does the cyanide ion when initiating this reaction. Mayo (private communication) reports that the cyanide is more reliable at the micro scale. It can be used in Experiment 55C where the amount used is so much reduced. The use of thiamine hydrochloride is illustrated in this experiment.

Reagent Preparation   Dissolve 5.6 g of potassium hydroxide in 100 mL of 95% ethanol to make about 1M alcoholic KOH, which should serve 30 students. *Fresh* or freshly distilled or purified benzaldehyde should be used. Benzaldehyde can be purified by extraction with bicarbonate solution, followed by drying and distillation. The thiamine hydrochloride must also be freshly opened and kept refrigerated.

Procedure

Reflux the reagents to initiate the reaction. In a 25-mL flask equipped for reflux and addition (Unit II, Figure 1.7), place 0.5 g of thiamine hydrochloride and a magnetic stirrer or boiling stone, and dissolve the salt in 3 mL of water with stirring. Cool the flask, and add 1.5 mL of pure benzaldehyde through the addition funnel. Cool as needed; then stir, and add 3 mL of ethanolic KOH dropwise through the funnel, slowly so that the mixture does not boil. Do not let it become too hot. When the addition has been completed, initiate gentle reflux (do not overheat), and reflux for 30–90 minutes.

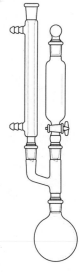

Figure 1.7, Unit II (page 24).

See Figures 2.6A, B; 2.7 (pages 52–55).

Isolate and purify the product. Cool to room temperature and then cool thoroughly in ice and scratch. If an oil separates, add a few drops of ethanol, and heat to redissolve it before recooling slowly. Collect the crystals on a Hirsch or Büchner funnel, and wash them with three 10–15-mL portions of ice water. Recrystallize from a minimum of hot 95% ethanol (about 6–8 mL per gram). When dry, determine the weight, yields, melting range (near 135°C), and infrared spectrum (KBr pellet).

Summary

1. Dissolve the catalyst in water, cool, and add the aldehyde.

2. With cooling, add the alcoholic base dropwise.

3. Reflux gently.

4. Cool gradually, then thoroughly to crystallize the product.

5. Collect, wash, and recrystallize the product.

6. Analyze the product as directed.

Questions

1. What is the function of the alcoholic potassium hydroxide?

2. Write the equation for the acyloin condensation of furfural.

3. Write the mechanism of this reaction catalyzed by CN^- and by thiamine.

$$R-\overset{+}{N}-C-CH_3$$

(The single hydrogen shown is acidic.)

4. Why are extensive washings and special disposal of the filtrate and washings necessary in the case of cyanide ion catalysis?

5. Should the oxidation of benzoin be easy? Compare it with the oxidation of fructose. Explain and give the product of the oxidation of benzoin.

6. Benzoin has a chiral center. Will the product of this experiment be optically active? Explain.

7. Analyze the infrared spectrum of benzoin and compare it with that expected for benzaldehyde.

Disposal of Materials

alc. KOH (D, N, I)	Benzaldehyde (OW, I)
Thiamine hydrochloride (S)	Wash water (N, I)
Ethanol (OS, D, I)	Product (I)

Additions and Alternatives

If you encounter difficulty in the initial crystallization, evaporate a drop of solution on a stirring rod to give a seed crystal. Place this crystal in the solution and scratch to facilitate crystallization. Alternatively, omit extensive cooling and allow the solution to stand until the next period.

See Figures 4.6A, B, C (pages 154–157).

Analyze the product and the refluxing mixture by thin layer chromatography. The development solvent might be made with more ethyl acetate than hexane. This analysis could also be used to indicate the needed length of reflux. Record the infrared spectrum of benzaldehyde and compare it with that of the product.

It would be interesting to explore the use of vitamin B_1 from the drug store (try extracting the thiamine) for this preparation.

The directions for the rearrangement of benzoin and benzil to benzylic acid are found in Experiments 46A and B.

Try treating the product with Fehling's or Tollens' reagent (Chapter 17, Exercises 8 and 9) for qualitative results.

Recent papers of Breslow and coworkers have pointed to the importance of hydrophopic effects for this reaction in water. Try double or half the amount of water, or try 1.5 mL of water, 5 mL of ethanol, and 1.5 mL of 2N NaOH.

See Experiment 55B and C.

E X P E R I M E N T 5 5 B

Preparation of Benzoin (M)

References: Experiment 55A and its references

Reagent Preparation Benzaldehyde should be used *fresh*, but it can be purified by distillation or by extraction with bicarbonate, drying, and distillation. The thiamine hydrochloride must also be *fresh* and refrigerated after opening. Dissolve 5.6 g of potassium hydroxide in 100 mL of 95% ethanol to make about 1M alc. KOH, which should serve 50 students or more.

Procedure

Mix the reagents and initiate the reaction with reflux. In a 5-mL vial or equivalent vessel with magnetic spin vane or stirrer and equipped with a reflux condenser (Unit IA or B, Figure 1.8A; use extra care with an air condenser), place 0.2 g of thiamine hydrochloride. Add 1 mL of water and stir to dissolve the solid. Maintain the reaction vessel at room temperature and continue stirring during both of the following additions. Add 0.5 mL of benzaldehyde dropwise with a graduated pipet so the drops reach the

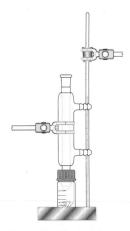

Figure 1.8 (page 26).

See Figures 2.8A, B, C
(pages 56–58).

vessel *directly*. Then add dropwise 1.2 mL of alc. KOH so the mixture does not boil. Do not let it become too hot. When the addition has been completed, initiate *gentle* reflux (do not overheat), and reflux for 20–60 minutes.

Isolate and purify the product. Cool the reaction vessel to room temperature. Then cool thoroughly in ice and scratch and/or seed to promote crystallization. If an oil separates, add a few drops of ethanol and heat to redissolve it before recooling slowly. Collect the crystals with suction on a Hirsch funnel and wash with three 3–5-mL portions of ice water. Recrystallize from a minimum of hot 95% ethanol (about 6–8 mL per gram). When the product is dry, determine the weight, yields, melting range (near 135°C), and infrared spectrum (KBr pellet).

Summary

1. Dissolve the catalyst in water, cool, and add the aldehyde.

2. With cooling, add the alcoholic base dropwise.

3. Reflux gently.

4. Cool gradually, then thoroughly to crystallize the product.

5. Collect, wash, and recrystallize the product.

6. Analyze the product as directed.

Questions

ANSWER QUESTIONS 1–7 FROM EXPERIMENT 55A.

Disposal of Materials

alc. KOH (D, N, I)	Benzaldehyde (OW, I)
Thiamine hydrochloride (S)	Wash water (N, I)
Ethanol (OS, D, I)	

Additions and Alternatives

Try using 0.3 g of thiamine with 1.8 mL of alc. KOH. A thin layer chromatographic analysis could be used to establish the length of reflux.

Try oxidizing the benzoin with Fehling's or Tollens' reagent. The oxidation to benzil is commonly done with concentrated nitric acid (3.5 mL per gram benzoin) at 90–100°C for 10–30 minutes (see Experiment 26). The resulting benzil can be rearranged to the salt of benzylic acid with strong aqueous base (Experiment 46B). Alternatively, the benzoin can be acetylated with acetic anhydride (*Organic Synthesis* Coll. Vol. II, Wiley, New York, 1943, p. 69).

See Experiment 55A.

E X P E R I M E N T   5 5 C

Preparation of Benzoin (M)
References: Experiment 55A and its references

Reagent Preparation Benzaldehyde can be purified by extraction with bicarbonate solution, by drying and distillation. The cyanide solution (about 0.62M) is prepared by dissolving 1 g of potassium cyanide in 7 mL of water and 17 mL of 95% ethyl alcohol (will serve 18 to 20 students). This reagent should be prepared and dispensed by the instructor.

Procedure

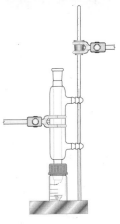

Figure 1.8 (page 26).

Initiate the reaction by refluxing the reagents. Equip a 3-mL or 5-mL vial or equivalent vessel with a spin vane or stirring bar and a reflux condenser (Unit IA or B, Figure 1.8A; extra care should be employed if an air condenser is used). In the vessel, place 0.4 mL of purified benzaldehyde, and add (*carefully!*) 1.20–1.25 mL of cyanide solution in the hood. Carefully, with the condenser functioning effectively and with stirring, establish a gentle reflux. If stirring is not used, add a boiling stone *before* heating. Maintain the reflux for 20–30 minutes. Allow the vessel to cool to room temperature; stop stirring and thoroughly cool it in an ice bath.

Collect and purify the solid product. Collect the crystals with care on a Hirsch funnel with suction. Wash them with a few drops of ice-cold ethanol and then several times with ice-cold water to remove any trace of potassium cyanide. Dispose of the filtrates as directed by your instructor. If desired, recrystallize the product from a minimum of hot 95% ethanol (ca. 0.7 mL/100 mg). The Craig tube should be appropriate or a test tube and Hirsch funnel can be used. When the product is dry, determine weight, yields, melting range, and infrared spectrum (KBr pellet).

See Figures 2.8A, B, C (pages 56–58).

Summary

1. Add cyanide solution to the aldehyde with care and then reflux.

2. Cool with stirring to room temperature, then thoroughly cool.

3. Collect the crystals with suction and wash thoroughly.

4. Recrystallize if directed to do so.

5. Analyze the product as directed.

ANSWER QUESTIONS 2–7 FROM EXPERIMENT 55A.

Q u e s t i o n s

Disposal of Materials

Potassium cyanide (HW, I, see comment for cyanide solutions.)

Cyanide solutions (**Toxic**) (Treat with bleach. Test a portion of the solution for cyanide as described in Experiment 72. If cyanide is still present, use more bleach until the test for cyanide is negative. Evaporate (care, no acid); HW, I)

Water and ethanol washes (Treat as cyanide solutions.)

Ethanol from recrystallization (Treat as cyanide solution.)

Ethanol (OS, D, I)

Product (I)

Additions and Alternatives

Use a thin layer chromatographic analysis to establish the length of reflux.

Try oxidizing the benzoin with Fehling's or Tollens' reagent. The oxidation to benzil is commonly done with concentrated nitric acid (3.5 mL per gram benzoin) at 90–100°C for 10–30 minutes (see Experiment 26). The resulting benzil can be rearranged to the salt of benzylic acid with strong aqueous base (Experiment 46B). Alternatively, the benzoin can be acetylated with acetic anhydride (*Organic Synthesis* Coll. Vol. II, Wiley, New York, 1943, p. 69).

See Experiment 55A.

E X P E R I M E N T   5 6

Preparation of Cinnamic Acid (Perkin Reaction) (SM)
References: Sections 1.3E–H, 2.3, 4.1, and 5.1

$$C_6H_5CHO + (CH_3CO)_2O \xrightarrow{K_2CO_3} C_6H_5CH = CHC(O)OC(O)CH_3 + H_2O$$

$$C_6H_5CH = CHC(O)OC(O)CH_3 + H_2O \longrightarrow$$

$$C_6H_5CH = CHC(O)OH + CH_3COOH$$

Aromatic aldehydes react with a variety of compounds characterized by having active hydrogens alpha to a carbon-to-oxygen double bond. Such modified aldol condensations are facilitated by the absence of α-hydrogens in the aldehyde, which will allow no condensation between two molecules of the aldehyde. The condensation products are not isolated when they dehydrate spontaneously. This loss of water in the current experiment may be partly explained by the enhanced stability of a double bond that is alpha to both a carbon-to-oxygen double bond and to an aromatic ring.

One of the more familiar condensations is the Perkin reaction, which involves the condensation of an aromatic aldehyde, such as benzaldehyde, with an acid anhydride in the presence of a salt of the acid. Other bases have been shown to be effective, and their use may save considerable time, as has been shown by Kalnin (*Helv. Chim. Acta*, 1928, *11*, 977). It may well be that the interaction of the anhydride with the base provides some acid salt to be used in the reaction. The present preparation utilizes such a procedure to prepare cinnamic acid from benzaldehyde, acetic anhydride, and potassium carbonate.

Reagent Preparation The benzaldehyde (see Experiment 55) and acetic anhydride should be freshly purified and/or distilled.

Procedure

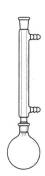

Assemble the apparatus with the reagents and initiate the reaction. In a 100-mL flask equipped for reflux (Figure 1.7, Unit I) with a drying tube, place 3.5 mL of benzaldehyde and 6 mL of acetic anhydride. Carefully add 3.5 g of anhydrous potassium carbonate through the condenser so that all of the solid reaches the flask. Heat the flask gently as long as foaming and frothing continues.

Figure 1.7, Unit I (page 24).

1. What gas is responsible?

When the foaming subsides, heat the mixture to reflux (180–200°C) (aluminum foil wrap may be needed) and continue for 50–60 minutes.

Purify the product mixture. Allow the flask to cool until it can be handled easily. Clean the joint. Pour and wash the contents into a 250-mL flask or beaker with 40 mL of water and 20 mL of 6N aqueous sodium hydroxide. Most or all of the sodium salt should dissolve, though additional water and heating may be needed. Insoluble tars may be removed by filtration, decantation, or subsequent extractions. Extract the cool, clear, basic solution with two 10–20-mL portions of methylene chloride. If a 60-mL separatory funnel is used, divide the aqueous solution in half and extract each half with two 12-mL portions of methylene chloride.

Question

Figure 4.1 (page 129).

Question

2. Why should the mixture be cool? Which layer is which? How could the aqueous layer be tested?

Extract the combined nonaqueous layers with 10–20 mL of water, and add the water to the basic solution(s).

See Figures 2.6A, B; 2.7
(pages 52–55).

Prepare, isolate, and purify the product. Cool the combined aqueous layers to ice temperature in a large Erlenmeyer flask. Acidify the aqueous solution with 20–30 mL of concentrated hydrochloric acid or until the solution tests acidic with external indicator paper (ca. pH = 3 or 4). When the precipitation is complete, collect the crystals by suction filtration. The product may be recrystallized from a minimum of boiling water (activated carbon may be needed). When the crystals are dry, determine yields, melting range (below 135°C), and other properties as directed, including an infrared spectrum.

Summary

1. Assemble the apparatus with the aldehyde and the anhydride.
2. Add K_2CO_3.
3. Carefully heat and then reflux.
4. Cool, transfer, and combine with aqueous base.
5. Extract the aqueous solution twice with methylene chloride.
6. To ensure that no product is lost, extract the organic layers with water.
7. Cool the combined aqueous layers, acidify, and cool again.
8. Collect with suction; recrystallize from a minimum of hot water.
9. Analyze the product as directed.

Questions

QUESTIONS 1 AND 2 ABOVE.

3. What is the purpose of the extraction with methylene chloride?
4. Why must the solution be basic for this extraction?
5. How is the acetic acid removed from the product?
6. What is the structure of the unisolated aldol condensation intermediate?
7. How could propanoic anhydride be used in this preparation? What would the product be?
8. Contrast the infrared spectrum of the product with that of benzaldehyde.

Disposal of Materials

Benzaldehyde (OW, I)	Acetic anhydride (D, N)
aq. NaOH (D, N)	Methylene chloride (OS, HO, I)
conc. HCl (D, N)	Methylene chloride (HO, evaporate, I)
Product (I)	Aqueous filtrates (I, D, N)

Additions and Alternatives

You could do this preparation on the micro scale at one-tenth the scale used here. Extractions would be done with a pipet.

Cinnamic acid can be photodimerized to give a cyclobutane.

If you carry out this reaction with salicylaldehyde, the *o*-hydroxycinnamic acid, which should be produced, cyclizes to give coumarin. See Cheronis, *Semimicro Experimental Organic Chemistry*, John de Graff, New York, 1958, pp. 312, 314. Alternative methods of preparing cinnamic acid are given in *J. Chem. Educ.*, 1950, *27*, 210; 1964, *41*, 565; 1990, *67*, A304 (M). Note that this last reference is to a microscale experiment. For an exploration of the aldol condensation, see *J. Chem. Educ.*, 1987, *64*, 367. See also Experiment 66.

EXPERIMENT 57

Preparation of Diphenylcarbinol (Grignard Reaction) (SM)
References: Sections 1.3E–H, 2.3, 4.1, and 5.1.

$$C_6H_5Br + Mg \xrightarrow{\text{ether}} C_6H_5MgBr$$

$$C_6H_5MgBr + C_6H_5CHO \longrightarrow (C_6H_5)_2CHOMgBr$$

$$(C_6H_5)_2CHOMgBr + H^+ \xrightarrow{H_2O} (C_6H_5)_2CHOH + Mg^{2+} + Br^-$$

The Grignard reagent, as generally understood, is an organomagnesium halide–ether complex that is formed by the action of magnesium with an organic halide in a solvent such as ether. It can react with some elements and a variety of compounds to yield (when not involving simple metathesis) reaction intermediates or complexes. These complexes or the reagent itself can be hydrolyzed with dilute acids or saturated ammonium chloride solution to yield compounds in which the magnesium halide has been replaced by hydrogen. Compounds containing the carbonyl grouping, such as aldehydes, ketones, esters, and acyl halides, are the most widely used with the Grignard reagent.

1. Why not carboxylic acids?

Question

These reactions followed by hydrolysis find synthetic use in many preparations, particularly of alcohols.

The restrictions on the choice of alkyl or aryl halides used with the Grignard reagent are that the carbon-to-halogen bond must not be too strong for the magnesium to break and must not be too weak lest a coupling reaction predominate. The reactivity of the halides increases in the order: chloride < bromide < iodide. The aryl chlorides, for example, are not usually affected by magnesium in ether. Since the yields are said to be inversely related to the reactivities, the bromide is often the halide of choice. See *J. Chem. Educ.*, 1989, *66*, 586. All equipment and all reagents used in this preparation must be scrupulously dry.

2. Why must everything be dry?

The ether *must* be anhydrous.

Procedure

Assemble the apparatus and activate the magnesium. Place one or two crystals of iodine and 0.5 g of bright magnesium filings or turnings (in that order) in a 100-mL flask equipped for addition and reflux (Unit II, Figure 1.7; if possible the glassware should be oven-dried and *cooled*), with the funnel and condenser fitted with drying tubes. Remove the flask from the apparatus, hold it with a clamp, and, with a micro burner, gently flame the flask. Do not push the purple cloud of iodine vapors beyond the metal, because the purpose is to deposit the iodine on the metal. This step may not be needed if a magnesium alloy requiring no activation is used. Turn off the flame and allow the apparatus to cool to room temperature. Do NOT GO BEYOND THIS POINT UNTIL ALL FLAMES HAVE BEEN EXTINGUISHED!

Initiate the formation of the Grignard reagent. Add about 1.5 mL of anhydrous ether to the contents of the cooled flask. Mix 2 mL of anhydrous bromobenzene and 5 mL of anhydrous ethyl ether in a small, dry flask and place the mixture in the addition funnel. With water flowing in the condenser, add about 1 mL of the ether–halide *mixture* to the magnesium in the flask. If everything is dry and free of contamination, the ether mixture may start to boil from the heat of reaction. Discolored spots, which may appear on the metal, are among the signs of reaction. Be prepared to cool the reaction if it is too vigorous, or heat it gently if it is reluctant to start. (If the reaction does not become *self-sustaining* after brief reflux, cool the flask and mixture, add a crystal of iodine directly to the metal, rub the crystal gently into the metal, reassemble, add a few drops of ether–halide mixture, and try again.) As the reaction subsides, add more of the ether–halide mixture from the funnel at such a rate that the reaction is maintained without resorting to cooling or heating.

When the reaction becomes sluggish, swirl or gently shake the mixture to ensure complete mixing. Briefly reflux the reaction mixture. When there appears to be no more reaction and much or most of the magnesium has been consumed, cool the flask in a cold water bath.

React the Grignard reagent with the aldehyde. Add 1.9 mL of benzaldehyde dropwise from the funnel with frequent shaking or swirling to mix the reactants. When the addition is complete, allow the mixture to warm to room temperature, and, with a clean, dry stirring rod, stir or probe the

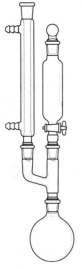

Figure 1.7, Unit II (page 24).

semisolid mixture to expose portions of the mixture that, because of viscosity, have not mixed and reacted. Quickly reassemble.

Break up the complex. Pour 10 mL of 3M sulfuric acid on an ice cube to thoroughly cool it. Cautiously add the cold acid solution to the reaction mixture from the funnel. Shake, swirl, and mix. If necessary, probe the mass with the stirring rod to keep the reagents in contact with each other. If the upper layer is not fluid or all of the magnesium has not dissolved, add a bit more 3M acid to achieve this. This acid need not be cooled.

Isolate the product. Transfer the *liquid* contents of the flask to a separatory funnel, and rinse the flask with a little *solvent grade* ether, adding this ether to the funnel. All of the product should be dissolved in the ether. Repeat this procedure with a little water. Separate the aqueous layer.

Figure 4.1   (page 129).

Q u e s t i o n

3.   Which is the aqueous layer? How could you test it?

Wash the ether layer with an equal volume of water and retain the ether layer.

Heat about 40 mL of water in a 250-mL beaker to its boiling point, and take it to the hood. NO FLAMES HERE. Slowly pour the ether layer into this hot water to evaporate the ether. When the ether has been evaporated, boil the water vigorously for a few minutes to steam distill out any unreacted bromobenzene or benzaldehyde. Stir the mixture to disperse the droplets and prevent bumping. Cool and add an ice cube to solidify the crude product. Decant the water or filter.

See Figures 2.6A, B; 2.7 (pages 52–55).

Recrystallize the product from heptane or another hydrocarbon solvent of similar boiling point. *No Flames.* Use a water or steam bath. If water is observed after the product has dissolved, decant the solution carefully into a dry flask, leaving the water behind. If the product oils, extract the oil with fresh solvent and retain all portions of solvent. Cool the solvent portions thoroughly and collect the flaky crystals on a Büchner or Hirsch funnel, using suction. During the suction, cover the funnel with a watch glass to prevent "top evaporation." Wash the crystals once with a little cold solvent, and allow them to air-dry. When dry, determine yields, melting range (below 70°C), and other properties, including the infrared spectrum, as directed.

Summary

1.   Activate the magnesium with iodine and assemble the apparatus.

2.   Cool, add ether, and then add a bit of the mixture.

3.   Continue the addition to just sustain the reaction.

4.   Reflux briefly and cool.

5.   Add benzaldehyde and mix thoroughly.

6. Acidify the mixture, destroying any excess metal.

7. Separate the organic layer and wash it with water.

8. Evaporate the ether (in the hood) by pouring it into boiling water.

(8. Alternate—Use Section 1.3I and then add a little water for steam distillation.)

9. Steam distill the impurities, if any.

10. Crystallize, and collect by decanting through a filter.

11. Recrystallize from heptane.

12. Analyze the product as directed.

Questions QUESTIONS 1–3 ABOVE.

4. Explain the method for removing ether from the product.

5. Explain why more H_2SO_4 is added if some magnesium remains. What is the reaction?

6. Write the equations for preparations discussed in Additions and Alternatives.

7. How could *p*-chlorobenzoic acid be produced by Grignard reaction from *p*-chlorobromobenzene? (Equation and explanation)

8. Analyze the infrared spectrum of the product. Contrast this spectrum to that expected for the starting materials.

Disposal of Materials

Iodine (*Caution*, HW, S, I)   Magnesium (S)

Bromobenzene (HO, OS, I)   Ethyl ether (OS, I)

Benzaldehyde (OS, OW, I)   3M H_2SO_4 (D, N)

Aqueous solutions (N, I)   Heptane, etc. (OS, OW, I)

Heptane solutions (OW, I)   Product (I)

Additions and Alternatives

Try this preparation on the micro scale at one-tenth the scale used here, with some modification.

For reluctant reactions, make a starter mix with iodine, magnesium, bromobenzene, and ether in a small test tube. With a dry stirring rod, rub the magnesium and iodine together. Warm the test tube with a warm water bath. When this starter reaction is well started, add all the contents of the tube to the flask through a wide-stem funnel to avoid contact of the contents with the joint.

Essentially the same procedure can be used if acetone, benzophenone, or methyl (or ethyl) benzoate is added in place of benzaldehyde. You could prepare valeric acid (boiling point near 185°C) in much the same manner. Form the Grignard reagent by adding, to 1 g of magnesium, a mixture of 3.7 mL of purified *n*-butyl bromide (student preparation can be used) in 13 mL of dry ether. Pour the mixture containing the Grignard reagent on 15 g of dry ice and hydrolyze the new complex with 25 g of ice and 5 mL of concentrated hydrochloric acid. Wash the aqueous layer and apparatus twice with 5–10 mL of ether. Fractionally distill the combined nonaqueous layers (above 140°C, air should circulate through the condenser, not water). The acidity of the product can be demonstrated by Exercise 6, Chapter 17.

Other examples of Grignard reactions and extensions can be found in *J. Chem. Educ.*, 1974, *51*, 57; 1976, *53*, 457; 1990, *67*, 271; and 1991, *68*, 71.

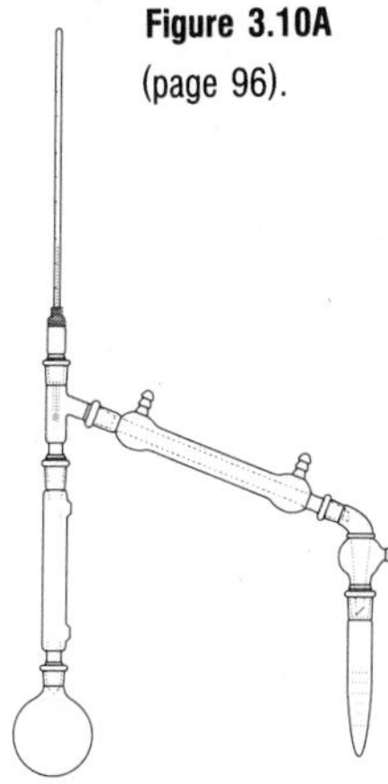

Figure 3.10A
(page 96).

CHAPTER 13

Natural Products

13.1 Introduction

It may seem strange to have a special chapter devoted to the chemistry of natural products, but it may well be that the source of all organic materials (even crude petroleum) is the biosphere. Many of the experiments in this book illustrate this idea by using naturally occurring substances as starting materials (see, for example, Experiments 7A and B, 9A and B, 10B, 28B, 69, 70, and the alternatives for 10 and 11).

The experiments in this chapter focus on the properties of three types of biomolecules—proteins (and amino acids), carbohydrates, and triglycerides. Each experiment has two or more parts that can be mixed and matched to produce experimental variations. For example, you could choose parts that lead to an emphasis on analysis or on unknowns.

For an illustration of the use of the polarimeter or of naturally occurring optically active substances in the resolution of racemic mixtures, consult *J. Chem. Educ.*, 1965, *42*, 269; 1986, *63*, 646; and 1989, *66*, 608, or Experiment 69.

13.2 Amino Acids and Proteins

Proteins are a necessary component of the human diet. Without the ingestion of some protein as found in meat, poultry, fish, cheese, beans, and other foods, the human being would cease to function.

Proteins are polymers made of α-aminocarboxylic acids, i.e., amino acids. The amino acids are chemically bonded to each other by means of an amide (or peptide) linkage:

$$-\overset{\overset{\displaystyle O}{\|}}{C}-N\big\langle$$

There are more than 20 naturally occurring amino acids, which can bond to each other in a tremendous variety of sequences. These long chains of amino acids, or proteins, may contain thousands of units.

397

The two preparations that follow illustrate a method used to form an α-amino carboxylic acid, glycine, and a method used to form the amide linkage, the preparation of hippuric acid by the Schotten–Baumann technique.

Glycine is prepared by the ammonolysis of the readily available α-chloroacetic acid. The α-chloroacetic acid could be prepared by reaction of acetic acid under Hell–Volhard–Zelinsky conditions.

Hippuric acid, which is prepared from glycine, is not a dipeptide, but the benzoylation of glycine suggests a method of forming the peptide linkage. The Schotten–Baumann technique utilizes sodium hydroxide both to promote the reactions of benzoyl chloride, which is less reactive than aliphatic acid halides, and to neutralize the hydrogen chloride being generated, to avoid protonation of the amine group in glycine.

As indicated in the directions, these two preparations may be done separately or as a unit. The third section of Experiment 58 is included to acquaint you with some of the simple tests used with proteins. Careful planning and execution should permit the completion of all three sections in one period or a little more. Any two sections could be used together and completed easily; this may be a better plan if two or more substances are used for each test.

E X P E R I M E N T 5 8 A

Part One: Preparation of Glycine (SM)
References: Sections 1-3l, 2.3, and 3.5.

$$ClCH_2COOH + 2NH_3 \longrightarrow ClNH_3CH_2COONH_4 \longrightarrow NH_2CH_2COO^-$$

Start the reaction one week ahead. At the beginning of the laboratory period prior to the one for which this experiment is assigned, add 3 g of chloroacetic acid in 5 mL of water cautiously to 40 mL of concentrated ammonium hydroxide in a 125-mL Erlenmeyer flask. Then set the labeled, well-stoppered container in the hood.

See Figure 1.9 (page 30).

Isolate the glycine. If you are isolating the glycine, reduce the solution to about one-half of its original volume, and add a solution of 10 g of $(NH_4)_2CO_3$ in 10 mL of warm water to it. Continue the evaporation until the volume is about 5–6 mL. Cool the solution; stir it with 25 mL of methyl alcohol, and cool it to well below 5°C. Decant the liquid through filter paper, then stir the crystals with 10 mL of methanol (to remove the ammonium salts), and collect them by suction filtration. The crystals may be further purified by dissolving them in a minimum of hot water, decolorizing with charcoal pellets, filtering off the charcoal, cooling the solution, and precipitating the crystals by the addition of methanol.

See Figures 2.6A, B; 2.7 (pages 52–55).

Prepare the glycine solution for Part Two. If the entire original solution is to be converted to hippuric acid without the isolation of glycine, add

18 mL of dilute NaOH and a few boiling stones to the solution; concentrate the solution by evaporation in the hood until the volume is about 15 mL. If the odor of ammonia persists, continue boiling with the addition of enough water to maintain a volume of 15 mL. If evaporation is not done in the hood, you can concentrate the solution by using the setup for simple distillation, with a rubber tube from the vacuum adaptor's outlet that dips into a trap containing water. If your instructor wishes you to isolate some glycine while converting most of it to hippuric acid, withdraw one-fifth of the total volume before adding the NaOH, and place it in a test tube. Handle the smaller portion as described in the preceding paragraph but on one-fifth the scale.

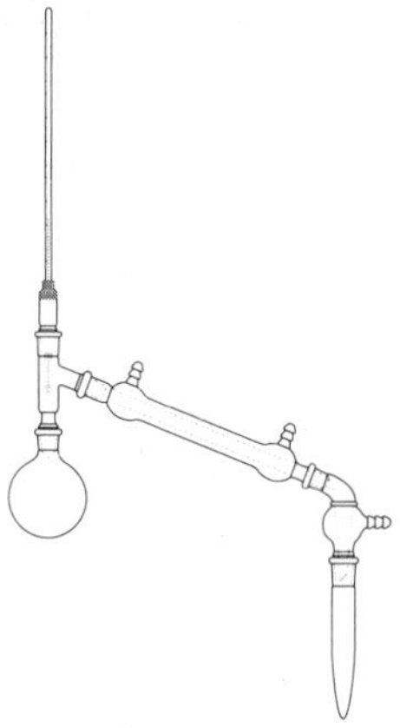

Part Two: Preparation of Hippuric Acid (SM)
References: Sections 2.3, 4.1, and Experiment 58A, Part One.

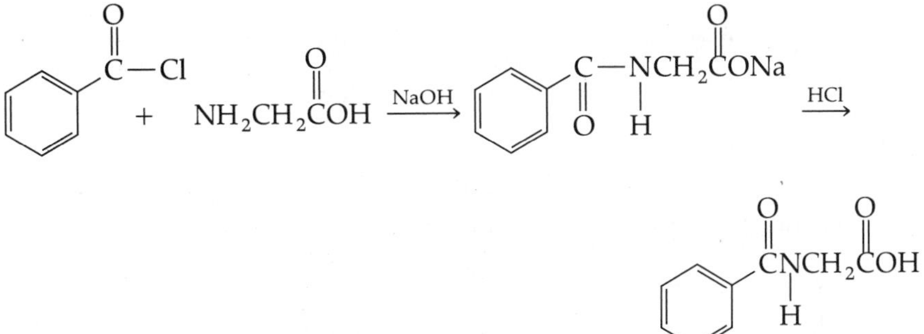

Figure 3.8 (page 87).

See Figure 4.1
(page 129).

Initiate the first reaction in a separatory funnel. Place the basic solution from Part One or a solution of 2 g of glycine in 15 mL of dilute (6M) NaOH in a separatory funnel, and cautiously add 3.5 mL of benzoyl chloride at room temperature. Stopper and shake the flask; release the pressure by carefully opening (invert!) the stopcock occasionally. Shake the mixture and vent for 15 minutes or until the odor of benzoyl chloride is no longer detectable, whichever comes first.

Acidify the solution, collect and purify the product. Pour the solution into a mixture of 8 mL of concentrated hydrochloric acid and 25 mL of cracked ice. After thorough stirring, the solution should test acidic on an external indicator paper (ca. pH = 3) (if not, add more acid). Collect the precipitate by suction filtration; then break the suction. Wash the precipitate by stirring it thoroughly with 6 mL of cold water; reapply the suction. Wash the precipitate a second time with 6 mL of ether. The hippuric acid may be further purified by recrystallization from a minimum amount of hot water. When the product is dry, determine the yields and the melting range (near 190°C).

Summary

Part One:

1. Mix the reagents and allow them to stand in a sealed flask for one week.

For isolation:

2a. Evaporate about one-half of the water.

3a. Add ammonium carbonate solution and evaporate again.

4a. Cool, stir in methanol, and cool well.

5a. Decant the solvent and wash the crystals again with methanol.

6a. Collect by gentle suction and purify if so directed.

For direct continuation to Part Two:

2b. Add base and boiling stones.

3b. Evaporate off the water to reduce the volume to 15 mL.

4b. If the smell of ammonia persists, add water and repeat.

Part Two:

7. Place in a separatory funnel either the basic solution from (4b) or 2 g of solid from (6a) dissolved in base.

8. Cautiously add benzoyl chloride; shake and release the pressure.

9. When the reaction is completed, pour the solution into iced HCl.

10. Collect the product by suction filtration from the acidified mixture.

11. Wash the product twice and recrystallize if so directed.

12. Analyze one or both products as directed.

Questions

1. Suggest how this method could be used to make phenylalanylglycine from glycine and β-phenylpropionic acid.

2. Suggest a synthesis of hippuric acid from benzaldehyde and acetic acid.

3. How would the Schotten–Baumann technique be used in making ethyl benzoate?

4. Half as much ammonium carbonate may be added in place of the ammonia in the preparation of glycine. How does it act? What precautions would have to be taken?

5. Why must all of the ammonia be removed before the addition of benzoyl chloride?

6. The principal impurity in this hippuric acid is benzoic acid. How is it formed and how is it removed?

7. The melting point of hippuric acid is determined as a check for the purity of the product. Why isn't the melting point of glycine taken?

Disposal of Materials

Chloroacetic acid (HO, HW, I, let stand with a stoicometric 2 : 1 ratio of base, evaporate)

Ammonium hydroxide (D, N)   Ammonium carbonate (S, I)

Methanol (OS, I, D)   Methanol solutions (evaporate, I)

Charcoal (evaporate, S, I)   Glycine (I)

dil. NaOH (N)   Benzoyl chloride (*Caution*, let stand with base, evaporate, I, S)

conc. HCl (D, N)

Wash water (N)   Wash ether (evaporate, I)

Hippuric acid (I)

Part Three: Test Reactions for Proteins and Their Hydrolysis Products

This portion of the experiment will introduce you to some of the chemical and physical characteristics of proteins and amino acids. Start the hydrolysis at the beginning of the period, using a colloidal solution of casein or a soluble egg albumin solution.

Hydrolysis of Proteins   Heat the mixture of 22 mL of dilute hydrochloric acid and 0.5 g of casein in a large boiling tube (22 or 25 × 200 mm) or a 125-mL flask (or use 15 mL of concentrated hydrochloric acid and 30 mL of casein solution or 5 mL of concentrated hydrochloric acid and 50 mL of 1% egg albumen solution) in a beaker of boiling water for 60–75 minutes.

8. Write the general equation for this hydrolysis reaction.   **Question**

Note: The following reactions should be conducted with the hydrolyzed material as well as with the original colloidal casein or soluble egg albumin solution. If casein solution was used in hydrolysis, increase the protein solution in each test by 1 mL; if egg albumen was used, increase by 2 mL.

Van Slyke Reaction

Just as nitrous acid is able to convert aromatic amines into diazonium salts (Section 9.1) that are subsequently hydrolyzed into phenols and nitrogen gas,

$$\underset{}{\text{C}_6\text{H}_5\text{NH}_2} \xrightarrow{\text{HNO}_2} \underset{}{\text{C}_6\text{H}_5\text{N}_2{}^+\text{Cl}^-} \xrightarrow{\text{H}_2\text{O}} \underset{}{\text{C}_6\text{H}_5\text{OH}} + \text{N}_2$$

so nitrous acid reacts in a similar fashion with amino acids.

To 5 mL of distinctly acidic protein solution (to the solutions that are neutral or have been neutralized with hydrochloric acid, add 0.5 mL of dilute hydrochloric acid), add 2 mL of freshly prepared 10% $NaNO_2$ solution. Record your observations.

Biuret Test

If urea is heated briefly above its melting point, ammonia is liberated and the residue that solidifies is **biuret**.

$$\underset{\text{O}}{\overset{\text{O}}{\underset{\|}{\text{NH}_2\text{CNH}_2}}} \xrightarrow{\Delta} \text{NH}_3 + \text{H}-\text{N}=\text{C}=\text{O}$$

$$\text{H}-\text{N}=\text{C}=\text{O} + \underset{\|}{\overset{\text{O}}{\text{NH}_2\text{CNH}_2}} \longrightarrow \text{NH}_2-\overset{\text{O}}{\underset{\|}{\text{C}}}-\underset{\text{H}}{\overset{\|}{\text{N}}}-\overset{\text{O}}{\underset{\|}{\text{C}}}-\text{NH}_2$$

biuret

When biuret is dissolved in water and a few drops of dilute cupric sulfate solution, and a few drops of 10% sodium hydroxide solution are added, a pink-violet color appears. The same results are obtained with all compounds containing two or more peptide bonds. This test may be used to determine when the hydrolysis reaction of a protein is complete.

To 2 mL of distinctly basic protein solution (to the solutions that are neutral or have been neutralized with sodium hydroxide, add 1 mL of 10% sodium hydroxide) add 6 mL of water and 2 drops of 2% $CuSO_4$. Record your observations.

Place 0.5 g of urea in a clean, dry test tube, and heat it until the evolution of gas has practically stopped. Test the gas by holding a piece of moist, red litmus paper at the mouth of the test tube. Dissolve the product in 5 mL of water, and filter the solution. Then add 1 mL of 10% sodium hydroxide and 2 drops of 2% $CuSO_4$. Record your observations.

Xanthoproteic Test

In the xanthoproteic test, aromatic rings of proteins are nitrated by heating the protein with concentrated nitric acid to give a yellow solution. The yellow color is intensified by making the solution basic.

Cautiously warm 1 mL of the protein solution to be tested with 2 mL of concentrated nitric acid. The cooled solution may be neutralized with 10% sodium hydroxide. Record your observations.

Questions

9. Tabulate the results of each of the previous tests on both hydrolyzed and unhydrolyzed protein. Explain your observations.

10. Why might amino acids exist as dipolar ions $[^+NH_3CH_2C(=O)O^-]$?

11. What effect should the protonated amine that is α to the carboxylic acid group have on the K_a (or pK_a) of the carboxylic acid group? Check your answer by looking up the K_a (or pK_a) value(s) for glycine and acetic acid.

12. How can the biuret test be used to determine when hydrolysis of a protein is complete?

13. If the biuret test were positive with hydrolyzed casein, what would it indicate?

14. How could the Van Slyke test be used to determine the amounts of protein and urea in the blood?

15. Which amino acids are responsible for a positive xanthoproteic test?

Disposal of Materials

Protein solutions (N or D, I)

aq. $NaNO_2$ (I, acidify in hood or treat with dilute bleach, D)

Urea (S)

conc. HNO_3 (D, N)

aq. HCl (D and/or N)

aq. $CuSO_4$ (D)

Urea solution (N or D)

Nitric acid solutions (N)

Solutions

Colloidal casein is obtained by shaking together 2 g of casein, 100 mL of water, and 25 mL of 10% NaOH.

Soluble egg albumin solution may be prepared in at least two ways. Dried egg albumin may be used to make a 1% aqueous solution, or a 1 : 10 solution of fresh egg white and water may be beaten and filtered through a porous medium such as cheesecloth.

E X P E R I M E N T   5 8 B

Part One: Preparation of Glycine (M)
References: Experiment 58A, Part One

See Figure 1.9 (page 30).

Start the reaction one week ahead. At the beginning of the lab period prior to the one for which this experiment is assigned, dissolve 225 mg of α-chloroacetic acid in 0.4 mL of H_2O. Cautiously add this solution to 3 mL of concentrated ammonium hydroxide in a 5-mL conical vial. Set the labeled, stoppered container in the hood until the next period.

Isolate the glycine. If you are isolating glycine, concentrate the solution to about one-half of its original volume with magnetic stirring and gentle heating. Then add a solution of 0.75 g of $(NH_4)_2CO_3$ in 1 mL of warm water to the reaction mixture by pipet. Continue the evaporation until the volume is reduced to about 0.5 mL. Cool the solution; stir it with 2.5 mL of methanol, and cool it below 5°C. Crystals of glycine form. Remove as much of the liquid as possible with a plugged pipet (Section 1.3E), and add 1 mL of methanol, with stirring, to the crystals (to remove the ammonium salts). Collect the crystals by suction filtration in a Hirsch funnel.

Prepare the glycine solution for Part Two. If the entire reaction mixture (3.5 mL) is to be converted to hippuric acid without the isolation of glycine, add, in two or three portions, a total of 1.6 mL of dilute NaOH (6M) to the solution. Evaporate until the volume is reduced to about 1.5 mL. If the odor of ammonia persists, continue boiling with the addition of enough water to maintain a volume of 1.5 mL.

Part Two: Preparation of Hippuric Acid (M)
References: Experiment 58A, Part Two

Initiate the reaction. To the basic solution from Part One or to a solution of 150 mg of glycine in 1.2 mL of dilute NaOH (6M) in a 5-mL conical vial, cautiously add 0.3 mL of benzoyl chloride. Stopper the flask, and magnetically stir its contents. From time to time release the pressure by careful removal of the stopper. Continue stirring until the odor of benzoyl chloride is no longer detectable.

Isolate and purify the product. Transfer the solution via a pipet to a mixture of 1 mL of concentrated hydrochloric acid and 2.5 mL of ice in a 10-mL beaker. After thorough stirring, the solution should test acidic to an external indicator paper (ca. pH = 3). If not, add more acid and test the solution after each addition. Collect the precipitate by suction filtration on a Hirsch funnel, and wash it with 1 mL of cold water and then with 4 mL of ether. It is important to stir the crystals thoroughly with the ether while discontinuing the suction. Determine the melting range, which should be near 190°C. The hippuric acid may be further purified by recrystallization from hot water (0.3–0.5 mL) in a Craig tube.

See Figures 2.8A, B, C (pages 56–58).

Summary

See Experiment 58A.

Part Three: Test Reactions for Proteins and Their Hydrolysis Products

See Experiment 58A.

SEE THOSE FOR EXPERIMENT 58A.

Questions

Summary

See Experiment 58A.

Disposal of Materials

Protein solutions (N or D, I)	aq. HCl (D and/or N)
aq. NaNO₂ (I, acidify in hood or treat with dilute bleach, D)	aq. CuSO₄ (D)
Urea (S)	Urea solution (N or D)
conc. HNO₃ (D, N)	Nitric acid solutions (N)

13.3 Carbohydrates

Sugars or carbohydrates are either aldehydes or ketones containing several hydroxyl groups or compounds that can be converted into such aldehydes or ketones. **Monosaccharides** contain one sugar molecule; that is, they cannot be hydrolyzed into simpler sugar molecules. **Disaccharides** can be hydrolyzed into two monosaccharides. **Polysaccharides** can be hydrolyzed into many monosaccharides. Aldoses are monosaccharides containing an aldehyde group, and ketones are monosaccharides containing a ketone group.

The fermentation of sugar to ethanol is of practical importance. This fermentation with the enzymes in yeast is illustrated in Experiment 60A. The starting material is a disaccharide. Experiment 60B illustrates the fact that these enzymes can be used to carry out other reactions.

Part Two of Experiment 59 consists of a group of test reactions commonly used with sugars. They encourage more familiarity with the chemistry of carbohydrates and can be used to identify an unknown.

The synthesis and purification of a sugar presents an exceedingly difficult problem, but certain representatives of the class occur in natural products and can be isolated from them. Experiment 59 is typical of such an isolation procedure. The sugar occurring in milk is called lactose. It is the only disaccharide synthesized by a mammal. Lactose is 4,O-(β-D-galactopyranosyl)-D-glucopyranose);

it can be isolated from milk after the fat and protein have been removed. Skimmed milk, i.e., the fat has been removed, is available commercially as a liquid or as a powder to be reconstituted. Skimmed milk contains the proteins casein and albumin as well as minerals and about 5% by weight of lactose. Nonfat powdered milk may be mixed and used for Experiment 59, Part One.

Casein exists in milk as the soluble calcium salt. The salt can be reacted with acid and warmed to produce the insoluble protein. Excessive acid and heating should be avoided because it may lead to the hydrolysis of lactose.

Albumin is then precipitated by boiling the solution, which has been neutralized. The albumin is denatured, i.e., precipitated irreversibly, because of what is believed to be a change in the stereochemistry of the protein, brought about by heat. After the albumin has been removed, the filtrate is concentrated. Ethanol is added to the concentrated solution. Lactose is insoluble in ethanol and when ethanol is added to the aqueous solution, the lactose crystallizes. The filtration of this alcoholic solution and the complete removal of suspended material will be assisted by adding active carbon and covering the filter with a layer of wet silica filter aid. A clear filtrate is essential because even traces of impurities may jeopardize the crystallization of the lactose. The clear, cool filtrate should yield crystalline lactose on standing.

E X P E R I M E N T 5 9

Isolation of Sugar; Test Reactions for Carbohydrates (SM)
References: Sections 2.3 and 13.3

Part One: Lactose from Milk

Remove casein and albumin by coagulation. In a 250-mL beaker, warm 100 mL of skimmed milk to 40°C, and add a solution of acetic acid (1 part glacial acetic acid to 2 parts water) dropwise with continuous stirring as long as the casein separates. From 0.5 to 1.0 mL of the acid is usually required. The casein may be removed after it has been congealed, by rubbing it into one aggregate with a stirring rod. If the casein is not easily removed with the stirring rod, you may decant the liquid into a beaker of the same size or larger. Then add 2.5 g of powdered calcium carbonate immediately, and stir the mixture well; heat it at its boiling point for ten minutes. The heating coagulates the albumin.

Remove the precipitated albumin by suction filtration of the hot solution, and concentrate the filtrate by evaporating it in a beaker to a volume of about 15 mL. Next add 90 mL of ethyl alcohol to the solution and about 0.5 g of activated charcoal pellets to it after it has cooled. Reheat the solution for a few minutes, stir it well, and then filter it *warm* by gentle suction through a layer of filter aid or Celite (face powder also works well). (Mix the filter aid with some solvent to form a thin paste, and then filter it through a Büchner funnel by suction to give a layer about 3-mm deep. Clean the filter flask, and return it to the setup to be used for filtration.)

See Figure 1.9 (page 30).

Crystallize, collect, wash, and dry the lactose. When the clear filtrate is allowed to stand one week in a tightly stoppered Erlenmeyer flask, hard crystals of lactose will separate. Collect these by suction filtration, and wash them with a little dilute ethanol (1 : 4 ethanol to water); dry them, and weigh them. Weigh the product after it is dry, and determine its melting range (literature, value, α isomer 222°C, β isomer 253°C). Calculate the weight percent yield of the lactose by assuming that the density of milk is 1.03 g/mL.

Summary

1. Acidify warm skimmed milk with acetic acid.

2. Remove the casein.

3. Add $CaCO_3$ and heat at boiling.

4. Remove the albumin with suction filtration.

5. Concentrate the solution, cool, add alcohol and carbon, and reheat.

6. Gently suction filter warm.

7. Allow to stand, stoppered, for one week.

8. Collect, wash, and allow the crystals to dry.

9. Analyze the product as directed.

Questions

1. What does heat and (or) acid do to proteins like casein?

2. What is the purpose of adding calcium carbonate after removal of the casein?

3. Explain the purpose and function of the addition of ethanol to the aqueous solution.

Disposal of Materials

Skimmed milk (D)	Glacial and diluted acetic acid (D, N)
Casein (wash, S)	Albumin (S)
Ethyl alcohol (D)	Activated charcoal (S)
Filter aid (S)	Lactose (I)

Part Two: Test Reactions for Carbohydrates

This experiment is intended to familiarize you with some of the common tests used to characterize mono-, di-, and polysaccharides. You should tabulate the results of all the tests. A sample table is given below. A system of plus and minus signs may be used to indicate positive or negative reactions. The times required for osazone formation should also be noted in Table 13.1. These times and the appearance of the osazones under a microscope are sometimes used to identify specific sugars. The Benedict's, Tollens', and osazone tests will be tried with a starch solution and with 2% solutions of glucose, fructose, sucrose, and lactose.

Table 13.1 Results of Tests

Compound	Benedict's	Tollens'	Osazone Formation Time	Iodine
Starch (hydrol.)				
Starch				
Glucose				
Fructose				
Sucrose				
Lactose				

Hydrolysis of Starch

This experiment should be started at the beginning of the period so that some of the tests may be done concurrently. Add 5 drops of concentrated hydrochloric acid to 20 mL of starch solution in a 22 or 25 × 200-mm test tube or 125-mL Erlenmeyer flask (or see Figure 1.7, Unit I), and heat this mixture in a water bath for 40–45 minutes. When cool, neutralize the solution with 10% NaOH. Test it with external indicator paper. Filter the solution by gravity if necessary so that it is clear. Record all observations.

Benedict's Test

Benedict's test is used to determine whether a sugar is a reducing or nonreducing sugar.

A **reducing sugar** is one that exists as an actual or potential aldehyde or α-hydroxyketone. The potential aldehyde or ketone is present as the **hemiacetal** or **hemiketal**,

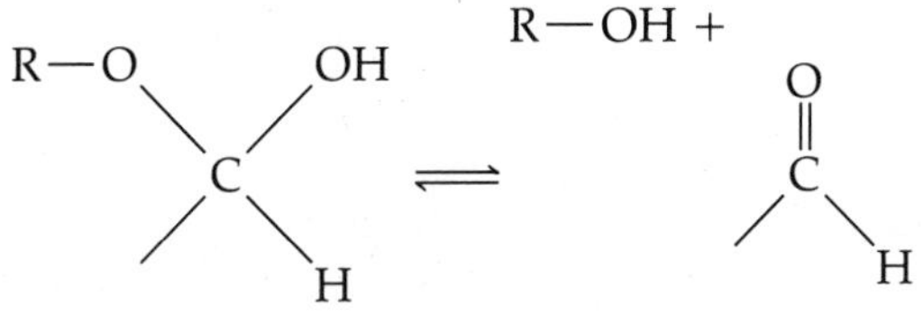

hemiacetal

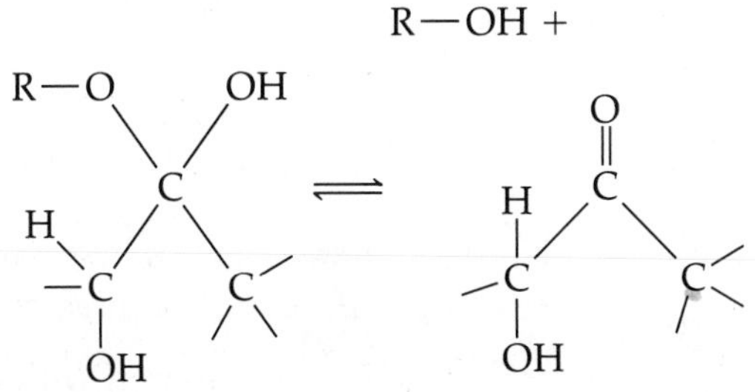

hemiketal

which is in equilibrium with the aldehyde or α-hydroxyketone in solution.

The aldehyde or α-hydroxyketone is oxidized by Benedict's reagent, a basic solution of cupric ion complexed with citrate ion, to give an acid salt while the reagent is reduced to form a red precipitate of cuprous oxide.

$$-\overset{\overset{\displaystyle O}{\|}}{C}H + 2\,Cu^{+2} + 5\,OH^- \longrightarrow -\overset{\overset{\displaystyle O}{\|}}{C}O^- + \underset{red}{Cu_2O} + 3\,H_2O$$

$$-\overset{\overset{\displaystyle H}{|}}{\underset{\underset{\displaystyle OH}{|}}{C}}-\overset{\overset{\displaystyle O}{\|}}{C}-H \rightleftharpoons -\overset{}{\underset{\underset{\displaystyle OH}{|}}{C}}=\overset{}{\underset{\underset{\displaystyle OH}{|}}{C}}-H \qquad [\text{TAUTOMERIZATION}]$$

$$-\overset{}{\underset{\underset{\displaystyle OH}{|}}{C}}=\overset{}{\underset{\underset{\displaystyle OH}{|}}{C}}-H + 4\,Cu^{+2} + 9\,OH^- \longrightarrow -\overset{\overset{\displaystyle O}{\|}}{C}-\overset{\overset{\displaystyle O}{\|}}{C}-O^- + 2\underset{red}{Cu_2O} + 6\,H_2O$$

A **nonreducing sugar** is one that exists as either an acetal or a ketal that is not in equilibrium with the aldehyde or α-hydroxy ketone and alcohol in solution.

acetal   ketal

Since neither the aldehyde nor the α-hydroxy ketone is present, negative results (no red precipitate) will occur with Benedict's reagent.

Procedure

In each of the five test tubes, bring 5 mL of Benedict's solution to a gentle boil, and add 3 drops of 2% carbohydrate solution (hydrolyzed starch, glucose, fructose, sucrose, lactose). After a minute of continued boiling, add 2 or 3 more drops of carbohydrate solution and continue heating if the blue color persists. A red, brown, or yellow precipitate constitutes a positive test for a reducing sugar. A color change is not a positive test; a precipitate must result.

Tollens' Test

The Tollens' test can also be used to determine whether a sugar is a reducing or nonreducing sugar. The Tollens' reagent consists of the silver ammonia ion, $Ag(NH_3)_2^+$, which is capable of oxidizing aldehyde groups to the salt of the carboxylic acids and concomitantly produces the free silver, Ag, in the form of a mirror. An α-hydroxy ketone can also be oxidized by Tollens' reagent.

$$\overset{O}{\overset{\|}{-CH}} + 3OH^- + 2Ag(NH_3)_2^+ \longrightarrow \overset{O}{\overset{\|}{-CO^-}} + 2H_2O + 4NH_3 + 2\,Ag\,(mirror)$$

Procedure

TOLLENS' SOLUTION, USED OR UNUSED, SHOULD NOT BE WASHED DOWN THE DRAIN. IF THESE SOLUTIONS DRY OUT, EXPLOSIVE COMPOUNDS COULD FORM. Such solutions should be collected for silver recovery. Vessels that will contain or have contained this solution, especially the test tubes, should be cleaned with hot concentrated nitric acid (*Caution!*) before use to encourage mirror formation and with cold dilute nitric acid after use to destroy any fulminating silver. The cleanings should be followed by a thorough rinsing with distilled water. All these washes should be added to the silver recovery container.

In each of five clean test tubes, place 2 mL of 5% aqueous $AgNO_3$ and one drop of 10% NaOH. Dissolve the precipitate by the dropwise addition of a dilute NH_4OH solution (1 : 10 dilution) with shaking. Add 3–5 drops of carbohydrate solution, and gently heat in the hot water bath. If a black sludge results instead of a silver mirror, repeat the test with a cleaner test tube.

Osazone Test

Carbohydrates will react with phenylhydrazine to yield crystalline products called osazones.

$$\begin{array}{c} \overset{O}{\overset{\|}{C-H}} \\ | \\ CHOH \\ | \\ R \\ \text{aldose} \end{array} \xrightarrow{\text{3 PhNHNH}_2} \begin{array}{c} HC=NNHPh \\ | \\ C=NNHPh \\ | \\ R \\ \text{osazone} \end{array} + PhNH_2 + NH_3$$

Osazone formation is observed with both α-hydroxy aldehydes and α-hydroxy ketones.

Because sugars do not crystallize easily and prefer to form syrups, identification and purification of a sugar through formation of a solid derivative such as an osazone is of great help to the carbohydrate chemist.

Procedure

Add, to 5 mL of each carbohydrate solution, 3 mL of phenylhydrazine solution POTENTIALLY TOXIC! and 3 drops of saturated NaHSO$_3$ solution. Keep the test tubes in a boiling water bath for 30 minutes or until a precipitate forms (record the time needed). Those solutions that have not formed precipitates within 30 minutes may be induced to do so by cooling and scratching the test tube walls.

Iodine Test

Starch is a polysaccharide of glucose that generally contains about 20% of a water-soluble fraction, amylose, and about 80% of a water-insoluble fraction, amylopectin. Starch will form a blue color when tested with the iodine reagent. The blue color is believed to be due to the amylose chain wrapping around the iodine. Mono-, di-, and highly branched polysaccharides do not give a blue color.

Procedure

Add, to 5 mL of starch solution, one drop of very dilute I$_2$-KI solution, and observe the color. What changes, if any, take place when the solution is boiled and cooled? Compare the results with glycogen, hydrolyzed starch, glucose, fructose, sucrose, and lactose with those for starch.

Questions

1. What are the structures for glucose, fructose, sucrose, and lactose?
2. Write a generalized structure for amylose and amylopectin.
3. Write a generalized equation for the hydrolysis of starch.
4. Write an equation for each positive test recorded in the table *Results of Tests*.
5. Should glycogen be expected to react with I$_2$ as starch does?
6. What are the structural differences between reducing and nonreducing sugars?
7. Why does the hydrolyzed starch solution give a positive test with Benedict's reagent?
8. What results are observed for the reaction of a hydrolyzed starch solution with the iodine reagent?

9. What problems could result if not all of the nitric acid used to clean the test tubes in the Tollens' test was removed before the test?

Disposal of Materials

Starch solutions (filter, D)

aq. NaOH (N)

aq. Carbohydrate solutions (D)

aq. Ammonia solution (N)

aq. Bisulfite solution (save for disposal)

conc. HCl (D, N)

used Benedict's solution (filter, N, D)

any Silver solutions (container for recovery, I)

used Phenylhydrazine (*HW*, I)

dil. I_2-KI solution (treat with bisulfite, I, D)

Summary

Starch solution: Triturate 1 g of soluble starch with 10 mL of water, and add 100 mL of boiling water to this mixture. Cool and filter.

Benedict's solution: Dissolve 20 g of sodium citrate and 11.5 g of anhydrous Na_2CO_3 in 100 mL of boiling water. To the hot solution, add 20 mL of 10% aqueous $CuSO_4$ with constant stirring.

Phenylhydrazine solution: **PHENYLHYDRAZINE IS A SUSPECTED *CARCINOGEN*. HANDLE IT WITH CARE. AVOID INHALATION OR SKIN CONTACT.** Dissolve, with shaking, 12 g of phenylhydrazine hydrochloride and 18 g of sodium acetate in 125 mL of water to which a drop or two of glacial acetic acid has been added. Gentle heating may be needed to dissolve the solid. Stirring with a pinch of active carbon and filtering will clarify the solution if that is needed.

> **CAUTION**

Dilute iodine–potassium iodide solution: Dissolve 1 g of potassium iodide, KI, in 25 mL of distilled water. Add 0.5 g of iodine (*Caution!*), and shake the solution until the iodine dissolves. Dilute the solution to 50 mL with distilled water.

E X P E R I M E N T 6 0 A

Fermentation of Sucrose (SM)
References: Sections 3.4, 7; 4.2D; 13.3; and Experiment 59

Introduction   Fermentation of sugar was one of the first chemical reactions carried out and probably one of the first to be accidentally discovered. It has been studied and used throughout the centuries and was the subject of heated controversy in the nineteenth century when organic chemistry was moving away from vitalism. The work of Pasteur, Büchner, and others established that many enzymes in yeast catalyze a series of reactions of which the easily observable products are carbon dioxide and ethanol. The last enzyme catalyzed step is the reduction of acetaldehyde to ethanol. The enzyme that catalyzes this reaction also catalyzes the reverse reaction. This reverse reaction can create a severe hangover or poisoning if another alcohol is used. In fact, one first-aid treatment for methanol or ethylene glycol poisoning in many animals is to introduce large amounts of ethanol intravenously. The enzyme binds to ethanol better than the other alcohols, which will be excreted in aqueous solution if not metabolized.

Ethanol denatures enzymes if it is present in large enough amounts. Normal yeast cannot produce a solution of much more than 12% ethanol no matter how much sugar is present. The product of the fermentation is a very dilute aqueous ethanol solution (along with cell debris, etc.). It requires careful, fractional distillation if the azeotrope of water and ethanol is to be isolated.

Assemble the apparatus and introduce the reagents. In a 250-mL beaker, place 30–40 mL of warm water (30–35°C). In a 50-mL Erlenmeyer flask, dissolve 3 g of sucrose in 15 mL of warm water and 1 mL of nutrient solution. Place the flask in the beaker and stir. Suspend 0.2 g of dry baker's yeast in 4 mL of warm water in a test tube, with mixing. Place the test tube in the warm water bath. In another test tube, place 3–5 mL of saturated aqueous calcium hydroxide, and float about 1 mL of mineral oil on top of it.

Initiate the reaction. When the yeast has been suspended in warm water for at least five minutes, add the suspension to the sucrose solution and stir thoroughly. Close the flask with a one-holed stopper that has a short piece of glass tubing in it (just through the bottom of the stopper). Attach a piece of flexible, unreactive tubing (Teflon, tygon, polypropylene, etc.) to the glass tubing. The other end of the flexible tubing should extend 1–2 cm below the surface of the $Ca(OH)_2$ solution in the test tube. Place the test tube in the beaker. If you are using a magnetic stirrer, gently stir a bit more. Place the beaker and its contents in a constant temperature bath set at 33–35°C for about one week.

Prepare a density/composition graph for ethanol and water. Do this during the first period if so directed by your instructor. Assume that the density of ethanol is 0.789 g/mL and that of a 95% ethanol and water solution is 0.804 g/mL. Check the density of the 95% ethanol. Next take a small sample of 95% ethanol and add a much smaller sample of water. Record the volume. Calculate the weight composition of the new mixture and determine its density (Section 3.7). Assume that the density of water is 1 g/mL. (Suppose you took 5 mL of 95% ethanol and 0.5 mL of water. 5 mL × 0.804 g/mL = 4.02 g solution. 4.02 × .95 = 3.819 g ethanol and

0.201 g water + 0.5 g water = 4.52 g new mixture. 3.819 g ethanol/4.52 g mixture $\times$ 100 = 84.5% ethanol.) Repeat this procedure until you have four or five values evenly spaced between 70% and 95%. Graph your results with percent composition on the ordinate and density on the abscissa. You may also want to inject 95% ethanol and/or one or more of your mixtures into the gas chromatograph to see if quantitative analysis can be achieved.

Isolate and analyze the azeotrope. Using a pipet, equally divide the solution in the flask into two centrifuge tubes (leave behind as much solid as you can). Centrifuge for 3–5 minutes to get a good pellet. With a pipet (or plugged pipet or one equipped with a plugged segment of tubing) remove as much liquid as possible without the solid and transfer the liquid (after removing the plugged segment of tubing) to a 25-mL round-bottom flask equipped for fractional distillation (Figure 3.10A).

Alternatively, if you do not centrifuge the solution, filter a suspension of Celite or face powder (see Experiment 59, Part One) through a Hirsch or small Büchner funnel with *gentle* suction to give a 1–2-mm layer on the filter paper. With continued gentle suction, decant the solution through this filter aid pad. Transfer the filtrate to the 25-mL flask with a boiling stone or stir bar. The fractionating column should be a short condenser containing a small amount of unreactive metal wool or glass bead packing. Block the water attachments. Use a short path head if it is available. Use the shortest available condenser as the column.

See Figure 3.10A (page 96).

With water flowing in the condensing condenser, begin gentle reflux into the fractionating column. Gradually move the condensate up until the thermometer bulb is bathed with vapor. The distillation temperature may be erratic. Collect the fraction distilling near 78°C. Discontinue distillation and collection when the temperature threatens to reach or exceed 85°C.

Analyze the product. Determine the density of the collected distillate and compare it with your graph. Record the gas chromatography trace for the distillate and estimate the composition.

1. With these pieces of data and the distillation temperature, what do you conclude about your product?

Question

Summary

1. Prepare a sugar solution, yeast suspension, and fermentation air lock.

2. Mix the sugar solution and the yeast, and arrange so that gas can escape but air cannot enter.

3. Incubate and allow to ferment.

4. Prepare a graph of composition versus density for ethanol/water mixture rich in ethanol.

5. Separate the solution from any solids.

6. Fractionally distill the solution.

7. Analyze the distillate.

Questions

QUESTION 1 ABOVE.

2. Why must gas escape and air not enter during fermentation?

3. What is the white solid that appears, over time, in the air lock solution?

4. Detail your calculations for the composition graph.

5. How could you establish correction factors for weight composition from the gas chromatograph?

6. If the boiling point of ethanol is 78.4°C and that of water is 100°C, why did you collect at or below 78°C?

Disposal of Materials

Sucrose (S)   Yeast (S)

aq. $Ca(OH)_2$ (N, I)   Mineral oil (OW, I)

Nutrient solution (D)   Cell debris, filter aid, etc. (S, I)

aq. Ethanol solutions (D)   Product (I)

Additions and Alternatives

Try a Carbowax® column for the gas chromatography.

   Some instructors recommend saturating the solution to be distilled with an alkali carbonate. You may want to try this. You could also determine the refractive index before and after distillation. You could extend the density/composition graph to lower values and analyze the filtrate before distillation by determining its density. Try the filtration with well-packed glass wool and no suction. Other sugars such as glucose, fructose, or galactose could be used.

   You could do this experiment on a micro scale by reducing all reagents except water by a factor of 5. The water is reduced to 2.5 mL. A side-arm test tube or a 5-mL vial (with or without an air condenser) and an inlet adaptor could be used for the fermentation vessel. One centrifuge tube (balanced!) will hold the solution. Either a short path head or a Hickman still should be used on top of the fractionating column. You may have to combine samples if density is used for analysis.

EXPERIMENT 60B

The Reduction of Vanillin by Fermenting Yeast (SM)

References: Sections 1.3I, 2.3, 3.4, 4.2C; Experiments 29A and B; 59,
Part A; and 60A

Introduction   As noted in Experiment 60A, the last step in the production of ethanol is the reduction of an aldehyde. The enzyme that catalyzes this step is not totally specific for acetaldehyde; it has been found that other aldehydes and ketones can be reduced with the help of this enzyme. The fermenting yeast provides the cofactors as well as the enzyme needed for this reduction. The reduction of vanillin has been chosen to illustrate this property and as a comparison with Experiment 29B. Other suggestions are found in the Additions and Alternatives section.

Assemble the apparatus and introduce the reagents. Follow the approach of Experiment 60A with the following exceptions: Use a 125-mL Erlenmeyer flask and a 400–600-mL beaker. Make the sugar solution from 9 g of sucrose, 50 mL of warm water, and 3 mL of nutrient solution, and dissolve 0.3 g (304 mg) of vanillin in the solution. Stir gently, with a magnetic stirrer if possible. Use a 25-mL Erlenmeyer flask to disperse and incubate 1 g of dry baker's yeast in 15 mL of warm water.

Initiate the reaction. When the yeast has warmed, add the contents of the 25-mL flask to those in the 125-mL flask. Note the time. Close the larger flask, and connect it to the air lock. Arrange the flask and air lock in the beaker and incubate for a week or more.

Monitor the progress of reaction. If directed to do so, take 0.4–0.5-mL samples with a graduated pipet from the fermenting mixture every 20–30 minutes for the rest of the period (up to three samples; draw the last one 15–20 minutes before the end of class). Record the time for each sample. Add each sample to a different tube containing 0.5 mL of methylene chloride. Mix completely in the pipet. Separate the nonaqueous layer, and transfer it to a clean, dry, labeled test tube. Concentrate the samples (Section 1.3I) almost to dryness; only a few drops should remain. Take a similar sample the second week, before completing the experiment, and treat it the same way. Run a thin layer chromatography plate for each sample as suggested for Experiment 28B. The plate should have at least two spots for the sample and one for a vanillin standard. (If vanillyl alcohol is available, it can also be run as a standard on the plate.)

See Figures 4.6A, B, C (pages 154–157).

1. What do your results suggest about how long fermentation must go on to reduce most of the vanillin?

Isolate the product. The second laboratory period, place a layer of filter aid (see Experiment 59A) on the paper in a Büchner funnel with gentle

See Figures 2.6A, B; 2.7
(pages 52–55).

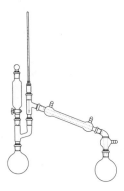

Figure 7.1 (page 257).

suction. Filter the fermentation mixture. (If a centrifuge is available, see Experiment 60A for another way to do this separation.) Assemble a simple apparatus for simultaneous addition and distillation (Figure 7.1) with a 25-mL round-bottom flask. Use a magnetic stirrer if possible or a boiling stone. Distill off as much alcohol and water as possible from the filtrate (down to a *few* milliliters) without heating the flask to dryness. Extract the remaining liquid with methylene chloride (Pasteur pipet). Gently evaporate the solvent and recrystallize the solid product from a minimum of hot ethyl acetate (see Experiment 29B). Determine the weight and the melting range of the dry product.

Summary

1. Prepare the sugar/vanillin solution, yeast suspension, and air lock.

2. Mix the yeast and the sugar solution and arrange for anaerobic fermentation.

3. Incubate and allow to ferment.

4. Take samples; prepare for and do thin layer chromatographic analysis if you are doing that part of the experiment.

5. Separate the solution from the solids.

6. Distill off almost all of the water and alcohol.

7. Extract and recrystallize the product.

Questions

QUESTION 1 ABOVE AND QUESTIONS 1 AND 2 FROM EXPERIMENT 60A.

3. Why not extract the product from the water in the distillation flask with ethyl acetate?

Disposal of Materials

Sucrose (S)	Yeast (S)
Nutrient solution (D)	Mineral oil (OW, I)
aq. Ca(OH)$_2$ (N, I)	Vanillin (S)
aq. Ethanol solutions (D)	Methylene chloride (HO, I)
Ethyl acetate (OS, OW, I)	Ethyl ether (OS, OW, I)
Product (I)	Cell debris, filter aid, etc. (S, I)

Additions and Alternatives

Try ethyl ether to extract the concentrated filtrate, but remember the difference in density.

Instead of concentrating the solution, saturate it with salt and extract three times with 25–30 mL of ethyl ether. Dry and evaporate or distill off the solvent.

Compare the results of this experiment with those of Experiment 29B.

Try cyclohexanone or some of the compounds suggested in Experiment 29B with this approach.

Try the reduction of ethyl acetoacetate to (S)(+)ethyl 3-hydroxybutanoate (*Organic Synthesis*, Vol. 63, Wiley, New York, 1984). Use double the scale of this experiment and use 2.6 g (or 2.6 mL) of ethyl acetoacetate in place of vanillin. Isolate after filtration by multiple extractions with methylene chloride, evaporation, and microcolumn chromatography with alumina and methylene chloride.

13.4   Fats, Oils, and Soap

Soap may be prepared from animal fat or vegetable oil. Both fats and oils are esters of the trihydroxy alcohol, glycerol.

$$
\begin{array}{c}
\text{H}_2\text{C}-\text{O}-\overset{\displaystyle\text{O}}{\overset{\|}{\text{C}}}-\text{R} \\[4pt]
\text{HC}-\text{O}-\overset{\displaystyle\text{O}}{\overset{\|}{\text{C}}}-\text{R}^{\text{I}} \\[4pt]
\text{H}_2\text{C}-\text{O}-\overset{\displaystyle\text{O}}{\overset{\|}{\text{C}}}-\text{R}^{\text{II}}
\end{array}
$$

fat or oil

The three R groups in the formula above may be the same or different from each other. The R group usually consists of long, straight carbon chains containing eleven to seventeen carbon atoms. Fats are more saturated than oils, which generally contain one to three carbon–carbon double bonds.

Soap is prepared by hydrolyzing the fat or oil under basic conditions to produce the salts of the carboxylic acids.

$$
\begin{array}{c}
\text{H}_2\text{C}-\text{O}-\overset{\text{O}}{\overset{\|}{\text{C}}}-\text{R} \\
\text{HC}-\text{O}-\overset{\text{O}}{\overset{\|}{\text{C}}}-\text{R}^{\text{I}} \quad + \text{ NaOH} \xrightarrow{\ \text{H}_2\text{O}\ } \\
\text{H}_2\text{C}-\text{O}-\text{C}-\text{R}^{\text{II}} \\
\overset{\|}{\text{O}}
\end{array}
\quad
\begin{array}{c}
\text{H}_2\text{C}-\text{OH} \\
\text{HC}-\text{OH} \\
\text{H}_2\text{C}-\text{OH} \\[4pt]
\text{glycerol}
\end{array}
\ + \
\begin{array}{c}
\text{R}-\overset{\text{O}}{\overset{\|}{\text{C}}}-\text{O}^-\text{Na}^+ \\
\text{R}^{\text{I}}-\overset{\text{O}}{\overset{\|}{\text{C}}}-\text{O}^-\text{Na}^+ \\
\text{R}^{\text{II}}-\overset{\text{O}}{\overset{\|}{\text{C}}}-\text{O}^-\text{Na}^+ \\[4pt]
\text{soap}
\end{array}
$$

Soaps consist of a mixture of salts of carboxylic acids, usually having 12, 14, 16, and 18 carbon atoms in a straight chain.

Soaps function as dirt removers because of their structure in which one end of the molecule, the salt of the carboxylic acid, is polar and soluble in water, while the other end of the molecule, the long hydrocarbon chain, is nonpolar and soluble in the oil or grease that contains the dirt. The soap aligns itself so that the hydrocarbon ends dissolve in the dirt and oil while the polar ends dissolve in water.

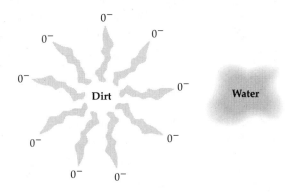

The "oil droplet" is removed by being strongly attracted to water via the polar end of the soap. The "oil droplet" is solvated by the water.

E X P E R I M E N T   6 1 A

Production of Soap (M)
References: Section 2.3 and Experiment 70

Hydrolyze the fat. Dissolve 0.5 g of sodium hydroxide pellets in a mixture of 2 mL of water and 2 mL of 95% ethanol in a 10-mL beaker or Erlenmeyer flask. Place a magnetic stirrer and 1 g of liquid shortening (vegetable oil) or fat, e.g., Spry or Crisco, in a 25-mL beaker and, using a Pasteur pipet, transfer the basic solution into the 25-mL beaker with stirring. Heat the beaker in a steam or sand bath at 100°C for 30 minutes with stirring. During the 30-minute period, add, in small portions, a mixture of 2 mL of water and 2 mL of ethanol to maintain the volume. Alternatively, reflux using a 10-mL flask, boiling stone or magnetic stirrer, and a water condenser.

See Figure 1.7, Unit I (page 24).

Isolate the soap. Dissolve 5 g of sodium chloride in 15 mL of water in a 50-mL beaker. Heat the solution if necessary to effect solution. Cool the salt mixture, and quickly pour the reaction mixture from the 25-mL beaker into this salt solution. If you used the flask, make this transfer with a pipet.

Do NOT pour this basic mixture over the ground joint. Stir the solution while cooling it in an ice bath. Filter the precipitate with suction through a Hirsch funnel, and wash it with two 1-mL portions of cold water. Let the soap dry until the next laboratory period when it can be weighed and the percent yield calculated.

Summary

1. Dissolve the NaOH in aqueous alcohol and add this solution to the fat.

2. Heat the mixture, maintaining the volume.

(2a.) Use simple reflux.

3. Prepare a concentrated salt solution and add the reaction mixture to it with cooling.

4. Collect, wash, and allow the product to dry.

Questions

1. What difference would there have been if potassium hydroxide had been used instead of sodium hydroxide in this experiment?

2. Why is an ethanol/water mixture rather than water used as the solvent for this reaction?

3. Why must the 50:50 mixture of water and alcohol be added to the reaction mixture during the 30-minute heating period if there is no reflux condenser?

4. Why does the soap precipitate when the reaction mixture is added to the sodium chloride solution?

Disposal of Materials

Fat or oil (OW, I)   Sodium hydroxide (D, N, I)

NaCl solution and filtrate (D or N, I)   Sodium hydroxide solution (N, I)

Product (I)

E X P E R I M E N T   6 1 B

Production of "Soap Hydrocarbons" (SM)
References: Sections 2.3; 3.4, 5; and 4.2D

$$RCOONa + NaOH \longrightarrow RH + Na_2CO_3$$

Prepare the pyrolysis mixture. Mix 10 g of a good powdered, flaked, or granulated soap or soap flaked from a cake of soap (not one of the so-called soap powders, which may contain very little soap and are synthetic detergents) with 10 g of soda lime, and grind the mixture to a fine powder in a mortar. Transfer this mixture to a 22 or 25 × 200-mm boiling tube equipped with a one-holed cork stopper bearing a glass delivery tube and a condenser (Figure 11.2). Support the assembly by clamps attached at the points indicated. The clamp that supports the pyrolysis tube must be made of iron and not of Castaloy or of white metal because these alloys have such low melting points that they will fuse under the conditions required in this experiment. Although this pyrolysis could also be done with assemblies like Units III and IV, Figure 1.7, it would be more advantageous to use the setup shown in Figure 11.2.

Pyrolyse the mixture. Whatever the setup, first apply heat at point A, using an ordinary Bunsen burner. A microburner does not deliver enough heat, and a Meker or Fisher grid-top burner may deliver too much. As the mixture in this region ceases frothing, move the flame very gradually toward point B. In this way the passage of froth into the delivery tube is prevented.

Question

1. How does this help to prevent froth from getting into the receiver? Why should the froth be prevented from passing into the receiver?

Finally, heat the entire test tube to drive out all volatile products.

Isolate the hydrocarbon layer. The condensed liquid consists principally of water and hydrocarbons. A little soap is carried along as a spray and dissolves in the water. Withdraw the bottom aqueous layer using a pipet, and add 2 mL of water to the hydrocarbon layer.

Question

2. Which one is the hydrocarbon layer? How do you know (or how could you test the aqueous layer)?

Allow the mixture to separate into layers after shaking it thoroughly. Withdraw the aqueous layer again, and shake the hydrocarbon layer with sufficient anhydrous magnesium sulfate to dry it.

Filter the dried material through a small pinch of fine glass wool placed in the stem of a funnel. Collect the filtrate in a clean, dry 25-mL round-bottom flask. Assemble the apparatus for simple distillation with an air condenser and a properly positioned, 360° thermometer (Unit III, Figure 1.7, or Figure 3.8). If there is less than 5 mL of product, use microdistillation.

Start the distillation. Record the temperature at which the first drop of *clear* distillate condenses. From this point on, change receivers after every 20° rise in temperature. Record all observations, including the highest temperature reached during distillation.

Allow the thermometer to cool in place. It must not be removed while hot and placed in contact with the cold table top or exposed to cold air. *Failure to observe this precaution may result in breakage of the thermometer.*

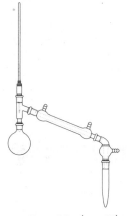

Figure 3.8 (page 87).

From the boiling points of hydrocarbons listed in a text or reference book, determine the hydrocarbons present in the distillate. From the relative amounts of distillate collected in each temperature interval, deduce the relative abundance of each of the fatty acids represented in the soap sample. Assume that fatty acids containing an odd number of carbon atoms are absent.

3. Why is this a reasonable assumption?

Deliver the hydrocarbons to your instructor, and make the appropriate report.

Summary

1. Grind the soap and base.

2. Spread the mixture in the pyrolysis tube and assemble the apparatus.

3. Pyrolyse the mixture.

4. Remove the aqueous layer and wash the organic layer.

5. Dry and filter the organic product mixture.

6. Distill and/or analyze the product mixture, as directed.

QUESTIONS 1–3 ABOVE.

4. What *hydrocarbons* might be present in your mixture, and which should be absent?

5. Characterize the fat (with respect to fatty acids) from which your sample of soap may have been made.

Disposal of Materials

Soap (S)   Soda lime (S, I)

$MgSO_4(S)$ Products (I)

Solids in the pyrolysis tube (cool, wash with dilute acid (carefully), N, D, I)

Additions and Alternatives

Develop a method to analyze this product mixture by gas liquid chromatography before and after the final distillation. If possible, employ hydrocarbon standards. This analysis could be done instead of distillation. The final distillation could be done at reduced pressure! The analysis would then be made by gas liquid chromatography with standards.

CHAPTER 14

Other Named and Type Reactions

14.1   Introduction

Many of the reactions encountered in the lecture portion of an organic chemistry course are remembered and referred to by their name, type, and technique. The five experimental approaches in this chapter are no exception. (A great many of the experiments in other chapters have similar associations.) Some of the methods used in this chapter are many decades old, whereas some have become important in the last decade. The Williamson ether synthesis (Experiment 62) involves nucleophilic displacement and so complements some of the experiments in Chapter 6. The other four provide various ways of making carbon-to-carbon bonds, and such processes are important in the synthesis of more complex molecules. Although Experiment 66 is not a true aldol condensation, it does introduce the use of phase transfer catalysis.

E X P E R I M E N T   6 2 A

Preparation of Methyl β-Naphthyl Ether (Williamson Synthesis) (SM)
References: Sections 1.3E–H and 2.3

In the Williamson synthesis, an alcohol, which has been converted to a salt, is reacted with an alkyl halide to produce an ether.

Phenol is like a relatively acidic alcohol and can be reacted with sodium hydroxide to form the stable salt, sodium phenoxide. Sodium phenoxide will function as a nucleophile in a reaction with an alkyl halide to generate the alkyl phenyl ether and sodium halide.

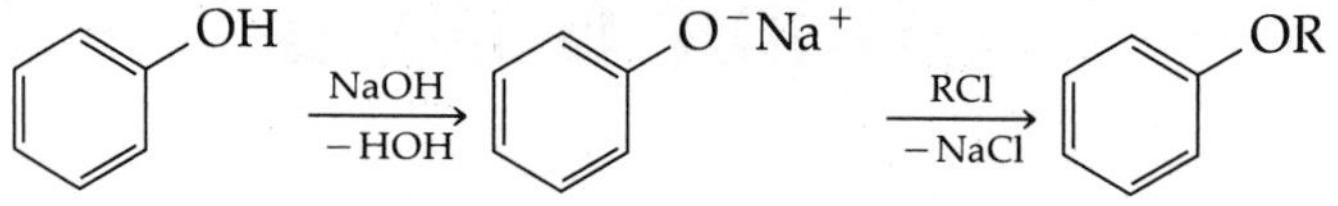

OH $\xrightarrow[-HOH]{NaOH}$ O⁻Na⁺ $\xrightarrow[-NaCl]{RCl}$ OR

For good yields of the ether (a substitution product) to result, the alkyl halide must be primary. If the alkyl halide were tertiary, the competing elimination reaction with the phenoxide ion reacting as a base would predominate. If the alkyl halide were secondary, both the elimination and desired substitution reactions would occur.

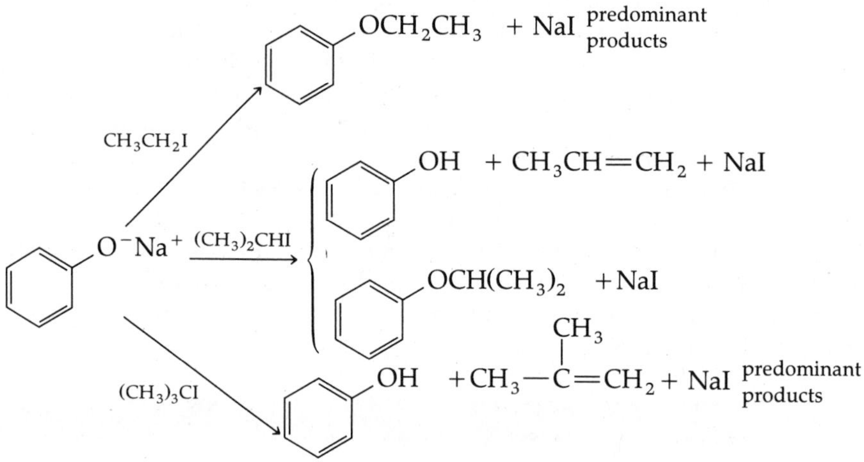

OCH₂CH₃ + NaI predominant products

CH₃CH₂I

O⁻Na⁺ (CH₃)₂CHI

OH + CH₃CH=CH₂ + NaI

OCH(CH₃)₂ + NaI

(CH₃)₃Cl

OH + CH₃—C=CH₂ + NaI predominant products
       |
       CH₃

In the following reaction, sodium hydroxide reacts with β-naphthol to produce the sodium salt of β-naphthol. The sodium salt reacts with methyl iodide to form methyl β-naphthyl ether. All of the species are brought into contact with each other by dissolving them in the solvent methanol.

Procedure

Figure 1.7, Unit I
(page 24).

CAUTION

Introduce the reagents and initiate the reaction in the hood. In a 50-mL round-bottom flask equipped with a reflux condenser (Unit I, Figure 1.7), place, in the following order and with mixing, 10 mL of methanol, 4.0 g (0.027 mole) of β-naphthol, 4 mL (0.025 mole) of 25% NaOH solution, and 2.5 mL (0.040 mole) of methyl iodide (d., 2.28 g/mL). VOLATILE, POTENTIAL CARCINOGEN; WORK IN A HOOD. Add a few boiling stones, and reflux the mixture for 30 minutes.

Isolate and purify the product. Pour the mixture quickly into 20 mL of cold water. Decant the water from the solid, add an additional 5 mL of cold water, and decant again. Transfer the wet solid into a 125-mL Erlenmeyer flask, and add 40 mL of 95% ethanol and a small portion (0.2–0.4g) of pelletized decolorizing charcoal. Boil the solution for a few minutes to dissolve the solid ether, and filter the hot solution by gravity through a *prewarmed* funnel containing *fluted* filter paper or a bit of glass wool.

See Figures 2.6A, B, C, 2.7 (pages 52–55).

Collect the crystals. Reheat the solution, and concentrate it to a little less than half of its original volume. Crystals should form upon cooling. If they do not form by the time the flask is cool enough to hold, the solution may be "seeded" by evaporating a little bit of it on a stirring rod and by using this rod, covered with "seed" crystals, to stir the solution and scratch the flask. Allow the crystals to form, cool in an ice bath, and then collect them by suction filtration. Wash the crystals with two 3-mL portions of cold ethanol. Allow the crystals to dry.

Analyze the product. Determine the weight of the crystals, their melting point range, and their percent yield. The melting point should be near 70°C.

Summary

1. Equip a round-bottom flask for reflux.

2. Add the reactants in the order given.

3. Reflux for 30 minutes.

4. Pour the reaction mixture into water and decant the water from the resulting solid.

5. Wash the solid again with water.

6. Dissolve the solid in hot ethanol with charcoal, and hot filter the solution.

7. Reheat and concentrate the alcoholic solution.

8. Cool to obtain crystals and collect them by suction filtration.

9. Wash the solid with cold ethanol.

10. Dry, weigh, and analyze the product.

Questions

1. Sodium hydroxide reacts with β-naphthol to form the sodium salt. Can sodium ethoxide, the sodium salt of ethanol, be formed in the same manner? If not, how is it made?

2. Explain the acidic nature of β-naphthol.

3. Why does the ether precipitate when the alcoholic solution is poured into water?

4. What else could have been used in place of methyl iodide to produce the same ether?

5. Outline the S_{N^2} mechanism for this reaction.

6. Why is sodium hydroxide used in the smallest molar amount?

7. Explain the "seeding" procedure.

Disposal of Materials

Methanol (OS, OW, I, D)

aq. NaOH (D, N)

Aqueous solutions (N, D, I)

Charcoal (evaporate, S, I)

β-Naphthol (HW, OW, S, I)

Methyl iodide (HW, HO, I)

Ethanol (OS, OW, D, I)

Ethanol solution (OW, evaporate, S, I)

Additions and Alternatives

You could use the ethyl iodide or the 1-iodopropane prepared in Experiment 22 in place of methyl iodide.

You could record the infrared spectrum and test the product with ferric ion solution for the presence of phenol.

Try the recrystallization with 75% or 85% ethanol.

E X P E R I M E N T 6 2 B

Preparation of Methyl β-Naphthyl Ether (Williamson Synthesis) (M)
References: Experiment 62A and Sections 1.3E–H and 2.3

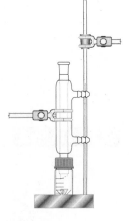

Introduce the reagents and initiate the reaction. In a 5-mL conical vial equipped with a reflux condenser and a spin vane, place, in the following order and with mixing, 1 mL of methanol, 400 mg (0.0027 mole) of β-naphthol, 400 μL (0.0025 mole) of 25% NaOH solution, and 250 μL (0.0040 mole) of methyl iodide (d., 2.28 g/mL). Use an automatic delivery pipet to dispense the liquids. Stir and reflux the mixture for 25–30 minutes in a sand bath or aluminum block.

Isolate and purify the product. Using a Pasteur pipet, quickly transfer the solution into a 5-mL conical vial (test tube, glass centrifuge tube, or small flask) containing 2 mL of cold water, and mix well. Using a pipet or plugged pipet, remove as much of the water as possible from the solid; add an additional 500 μL of cold water, and transfer the wet solid and slurry to the bottom portion of a 5-mL Craig tube. Insert the plug and

Figure 1.8 (page 26).

centrifuge in the usual manner. Alternatively, bend a piece of wire to create a loop and a handle. Put the loop of the wire around the bottom of the Craig tube; place it into a centrifuge tube, and centrifuge it right-side-up for two minutes. Make sure that the handle extends just out of the centrifuge tube. Remove the water by decanting it off or by using a pipet. Add about 1.7–1.8 mL of 95% ethanol to the solid in the Craig tube. Boil the solution gently (stir with a microspatula) until all of the solid is dissolved. It may be necessary to add small amounts of *hot* solvent to dissolve the crystals. Allow the solution to cool.

See Figures 2.8A, B, C (pages 56–58).

Collect the crystals. If crystals do not form by the time the tube is cool enough to hold, the solution may be "seeded" by evaporating a little of it on a spatula and by using this spatula, covered with "seed" crystals, to stir the solution. Cool the solution, and allow the crystals to form; then plug, invert, and centrifuge the Craig tube in the usual manner to collect the crystals.

Analyze the product. Allow the crystals to dry, weigh them, determine the melting range, and calculate the percent yield. The melting range should be near 70°C.

Summary

1. Equip a vial for reflux and add the reactants in the order given.

2. Reflux for 20–30 minutes, and pipet the reaction mixture into water.

3. Remove the water from the solid and wash it again with water.

4. Recrystallize from a minimum of hot 95% ethanol.

5. Dry, weigh, and analyze the product.

ANSWER THE QUESTIONS GIVEN IN EXPERIMENT 62A.

Questions

Disposal of Materials

Methanol (OS, OW, I, D)

aq. NaOH (D, N)

Aqueous solutions (N, D, I)

Ethanol (OS, OW, D, I)

β-Naphthol (HW, OW, S, I)

Methyl iodide (HW, HO, I)

Ethanol solution (OW, evaporate, S, I)

Additions and Alternatives

You can do the isolation and recrystallization in centrifuge or test tubes—centrifuged if necessary—and collect the solid on a Hirsch funnel with suction.

See Experiment 62A.

Preparation of Hexanoic Acid
(Malonic Ester Synthesis) (SM)
References: Sections 1.3E–I; 3.2, 4, 5; and 5.1

$$
\begin{array}{c}
\overset{\displaystyle O}{\overset{\|}{C}}OCH_2CH_3 \\
| \\
CH_2 \\
| \\
\underset{\displaystyle O}{\overset{}{C}}OCH_2CH_3 \\
\|
\end{array}
+ NaOCH_2CH_3 \longrightarrow
\begin{array}{c}
\overset{\displaystyle O}{\overset{\|}{C}}OCH_2CH_3 \\
| \\
CH^- Na^+ \\
| \\
\underset{\displaystyle O}{\overset{}{C}}OCH_2CH_3 \\
\|
\end{array}
+ HOCH_2CH_3
$$

$$
\begin{array}{c}
\overset{\displaystyle O}{\overset{\|}{C}}OCH_2CH_3 \\
| \\
CH^- Na^+ \\
| \\
\underset{\displaystyle O}{\overset{}{C}}OCH_2CH_3 \\
\|
\end{array}
+ CH_3CH_2CH_2CH_2Br \longrightarrow
\begin{array}{c}
\overset{\displaystyle O}{\overset{\|}{C}}OCH_2CH_3 \\
| \\
CH-CH_2CH_2CH_2CH_3 \\
| \\
\underset{\displaystyle O}{\overset{}{C}}OCH_2CH_3 \\
\|
\end{array}
+ NaBr
$$

$$
\begin{array}{c}
\overset{\displaystyle O}{\overset{\|}{C}}OCH_2CH_3 \\
| \\
CH-CH_2CH_2CH_2CH_3 \\
| \\
\underset{\displaystyle O}{\overset{}{C}}OCH_2CH_3 \\
\|
\end{array}
\xrightarrow{KOH}
\begin{array}{c}
\overset{\displaystyle O}{\overset{\|}{C}}O^- K^+ \\
| \\
CH-CH_2CH_2CH_2CH_3 \\
| \\
\underset{\displaystyle O}{\overset{}{C}}O^- K^+ \\
\|
\end{array}
\xrightarrow[\Delta]{H^+}
$$

$$
\begin{array}{c}
\overset{\displaystyle O}{\overset{\|}{C}}OH \\
| \\
CH_2-CH_2CH_2CH_2CH_3
\end{array}
+ CO_2
$$

The diethyl malonate synthesis takes advantage of the activation of hydrogen atoms attached to a carbon atom that is alpha to two carbon–oxygen double bonds. Such compounds in which the hydrogens bonded to the carbon atoms are acidic readily form their corresponding sodium salts when treated with equimolar amounts of anhydrous sodium ethoxide. The sodium salts react readily with alkyl halides to give the alkylated esters. All contact of the reaction mixture with water must be avoided until the alkylation reaction is complete. The ester is saponified and then acidified to produce the free acid. The acid will decarboxylate with heating to give an acid containing two more carbon atoms than the original alkyl halide, in other words, a substituted acetic acid. This general method may be used to synthesize mono- or disubstituted acetic acids. In the following reaction, hexanoic acid, the monosubstituted acetic acid will be synthesized from ethyl malonate and *n*-butyl bromide.

Procedure

Assemble the apparatus. Equip a 50-mL round-bottom flask with a magnetic stirrer and a Claisen adaptor to which is attached a reflux condenser and a separatory funnel. Connect calcium chloride drying tubes to both the funnel and the condenser in order to maintain anhydrous conditions for the reaction (Figure 1.7, Unit II, with two drying tubes).

Prepare the sodium ethoxide in a hood. Place 20 mL of absolute ethanol and 0.7 g of clean sodium (in the form of small pieces; add one piece at a time and wait) into the round-bottom flask. If the reaction becomes too exothermic, an external cold water bath may be used to control it. IF WATER COMES IN CONTACT WITH THE SODIUM, THERE IS A POSSIBILITY OF A FIRE OR EXPLOSION. If the reaction between the sodium and the alcohol does not go to completion, the mixture may be warmed gently.

Add the ester. When the reaction mixture temperature is just below 50°C, gradually add 4.7 mL (0.031 mole) of freshly distilled diethyl malonate (d., 1.05 g/mL) with stirring over a period of about five minutes. Complete the reaction by refluxing the solution for a few minutes.

React the mixture with the alkyl halide. Place 3.6 mL (0.033 mole) of *n*-butyl bromide in the separatory funnel, and gradually add it to the mixture, with stirring, over a 5–10-minute period. Let the reaction largely subside before each new addition is made. Cooling may be needed. Reflux the reaction mixture for 90 minutes. During this time, an inorganic salt will precipitate and cause bumping. The apparatus should be well clamped.

Isolate the ester. Rearrange the apparatus for simple distillation (Figure 6.2). Remove the ethanol by distillation from a boiling water bath. Cool the residue, and add 20 mL of water containing 2 drops of concentrated hydrochloric acid. Decant the mixture into a separatory funnel, and rinse the flask with 5 mL of water. Add the water rinse to the separatory funnel and shake. Separate the aqueous layer from the organic ester layer, and then wash the ester with a small portion of fresh water. Dry the ester over anhydrous magnesium sulfate.

Purify the ester by distillation. You and one to three other students should now combine your products at the direction of your instructor. (See also Additions and Alternatives.) Decant the dried ester into a distilling flask, and distill (Figure 6.2) off any material boiling below 100°C. Distill the remaining ester with reduced pressure, and collect the alkylated ester in the range of 130–135°C/20 mm. Weigh the product, and calculate the

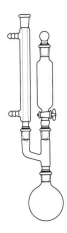

Figure 1.7, Unit II (page 24).

CAUTION

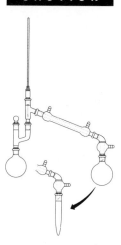

Figure 6.2 (page 237).

Figure 4.1 (page 129).

See Figure 3.11A (pages 104–106).

percent yield. The literature value is 235–240°C/760 mm for the boiling point.

Saponify the ester and distill out the resulting ethanol. NOTE: The amounts of chemicals in this part of the preparation must be adjusted appropriately if more or less than 8 mL of ester is used. Arrange addition–reflux equipment (Unit II, Figure 1.7). In the round-bottom flask (100-mL flask if the volume of the ester is less than 12 mL; 250-mL flask if the volume of the ester is more than 12 mL), place 8 g of potassium hydroxide and 8 mL of water. Dissolve the potassium hydroxide with stirring, and add drop-wise 8 mL of the alkylated ester. Reflux the mixture until saponification is complete and no oily layer remains (one to two hours). Convert to a distillation setup (Figure 6.2); add 8 mL of water, and then collect 8 mL or more of distillate to remove the ethanol that resulted from hydrolysis.

Acidify and reflux the mixture. Add a mixture of 12 mL of concentrated sulfuric acid and 8 mL of water. The rate of addition is governed by the behavior of the reaction mixture. The mixture should be strongly acidic. Reflux the mixture for one to three hours.

Isolate and purify the product. Cool the reaction mixture; separate the lower aqueous layer, and extract it four times with 25-mL portions of ethyl ether. Combine the ether extracts with the organic acid layer, and dry it over anhydrous magnesium sulfate. Decant off the ether solution into a beaker, and evaporate it on a steam bath to about 25 mL (or see Section 1.3I). Transfer the solution to a 50-mL round-bottom flask, and distill (Unit I, Figure 1.7, or Figure 3.8) the ether into a cold receiver. When all the ether has been removed, distill the hexanoic acid, collecting what distills in the range of 190–210°C.

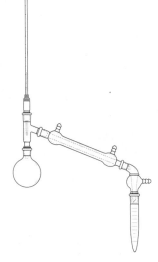

Figure 3.8 (page 87).

Analyze the product. Weigh the product, and calculate the percent yield. The literature boiling points are 205°C/760 mm, 99°C/10 mm, and 111°C/20 mm.

Summary

1. Equip a flask for reflux and addition, complete with drying tubes.

2. Add anhydrous ethanol and continuously add small sodium pieces.

3. Warm if needed to complete the reaction, but cool before the next step.

4. Gradually add the ester.

5. Reflux briefly.

6. Gradually add the alkyl bromide and cool as needed.

7. Reflux for about 90 minutes.

8. Distill off the ethanol and neutralize the mixture.

9. Isolate the nonaqueous layer and dry this substituted ester.

10. Distill the combined ester fractions at reduced pressure.

11. Saponify the ester by refluxing with aqueous KOH.

12. Distill off the ethanol that has formed.

13. Add acid, and reflux to produce the final product.

14. Isolate the product and extract any that remains in the water with ether.

15. Dry, evaporate, and distill the ether, and isolate the product by distillation.

Questions

1. Why must contact of the reaction mixture with water be avoided until the alkylation reaction is complete?

2. Which inorganic salt is precipitated that may cause bumping?

3. Why is it important to remove all of the ethanol from the reaction mixture before reaction with sulfuric acid?

Disposal of Materials

Sodium (I, return to storage or react cautiously with 2-propanol)

Ethanol (OS, OW, I)   Sodium ethoxide solution (I, D: *Caution*, no flames, N)

Diethyl malonate (OW, I)   *n*-Butyl bromide (HO, I)

aq. Ethanol (distillate) (D, I)   conc. HCl (D, N, I)

$MgSO_4$ (S)   aq. Acid solution (N, I)

KOH (I)   aq. KOH (N)

Ethyl ether (OS, OW, I)

Product (I)

Additions and Alternatives

You can employ chasers (acenaphthalene or phenanthrene) for the ester or acid distillation. Determine the infrared spectrum of the acid and of the ester. If benzyl chloride is used in place of *n*-butyl bromide, hydrocinnamic acid can be prepared (compare with Experiment 27). The use of 1-iodopropane (from Experiment 22) could produce valeric acid.

Individually, you can distill your alkylated ester at reduced pressure, using a large reaction vial or a very small round-bottom flask and a short path head. Quantities used from this point should be reduced to one-fourth of the scale (and proportionate to your ester), and flasks should be reduced in size as well. Distillation should take less time at this scale. The final distillation should be done with a 10-mL flask and, after removal of the ether, can be done at reduced pressure.

14.2 Wittig Reaction

The Wittig reaction may be used to convert a carbonyl group, $C=O$, into an alkene, $C=C$. A phosphorus ylide reacts with the carbonyl compound to form a betaine, which then forms the product alkene.

$$\mathrm{>\!\!C=O} \quad + R_3P\!=\!\overset{\overset{\displaystyle R'}{|}}{C}\!-\!R'' \longrightarrow -\overset{\overset{\displaystyle R'}{|}}{\underset{\underset{\displaystyle O^-}{|}}{C}}\!-\!\overset{\overset{\displaystyle |}{|}}{\underset{\underset{\displaystyle P^+R_3}{|}}{C}}\!-\!R'' \xrightarrow{\text{base}}$$

a carbonyl   an ylide   a betaine
compound

$$\mathrm{>\!\!C=C\!<^{R'}_{R''}} \quad + \quad R_3P\!=\!O$$

         an alkene   a phosphine
                               oxide

In order to visualize the formation of the phosphorous ylide, think about the reaction of ammonia, a basic substance, with hydrogen chloride, an acidic substance, to form an ammonium salt.

$$NH_3 + HCl \longrightarrow NH_4{}^+Cl^-$$

The ammonium salt can be reacted with base to regenerate the ammonia:

$$NH_4{}^+Cl^- + NaOH \longrightarrow NH_3 + HOH + NaCl$$

Now substitute an alkyl chloride for hydrogen chloride in the reaction with ammonia, and react the product, an alkyl ammonium salt, with sodium hydroxide to produce an amine.

$$NH_3 + RCl \longrightarrow \quad RNH_3{}^+Cl^-$$

                           alkyl ammonium
                               salt

$$RNH_3{}^+Cl^- + NaOH \longrightarrow RNH_2 + HOH + NaCl$$

                               amine

If the primary amine, RNH_2, is reacted with excess alkyl chloride, and then base, the tetraalkyl ammonium salt is eventually formed in the same stepwise fashion.

$$RNH_2 \xrightarrow{RCl} R_2NH_2{}^+Cl^- \xrightarrow[-HOH]{NaOH}{}_{-NaCl} R_2NH \xrightarrow{RCl}$$

$$R_3NH^+Cl^- \xrightarrow[-HOH]{NaOH}{}_{-NaCl} R_3N \xrightarrow{RCl} R_4N^+Cl^-$$

tetraalkyl
ammonium
salt

Phosphorus, being a member of the same family as nitrogen, is capable of reacting with an alkyl halide in the same fashion as does nitrogen. Triphenylphosphine can react with either a primary or a secondary alkyl halide and then base to form an ylide.

$$(C_6H_5)_3P \quad + R_2CHX \longrightarrow (C_6H_5)_3P^+{-}CHR_2\ X^-$$

triphenyl phosphine

$$(C_6H_5)_3P^+{-}CHR_2\ X^- \xrightarrow{Base} (C_6H_5)_3P^+CR_2{}^- + base{-}H^+ + X^-$$

ylide

Unlike nitrogen, phosphorus has empty *d* orbitals, which can accept electrons to form the ylene.

$$(C_6H_5)_3P^+CR_2{}^- \longleftrightarrow (C_6H_5)_3P{=}CR_2$$

ylide   ylene

The hybrid structures signal the stability for the ylide.

The ylide is usually not isolated but is reacted directly with a carbonyl compound to produce the betaine.

$$\begin{array}{c} R \\ \diagdown \\ \quad\ \ C{=}O + (C_6H_5)_3P^+{-}C^-R_2 \longrightarrow R{-}\overset{\displaystyle R}{\underset{\displaystyle \underset{O^-}{|}\ \underset{P^+(C_6H_5)_3}{|}}{\overset{|}{C}}}{-}CR_2 \quad\longrightarrow \\ R \diagup \end{array}$$

betaine

$$\left[R{-}\overset{\displaystyle R}{\underset{\displaystyle \underset{O}{|}\ \underset{P(C_6H_5)_3}{|}}{\overset{|}{C}}}{-}CR_2 \right]$$

One may *imagine* the negative oxygen reacting with the positive phosphorus to produce the four-membered ring, which can then react to produce the alkene and triphenylphosphine.

$$\left[R{-}\overset{\displaystyle R}{\underset{\displaystyle \underset{O}{|}\ \underset{P(C_6H_5)_3}{|}}{\overset{|}{C}}}{-}CR_2 \right] \longrightarrow \begin{array}{c} R \\ \diagdown \\ \quad\ \ C{=}CR_2 + O{=}P(C_6H_5)_3 \\ R \diagup \end{array}$$

alkene   triphenyl
phosphine

Prior to the introduction of the Wittig reaction in 1953, alkenes were generally synthesized by the dehydration of alcohols (Chapter 6). Frequently, not only would the desired alkene be synthesized but also isomeric alkenes. One advantage to the Wittig reaction is that the position of the double bond is known and isomers are not formed. A second advantage is that the reaction is carried out under mild basic conditions—unlike the dehydration of alcohols, which requires acidic conditions.

George Wittig introduced a new general method for the synthesis of alkenes and received the Nobel Prize in 1979 for his studies of the use of phosphorus in synthetic organic chemistry.

In this experiment, 9-anthraldehyde will be reacted with benzyltriphenyl phosphonium chloride, and then 50% NaOH to ultimately produce trans-9-(2-phenylethenyl) anthracene.

E X P E R I M E N T   6 4

Preparation of *trans*-9-(2-Phenylethenyl) Anthracene (Wittig Reaction) (SM)
References: Sections 1.3E–I, 3.3, 4.1, 5.1, and 5.2

9-anthraldehyde   benzyltriphenyl phosphonium chloride $\xrightarrow{50\% \text{ NaOH}}$

trans-9-(2-phenyl-ethenyl)anthracene

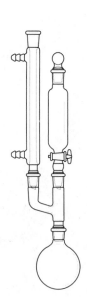

Figure 1.7, Unit II
(page 24).

Assemble the apparatus and introduce the reagents. Assemble a 25-mL round-bottom flask equipped with a stir bar, condenser, and an addition funnel (Unit II, Figure 1.7). Place 2 g of benzyltriphenyl phosphonium chloride, 1.25 g of 9-anthraldehyde, and 7 mL of methylene chloride in the flask. Stir the mixture completely for two to five minutes, and then add 2.5

mL of 50% sodium hydroxide from the addition funnel at a rate of 6–8 drops/minute. When the addition is complete, stir for half an hour.

Isolate and purify the product. Transfer the mixture to a separatory funnel. Wash the flask with 12 mL of methylene chloride and then with 12 mL of water. Add the washings to the separatory funnel. Shake the separatory funnel, and draw off the organic layer. Extract the aqueous layer with 7 mL of methylene chloride. Combine the organic layers. Dry the combined organic layers over 1–2 of anhydrous magnesium sulfate. Separate the solution from the drying agent. Evaporate the solvent by warming the solution in a water bath under the hood with a stream of dry nitrogen or air or aspiration (see Section 1.3I). Scratching, seeding, and/or standing for several days may be needed for crystals to form. Recrystallize the resulting waxy solid from a minimum amount of hot 1-propanol (20–30 mL or about 1 g/25 mL). Weigh the product; determine the percent yield and the melting range. The literature value for the boiling point is just over 130°C.

Determine the nuclear magnetic resonance and infrared spectra of the product. Proton nuclear magnetic resonance spectroscopy can be used to prove the formation of only the *trans* isomer. The compound is dissolved in $CDCl_3$. The infrared spectrum can be recorded as a KBr pellet. The infrared spectrum also suggests the trans isomer but it is ambiguous because of a monosubstituted benzene absorption that could imply the presence of the *cis* isomer. For a discussion of the use of spectroscopy in analyzing this product, see Silversmith, *J. Chem. Educ.*, 1986, *63*, 645. This report includes a microscale procedure.

Figure 4.1 (page 129).

See Figure 1.9 (page 30).

See Figures 2.6A, B; 2.7 (pages 52–57).

Summary

1. Equip a flask for reflux and addition.

2. Stir together all of the reactants except for the base.

3. Add the base dropwise and stir.

4. Transfer the mixture to a separatory funnel and add the washings.

5. Separate the organic layer and extract the aqueous layer.

6. Dry the organic layer, remove the drying agent, and evaporate the solvent.

7. Recrystallize the product from a minimum of hot 1-propanol.

8. Analyze the product as directed.

Questions

1. Why is vigorous stirring begun before the addition of base, and why is the base added slowly?

2. In the extractions/separations, which should be the bottom layer—the aqueous layer or the methylene chloride layer?

3. Explain why it would be highly unlikely for the *cis* isomer to form in this reaction.

4. Compare the coupling constants, J, for both a *cis* and a *trans* AX system. Explain how one can use nuclear magnetic resonance spectroscopy to show the formation of only the *trans* alkene.

Disposal of Materials

Benzyltriphenyl phosphonium chloride (irritant) (HW, I)

9-Anthraldehyde (OW, I)   Methylene chloride (OS, HO, I)

aq. NaOH (D, N)   Aqueous solution (N, I)

1-Propanol (OS, OW, I)   1-Propanol solution (evaporate, OW, I, S)

$MgSO_4$ (S, I, evaporate)

Additions and Alternatives

You can do this reaction on one-fifth to one-fourth the scale with microscale equipment.

E X P E R I M E N T   6 5 A

Preparation of 9,10-Dihydroanthracene-9,10-endosuccinic Anhydride (Diels–Alder Reaction) (SM)
References: Sections 1.3F–H and 3.3

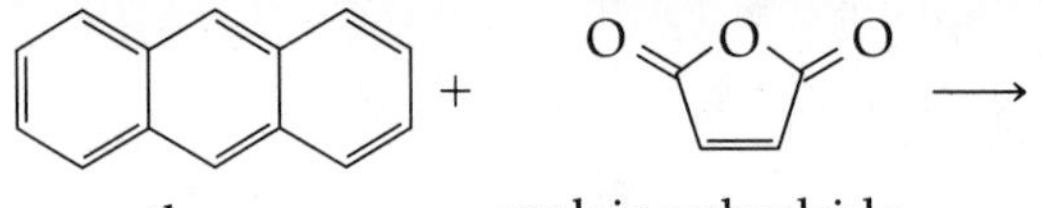

anthracene   maleic anhydride

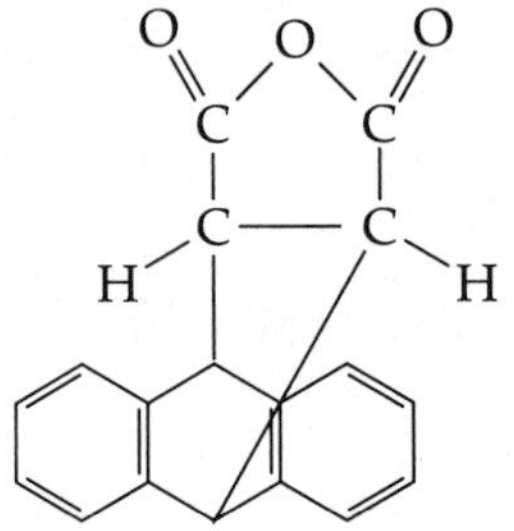

9,10-dihydroanthracene-
9,10-endo-α,β-succinic anhydride

The Diels–Alder reaction consists of the reaction of a *cis* **diene** with a substituted, activated alkene, the **dienophile**, to form a cyclic system.

diene   dienophile   Diels–Alder adduct

The reaction generally occurs very easily, requiring at most a moderate amount of heat.

In the following reaction, you will prepare 9,10-dihydroanthracene-9,10-endo-α,β-succinic anhydride. Maleic anhydride serves as the dienophile and the double bonds in the central ring of anthracene function as the diene.

Procedure

Assemble the apparatus and introduce the reagents. In a 50-mL round-bottom flask fitted with a reflux condenser and a magnetic stirrer (Unit I, Figure 1.7), place 2 g (0.011 mole) of anthracene, 1.1 g (0.011 mole) of maleic anhydride, and 25 mL of dry xylene.

Figure 1.7, Unit I (page 24).

Reflux the mixture with stirring for 20 minutes. Allow the mixture to cool well below its boiling point; add 0.2–0.4 g of pelletized decolorizing charcoal, and boil for another five minutes.

Isolate the product. Filter the hot solution by gravity filtration through a preheated, short-stemmed funnel containing fluted filter paper. Cool the solution completely, and collect with suction filtration, the crystals that form. Dry the crystals in a vacuum desiccator containing paraffin wax shavings (to absorb traces of xylene). Weigh the product; determine the percent yield and the melting range, which should be just over 260°C (with decomposition).

Reagents **Xylene** may be dried over anhydrous calcium chloride or over molecular sieves. Freshly opened **maleic anhydride** should be used because the anhydride is easily converted to maleic acid upon exposure to moisture in the air.

Summary

1. Equip a flask for reflux with stirring.

2. Add the reagents, and reflux with stirring for 20 minutes.

3. Cool, add pelletized carbon, and boil for five minutes.

4. Hot filter, cool, and encourage crystal formation.

5. Cool completely, suction filter, and dry in a desiccator.

6. Analyze the product as directed.

1. Why is it important that the xylene used in this reaction be dry?

2. Why is xylene used instead of toluene? How do they differ?

3. Why is it difficult to remove the last traces of the solvent, xylene, from the filtered crystals; that is, why are paraffin wax shavings placed in a vacuum desiccator with the crystals to "dry" the Diels–Alder adduct?

4. How are maleic and fumaric acid related? Their anhydrides?

Disposal of Materials

Anthracene (I, OW, S)   Maleic anhydride (I, OW, absorbs water)

Xylene (OS, OW, I)   Xylene solutions (OW, I)

Paraffin wax (S)   Decolorizing charcoal (S)

Product (I)

Additions and Alternatives

Record the infrared spectrum of the product in addition to or instead of the melting point.

If the reaction mixture is not highly colored, decolorizing carbon and hot filtration may be omitted.

EXPERIMENT 65B

Preparation of 9,10-Dihydroanthracene-9,10-endosuccinic Anhydride (Diels–Alder Reaction) (M)
References: Experiment 65A and its references

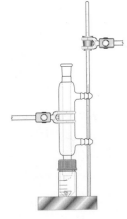

Figure 1.8 (page 26).

Assemble the apparatus and introduce the reagents. In a 50-mL conical vial fitted with a reflux condenser and a magnetic spin vane (Unit IA or B, Figure 1.8A), place 200 mg (0.0011 mole) of anthracene, 110 mg (0.0011 mole) of maleic anhydride, and 2.5 mL of dry xylene.

Reflux the mixture in a sand bath or aluminum block, with stirring, for 30 minutes.

Isolate the product. Cool the solution to room temperature and then to ice temperature. Filter the crystals by suction on a Hirsch funnel. Dry the crystals in a vacuum desiccator containing paraffin wax shavings (to absorb traces of xylene).

Analyze the product. Weigh the product; determine the percent yield and the melting range, which should be a bit over 260°C (with decomposition). Record the infrared spectrum if directed to do so.

Xylene may be dried over anhydrous calcium chloride or over molecular sieves.

Freshly opened **maleic anhydride** should be used because the anhydride is easily converted to maleic acid upon exposure to moisture in the air.

Summary

1. Equip a vial for reflux with stirring.

2. Add the reagents and reflux with stirring for 20–30 minutes.

3. Crystallize the product.

4. Collect the product by suction filtration and dry it in a desiccator.

5. Analyze the product as directed.

Answer those given for Experiment 65A.

Questions

Disposal of Materials

Anthracene (I, OW, S)   Maleic anhydride (I, OW, absorbs water)

Xylene (OS, OW, I)   Xylene solutions (OW, I)

Paraffin wax (S)   Decolorizing charcoal (S)

Product (I)

Additions and Alternatives

See Experiment 65A.

14.3   An Aldol-Like Reaction

An aldol condensation reaction produces a carbon–carbon bond by reacting an aldehyde or ketone with an aldehyde or ketone compound containing an α-hydrogen under acidic or basic conditions. A β-hydroxy aldehyde or a β-hydroxy ketone is produced, which can subsequently be dehydrated to an α,β-unsaturated aldehyde or an α,β-unsaturated ketone. For example, under basic conditions, acetaldehyde reacts to form the β-hydroxy aldehyde via the intermediate carbanion. The β-hydroxy aldehyde can be dehydrated to the α,β-unsaturated aldehyde.

$$
\begin{array}{c}
\text{O} \\
\parallel \\
\text{CH}_2\text{—CH} \\
| \\
\text{H}
\end{array}
$$

$$\downarrow \text{OH}^-$$

$$
\underset{\text{the aldehyde}}{\overset{\text{O}}{\underset{\parallel}{\text{CH}_2\text{CH}}}} \quad \longleftarrow \quad
\overset{\text{O}}{\underset{\parallel}{\text{—CH}_2\text{—CH}}} \quad \xrightarrow{\text{H}_2\text{O}} \quad
\text{CH}_3\text{—}\overset{\overset{\text{H}}{|}}{\underset{\underset{\text{OH}}{|}}{\text{C}}}\text{—CH}_2\text{—}\overset{\text{O}}{\overset{\parallel}{\text{CH}}}
$$

$$
\downarrow \underset{\text{CH}_2\text{=C—H}}{\overset{\text{O}^-}{\underset{|}{}}}
$$

the β-hydroxyaldehyde

$$
\underset{\text{the } \beta\text{-hydroxyaldehyde}}{\text{CH}_3\text{—}\overset{\overset{\text{H}}{|}}{\underset{\underset{\text{OH}}{|}}{\text{C}}}\text{—CH}_2\text{—}\overset{\text{O}}{\overset{\parallel}{\text{CH}}}} \quad \xrightarrow[\Delta]{\text{dilute HCl}} \quad
\underset{\text{the } \alpha,\beta\text{-unsaturated aldehyde}}{\text{CH}_3\text{—CH=CH—}\overset{\text{O}}{\overset{\parallel}{\text{CH}}}}
$$

There are many different condensation reactions that are similar to the aldol condensation. A carbanion, which is formed by the reaction of a base with a compound containing an acidic hydrogen, reacts with a carbonyl group to produce a β-hydroxy compound that in some cases may be dehydrated. In the following reaction, indene, which has acidic methylene hydrogens, will form the indene carbanion under basic conditions.* The carbanion can react with the benzophenone to generate a new carbon–carbon bond and form the hydroxyl group. The resulting intermediate is easily dehydrated (because of the possibility of extensive delocalization of electrons) to produce benzhydrylidene indene.

　* Aromatic compounds, which are unusually stable systems, contain $(4n + 2)\pi$ electrons where n = 0, 1, 2, etc. Indene forms a stable carbanion because, with the loss of one methylene hydrogen ion (H^+) and the subsequent rehybridization of the orbitals on the involved carbon atom from sp^3 to sp^2, the entire system contains 10π electrons, i.e., n = 2 in the formula $(4n + 2)\pi$ electrons. The indene carbanion is an aromatic system.

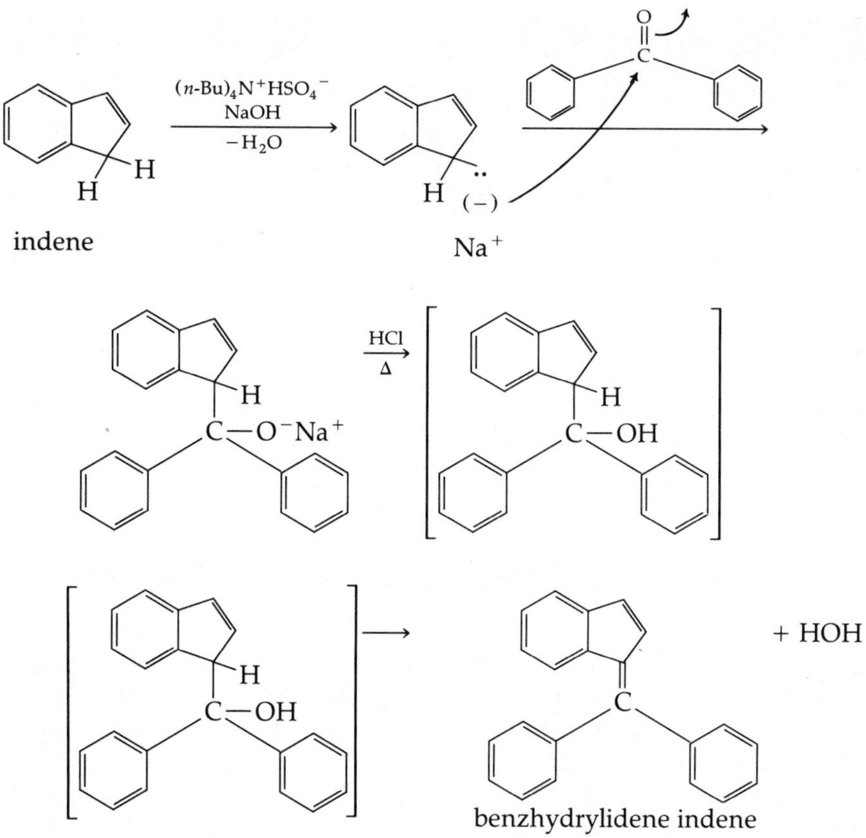

indene

Na⁺

benzhydrylidene indene

+ HOH

Indene is a nonpolar compound and is soluble in nonpolar solvents, such as xylene. Sodium hydroxide is an ionic compound and is soluble in polar solvents, such as water. In order for sodium hydroxide and indene to react to form the carbanion, they must come into contact with each other. The phase transfer catalyst, tetrabutyl ammonium hydrogen sulfate, the **"quat" salt,** is soluble in both polar and nonpolar solvents. Because it is ionic, the quat salt is soluble in polar solvents, and because nonpolar butyl groups surround the positive nitrogen atom, the quat salt is soluble in nonpolar solvents. The quaternary salt migrates from the polar to the nonpolar phase, carrying with it either the unreactive hydrogen sulfate ion or the highly reactive hydrogen ion. When indene reacts with hydroxide, it forms the sodium carbanion of indene, which reacts with benzophenone in the xylene to generate the sodium salt of benzhydrylidene indene. This organic ion may be carried into the polar phase as the negative counter ion of the quat salt. The sodium salt of benhydrylidene indene reacts in the aqueous phase with hydrochloric acid to generate benzhydrylidene indene, which is nonpolar and soluble in the xylene.

E X P E R I M E N T   6 6 A

Preparation of Benzhydrylidene Indene (an Aldol-like Condensation Reaction Using Phase Transfer Catalysis) (SM)
References: Sections 1.3E–H; 2.5; 3.3; 4.1, 4G; 5.1, and 14.3

indene + benzophenone $\xrightarrow[\text{xylene}]{\substack{(n\text{-Bu})_4N^+HSO_4^- \\ \text{NaOH}}}$ benzhydrylidene indene

Introduce the reagents and initiate the reaction. Dissolve 2.3 g of indene in 12 mL of xylene. Pour the solution into a 125-mL Erlenmeyer flask containing a magnetic stirrer. Add 2.4 g of sodium hydroxide pellets, 0.7 g of tetrabutyl ammonium hydrogen sulfate, 3.65 g of benzophenone, and 1 drop of water. Stir vigorously and heat the mixture. Discontinue heating as soon as an exothermic reaction begins. The solution will become deeply colored. Continue stirring until no more heat is evolved.

Isolate the product. Acidify the solution with 6N hydrochloric acid (litmus or Congo red paper); then separate and discard the aqueous layer. You might need a bright background to see the layers because of the intense color of both layers. Dry the organic layer. Remove the solvent by evaporation with reduced pressure (rotary evaporator, or distillation with vacuum pump or water aspirator).

See Figures 2.6A, B; 2.7 (pages 52–55).

Recrystallize the crude product from a minimum of hot 2-propanol or ethanol. Dry the crystals until the next period. Determine the percent yield and the melting range, which should be just under 115°C.

Summary

1. Mix the reagents as directed.

2. Stir vigorously.

3. Heat as needed to start the reaction but discontinue when heat evolves.

4. Stir as long as heat is evolved.

5. Acidify with dilute HCl.

6. Separate and discard the aqueous layer, using a separatory funnel.

7. Dry the organic layer with anhydrous $MgSO_4$.

8. Remove the xylene with reduced pressure distillation.

9. Recrystallize the resulting solid from a minimum of hot alcohol.

10. Analyze the product as directed.

1. Why is the solution of indene in xylene added to sodium hydroxide; that is, what is the purpose of the sodium hydroxide in the reaction? The drop of water?

2. Why must the reaction mixture be acidic before the organic solvent is separated and evaporated?

3. Why is the solvent removed at reduced pressure?

Disposal of Materials

Xylene (OS, OW, I)

Benzophenone (S, I)

dil. HCl (N)

2-Propanol (D, I)

Ethanol (D, I)

Recrystallization filtrate (I, evaporate, S or D)

NaOH (I, S, HW)

Tetrabutyl ammonium hydrogen sulfate (OW, S, I)

aq. HCl solutions (I, N)

Product (I)

Additions and Alternatives

Record the infrared and/or nuclear magnetic resonance spectra of the product. Fluorene may be used in place of indene, and other aromatic ketones (substituted benzophenones) or aldehydes (benzaldehyde only works with fluorene) that have no α-hydrogens could be tried. See *J. Chem. Educ.*, 1986, *63*, 916.

EXPERIMENT 66B

Preparation of Benzhydrylidene Indene (an Aldol-like Condensation Reaction Using Phase Transfer Catalysis) (M)
References: Sections 1.3E–I; 3.3, 5; 4.1; 5.1; 14.3; and Experiment 66A

Introduce the reagents and initiate the reaction. Dissolve 0.23 g of indene in 1 mL of xylene in a 12–15 mL centrifuge tube (glass) or 5-mL reaction vial containing a magnetic stir vane. Add 0.24 g of sodium hydroxide pellets, 80 mg of tetrabutyl ammonium hydrogen sulfate, 0.36 g of benzophenone, and 1 drop of water. Stir vigorously and heat the mixture in a sand bath or aluminum block. Discontinue heating as soon as the highly exothermic reaction begins. The solution will become highly colored. Continue stirring until no more heat is evolved.

Isolate the product. Acidify the solution by adding 6M hydrochloric acid (litmus or Congo red paper). Withdraw and discard the lower aqueous layer using a Pasteur pipet. Use a strong light source to distinguish the layers, because of the intense color of both layers. Transfer the organic layer to the bottom of a 5-mL Craig tube, and evaporate the solvent by heating under gently reduced pressure (or use a stream of dry nitrogen or a vacuum desiccator with paraffin wax shavings, see Section 1.3I). Recrystallize the crude fulvene derivative from a minimum of hot 2-propanol or ethanol in the Craig tube. Dry the crystals until the next period. Determine the melting range and the percent yield. The melting range should be just under 115°C.

See Figures 2.8A, B, C (pages 56–58).

Summary

1. Mix the reagents as directed.

2. Stir vigorously.

3. Heat as needed to start the reaction, then discontinue.

4. Stir as long as heat is evolved.

5. Acidify with dilute HCl.

6. Separate and discard the aqueous layer.

7. Evaporate the xylene.

8. Recrystallize the product from 2-propanol.

9. Analyze the product as directed.

Questions ANSWER THOSE GIVEN FOR EXPERIMENT 66A.

Disposal of Materials

Xylene (OS, OW, I)

Benzophenone (S, I)

dil. HCl (N)

2-Propanol (D, I)

Ethanol (D, I)

Recrystallization filtrate (I, evaporate, S or D)

NaOH (I, S, HW)

Tetrabutyl ammonium hydrogen sulfate (OW, S, I)

aq. HCl solutions (I, N)

Product (I)

Additions and Alternatives

You can do the evaporation and recrystallization in a centrifuge or test tube and use a Hirsch funnel with suction to collect the crystals.

See Experiment 66A.

CHAPTER 15

Kinetics

The study of reaction kinetics enables us to make a detailed analysis of the way a reaction takes place, and it is an important means for supporting the proposed mechanism. Many experiments have been suggested to illustrate the use of kinetics and the relationship between kinetics and reaction mechanisms. The three experiments in this chapter have been chosen because of their relative ease of observation. They also include suggestions for variations that will encourage you to explore various aspects of the kinetics of a reaction. These observations can then be compared with predictions from the mechanism proposed for the reaction.

15.1 Equations Derived for First- and Second-Order Reactions

For the reaction $A + B + C \rightarrow$ products, you would expect the rate to depend in some way upon the concentration of the reactants (i.e., rate $\alpha[A]^a[B]^b[C]^c$). The rate may be expressed as the appearance of a product ($\Delta P/\Delta t$) or the disappearance of a reactant ($-\Delta A/\Delta t$). It turns out that for the greatest accuracy, Δ must be infinitely small (denoting an instantaneous change), and the rate is better expressed as dP/dt or $-dA/dt$. When the two are combined and a proportionality constant is inserted, one form of the equation is $-dA/dt = k[A]^a[B]^b[C]^c$. The a, b, and c are determined experimentally, and each is called the order of the reaction with respect to that reactant. If $a = b = c = 1$, the reaction is third order overall and first order with respect to each component. If $a = b = 1$ and $c = 0$, the reaction is second order, as it would be if $a = 2$ and $b = c = 0$. For a first-order reaction, $a + b + c = 1$.

The rate equation for a first-order reaction is thus $-d[A]/dt = k[A]$, and this equation is rearranged ($d[A]/[A] = -kdt$) and integrated from $t = 0$ to some time, t.

$$2.303 \log \frac{[A]}{[A]_0} \text{ or } 2.303 \left(\log[A] - \log[A]_0\right) = -kt \qquad (1)$$

Rearranging to form the equation for a straight line:

$$\log[A] = -(k/2.303)t + \log[A]_0 \qquad (2)$$

In equation (2), [A] is the concentration at time t and $[A]_0$ (a constant) is the concentration at $t = 0$ or t_0 (NB: thus t equals $t - t_0$). A plot of log[A] versus t should yield a straight line, and k can be calculated from the slope.

The half life $(t_{1/2})$ of a reaction is the time needed for one-half of the starting material to react. Substituting in equation (1) and changing the signs gives:

$$2.303 \log \frac{A_0}{A_0 - .5A_0} = kt_{1/2} = 2.303 \log 1/.5 = 2.303 \log 2$$

or

$$t_{1/2} = 2.303 \log 2/k \text{ (for a first order reaction!)}$$

Similarly $t_{1/10}$ is the time required for one-tenth of the starting material to react and

$$2.303 \log \frac{A_0}{A_0 - .1A_0} = kt_{1/10}$$

or

$$t_{1/10} = (2.303/k)(\log 1/.9) = 0.1052/k$$

or

$$k = 0.1052/t_{1/10} \qquad (3)$$

Note that for a first order reaction $t_{1/10}$ should be independent of concentration.

If the reaction is second order where $-dA/dt = k[A]^2$ or $-dA/dt = k[A][B]$ with the initial concentration of A and B the same (i.e., they are the same throughout), both equations could be rearranged to $dA/A^2 = -kdt$. Similar integration then gives $1/A - 1/A_0 = kt$ or $1/A = kt + 1/A_0$, which when plotted should also be a straight line *if the reaction is second order*.

The half-life for a *second-order* reaction would then be:

$$1/(A_0 - .5A_0) - 1/A_0 = kt_{1/2}$$

or

$$1/.5A_0 - 1/A_0 = kt_{1/2}$$

or

$$t_{1/2} = 1/(kA_0) \qquad (4)$$

The equation for $t_{1/10}$ then becomes:

$$1/(A_0 - .1A_0) - 1/A_0 = kt_{1/10}$$

or

$$1/.9A_0 - 1/A_0 = kt_{1/10}$$

or

$$t_{1/10} = 0.1111/kA_0 \qquad (5)$$

Note that for a *second-order* reaction, $t_{1/10}$ will vary inversely with the initial concentration of the starting material.

15.2 The Effect of Temperature on Rate

For most reactions the rate will increase if the reaction temperature increases. In fact, one generalization is that the rate of many reactions approximately doubles for each $10°$ increase in temperature. The Arrhenius equation,

$$\log k = (-E_{act}/2.303R)(1/T) + C \qquad (6)$$

relates the specific rate constant (k) to the energy of activation (E_{act}) (C is a constant of integration). Note that the larger the E_{act} the more affected the rate will be by changes in temperature. This is the equation for a straight line if $\log k$ is plotted against $1/T$ (T is in K). The slope of this line, which can be determined experimentally, is equal to $-E_{act}/2.303R$, where R is the gas constant expressed in energy units (i.e., Joules or calories) (R = 1.99 cal/mole° or 8.3144 J/mole°). If a reaction is run at several temperatures so that k can be calculated at each temperature, this plot permits the determination of the energy of activation.

E X P E R I M E N T   6 7

The Solvolysis of *t*-Butyl Chloride

This experiment is a modified version of that described by Landgrebe (*J. Chem. Educ.*, 1964, *41*, 567). The rate is not directly measured; rather, the tenth life ($t_{1/10}$) is determined for each kinetic run. Runs at various temperatures permit the calculation of E_{act}, and the variations are designed to test the hypothesis that the mechanism is S_{N^1}. The reaction should be run at each of three temperatures until the times are consistent (at least in triplicate). These temperatures will allow you to determine the E_{act} and will give you times that will become the standards for comparison with other experimental variations. If the experiments span more than one period (you will do much better the second period), do a standard run each period for calibration and comparison.

$$(CH_3)_3CCl + H_2O \longrightarrow (CH_3)_3COH + HCl$$

The mechanism postulated for this reaction suggests that the alkyl halide ionizes in the rate step to form the chloride ion and a carbocation.

Question

1. What structural feature of halides and what type of solvent favor this?

The latter ion reacts rapidly with water, and the proton is subsequently lost.

Question

2. Write the steps of this mechanism.

The kinetic data obtained in this experiment must be consistent with the equations derived for first-order kinetics if the mechanism is to be supported. In order to determine $t_{1/10}$, one-tenth as much sodium hydroxide as alkyl halide is added to the reaction mixture, and the base is neutralized by the acid as the acid is produced. When one-tenth of the halide has been consumed, the acid–base indicator will change and the time ($t_{1/10}$) can be recorded. The rate constant, k, can be calculated from equation (3) (rearranged). Note that the solvent for the standard run is 70% water and 30% acetone.

Question

3. Explain how this composition of the solvent can be determined from the directions for the standard run below.

Standard Run

Set up a table of data. In your notebook, arrange and report all parts of this experiment in a table, including the following: run number, which variation you used, solvent, temperature, 1/absolute temperature, $t_{1/10}$, k, and log k.

Sample Table

Run No.	Run Type	Solvent Comp.	Temp. °C	1/T(in K)	$t_{1/10}$	k	log k

Measure the reagents and equilibrate them to temperature.. Take the temperature of a water bath that has been at ambient temperature for several hours and record it. From a 50-mL buret, add 3 mL of 0.1N *t*-butyl chloride in acetone to a 25-mL Erlenmeyer flask. Stopper it (cork), and suspend it by a wire in the bath if one is used. To another 25-mL Erlenmeyer flask, add 0.3 mL of 0.1N aqueous sodium hydroxide (using a pipet or 10-mL buret), 6.7 mL of deionized water from a 50-mL buret and 3 drops of bromophenol blue indicator solution. You may wish to stopper this flask. Suspend it in the bath if one is used.

Initiate the reaction. When the solutions have equilibrated to room temperature in the bath or on the bench top, mix them in the following manner. Noting the time, add the acetone solution quickly to the aqueous solution; swirl the mixture once, and pour it all back into the first flask. Stopper the flask, and return it to the bath if one is used. Record the time ($t_{1/10}$) when the blue color turns to straw yellow. Repeat this procedure two or three times or until two or more runs are within a few seconds of each other. Each run should start with clean flasks or those rinsed with acetone and drained for several minutes.

Summary

1. Set up a data table in your notebook.

2. Place the proper amounts of the acetone and of the aqueous solutions in the vessels.

3. Equilibrate them to temperature.

4. Mix and start timing as directed.

5. Record the $t_{1/10}$ when the color changes.

Temperature Variation (A)

The standard runs constitute the room temperature part of this variation.

Establish a cold water bath with ice chips, and regulate the temperature to about 10° *below* the temperature of the room temperature runs. Prepare the flasks for each run as directed in the standard procedure. Place the flasks in the bath; record the temperature, and allow five to ten minutes for the contents of the flasks to come to the same temperature as the bath.

Initiate the reaction. Mix as directed above; stopper and return the reaction flask to the bath. Record the time for the color change. Repeat two more times or until two or more runs are within 10–25 seconds of each other.

Establish a hot water bath. Now repeat a group of runs using a bath regulated to about 10° above that of the room temperature runs. Use the same procedure as above. Runs should agree within a few seconds.

Interpret the results. Plot $\log k$ versus $1/T$ using an expanded scale. From the plot, determine the slope of the line, and calculate E_{act} (with proper units!).

$$E_{act} = Slope(-2.303R)$$

Summary

1. Establish temperature baths 10° above and below room temperature.

2–5. Same as in the standard run. Be sure to return the reaction mixture to the bath after mixing.

Solvent Polarity Variation (B)

If the rate step involves ionization, solvents that can solvate ions will be expected to decrease the E_{act} and increase the rate of reaction.

Question

4. Explain fully.

The polarity of the reaction solvent used here can be varied by changing the proportions of water and acetone.

Question

5. Which is the more polar?

The percent water could be varied from 50–80%. The slight changes in concentration should not affect the results (see next variation).

Modify the amounts of reaction used. For runs using a solvent that is 80% water/20% acetone, make the following changes in the standard procedure. Use 2 mL of acetone solution, 0.2 mL of sodium hydroxide, and 7.8 mL of water.

Initiate the reaction. Do duplicate determinations; run them at room temperature or 10° above room temperature, but the temperature should be the same as that used for some standard runs (70/30).

Do another solvent composition. Determine the amounts of solutions needed for a 60% water/40% acetone solvent run and run these reactions in duplicate at the same temperature as the 80/20 runs.

Interpret the results. Determine the k values for these runs, and compare $t_{1/10}$ and k for runs with the three solvent polarities.

Question

6. Do the results support the mechanism proposed? Explain.

Summary

1. Measure the modified amounts of reagent for an 80/20 run.

2. Equilibrate the temperature of the solutions at a previously used temperature.

3. Proceed as in steps 2–5 of the standard run.

4. Measure the modified amounts of reagent for a 60/40 run.

5. Equilibrate the solutions to the same temperature.

6. Proceed as in steps 2–5 of the standard run.

7. Compare the k values at this temperature for the 80/20, 70/30, and 60/40 runs.

Variation of the Initial Concentration (C)

This variation will be run with half of all of the initial concentrations used in the standard reaction. According to the equations derived above (esp. 3, 4, and 5), a change in the initial concentration of substrate will not affect the $t_{1/10}$ of a first order reaction.

7. Should this change affect a second-order reaction? Why?

This experiment is designed to discriminate between these two types of reactions.

Prepare the reagents. Prepare the flasks as done for the standard procedure (70/30) except to the *water* (and NaOH) *flask* add 10 mL of a solution that is 70% water/30% acetone.

Initiate the reaction. Equilibrate duplicate runs at a standard temperature, and continue as in the standard run.

Interpret the results. Determine $t_{1/10}$ and k values, and compare them with their analogous standard values.

8. Is the $t_{1/10}$ observed consistent with a first- or second-order reaction? Explain.

Summary

1. Measure the modified amounts of reagents.

2. Equilibrate the temperature of the solutions to a temperature used above.

3. Proceed as in steps 2–5 above.

4. Compare the results with those of the standard run at this temperature.

Variations in Percent Completion (D)

The standard reaction is run to 10% completion. In theory, rates could be constant for a large proportion of a reaction. In practice, they tend to fall off as the substrate decreases and product increases. The directions can be modified for 5% ($t_{1/20}$), 15%, 20% ($t_{1/5}$), 25% ($t_{1/4}$), 30%, or 33% ($t_{1/3}$) completion. The lower percent completions are recommended. The modification that follows is for $t_{1/5}$.

Introduce the reagents and initiate the reaction. In the *water* flask, place 0.6 mL of 0.1N NaOH and 6.4 mL of water. Prepare the other flask as done in the standard run. Equilibrate duplicate runs at a standard temperature, and continue as above. Modify equation 3 to calculate k for these runs, and compare the k values calculated here with those for the analogous standard runs.

Question

9.   Is this consistent with a first-order reaction? Explain.

Summary

1.   Measure the reagents to find the $t_{1/20}$ or $t_{1/5}$, etc.

2.   Equilibrate and proceed as in steps 2–5 above.

3.   Calculate k by using a modified equation and compare with a standard run.

Structural Variations (E)

Prepare and/or use 0.1N acetone solutions of other secondary or tertiary alkyl or benzyl halides. These might include *t*-amyl chloride, isopropyl chloride, *sec*-butyl chloride, or benzyl chloride (or the corresponding bromides or iodides). Use these solutions as the acetone solution of *t*-butyl chloride was used in standard runs. If there is no color change after five to ten minutes, heat the reaction with a steam or water bath.

Question

10.   Postulate, on the basis of the mechanism, the change in $t_{1/10}$ or k that should be observed, and compare your prediction with your observation. Explain.

Summary

1.   Use solutions containing another halide for a standard run.

2.   Make qualitative comparisons.

Questions

QUESTIONS 1–10 ABOVE. THE COMPLETED DATA TABLE SHOULD BE PART OF YOUR REPORT AND CITED IN YOUR DISCUSSION. ALSO INCLUDE IN YOUR REPORT THE PLOT FROM PART (A).

Disposal of Materials

t-Butyl chloride in acetone (I, OS, HO, D)

aq. NaOH (N)

Reaction solutions (I, OS, D)

Solutions of other alkyl halides (I, OS, HO, D)

Additions and Alternatives

See Part E above. Use another halide prepared in Chapter 6. Collect enough data from Part B or D and use a computer graphics program to plot the data.

E X P E R I M E N T   6 8

The Formation of Alkyl Bromides

In this experiment, which was originally described by Cooley, et al. (*J. Chem. Educ.*, 1967, 44, 280), you will observe the rate of formation for primary and secondary alkyl bromides from their corresponding alcohols. The reagent is a mixture of concentrated hydrobromic acid and concentrated sulfuric acid. Two general mechanisms are usually suggested: an S_{N^1} process and an S_{N^2} process, both preceded by protonation.

$$ROH + H_2SO_4 \longrightarrow ROH_2^+ + HSO_4^- \text{ (protonation)}$$
$$S_{N^1}$$

$$ROH_2^+ \longrightarrow R^+ + H_2O \text{ (slow)}$$

$$R^+ + Br^- \longrightarrow RBr$$

$$Br^- \text{---} R \text{---} OH_2^+ \xrightarrow{S_{N^2}} RBr + H_2O$$

While different alcohols may react by different mechanisms, Cooley and coworkers report a lack of dependence on the concentration of hydrobromic acid for a limited number of cases. This seems to imply a predominance of S_{N^1} in the cases they studied.

1. Why does this absence of a concentration effect imply a first-order reaction?   **Question**

In this experiment, the initial mixture of alcohol and two acids should be soluble (only one layer). The alkyl bromide that is formed is insoluble in this mixture and forms its own layer, usually on top, in spite of densities of 1.2 g/mL or more.

2. What are the densities of the acids and what does this imply about the density of the aqueous layer?   **Question**

Either way, there is a change in the layer that is increasing. This change is referred to as the *change in length* for the rest of this discussion.

The standard procedure involves the use of alcohol to 48% (constant-boiling) hydrobromic acid to concentrated sulfuric acid in a volume ratio of 1 : 2 : 1. Descriptions of some different experimental setups will be given, along with recommendations for volumes, followed by a general description of the reaction and suggestions for variations. Depending upon the apparatus, the measurement of "length" may be volume, length, or degrees, but the treatment of the data is the same. If the apparatus is calibrated for volume (graduated receiver, centrifuge tube, eudiometer, etc.), the volume is used directly. If not, a wooden ruler (plastic softens) or a thermometer (report in degrees) can be used.

Question

3. How can a thermometer be used to measure "length"?

Experimental Setups

A. Use a graduated receiver with a water condenser (standard taper and horseshoe clamps). The ratio of reactants (alcohol : HBr : H_2SO_4) should be 3 mL : 6 mL : 3 mL for the standard run.

B. Use a graduated (15 mL) centrifuge tube attached to a 60–100-cm piece of glass tubing (as an air condenser) (cork or O ring and screw cap). Use the same volume ratio as in (A) or 3.5 : 7 : 3.5.

C. Use a 25 × 200-mm test tube with a cork and a 60–100-cm piece of glass tubing (as an air condenser). Use a volume ratio of 12 : 24 : 12.

D. Use a eudiometer tube with a 50–100-mL capacity. The total of the three volumes should be 12–15% of the available volume.

Standard Run

Make all determinations in duplicate if time permits, but allow only one to go to completion (infinity reading or Pc) to determine the amount of RBr formed.

Question

4. Why only one if the reactions are made up to the same concentrations? See the treatment of data below.

The time needed to reach completion will vary from 10–90 minutes, depending upon the alcohol used and the temperature.

Introduce the reagents and initiate the reaction. In the vessel to be used, place the alcohol and the hydrobromic acid, and cool them. Continue cooling this mixture while adding the sulfuric acid in small portions through a funnel.

Question

5. Why are these precautions required?

Set up a water bath maintained at 96–98°C in which the vessel can be placed (so that the height of the surface of the water will equal the height of the surface of the mixture). When the last addition of sulfuric acid has been made, *immediately* assemble the apparatus, and transfer it to the hot water bath.

Take readings. Note the time (t_0) when the vessel is placed in the hot water. With a small volume, take a reading after two minutes; then take ten more readings at 1–2-minute intervals and a few more at 2–3-minute intervals as long as there is an appreciable change in the amount of RBr. Record the elapsed time and "length" of RBr for each reading. For larger volumes, take the initial reading after three or four minutes, and take ten more readings at 2–3-minute intervals and a few more at 3–5-minute intervals while there is appreciable change. At each concentration, allow one run to go until there is no observable change for ten minutes (this volume is P_c). When making readings, use the bottom of the meniscus, and avoid errors of parallax. The initial and final few points may be a bit off.

Variations

a. *Use different alcohols.* Boiling point and solubility are important considerations. Possibilities for primary alcohols would include 1-butanol, 1-pentanol, 2-methyl-1-propanol, 3-methyl-1-butanol, 2,2-dimethyl-1-propanol, and benzyl alcohol. Secondary alcohols might include 2-butanol, 2-pentanol, 4-methyl-2-pentanol, cyclopentanol, cyclohexanol, and 2-methylcyclohexanol. If the reaction is too rapid, as reactions using some of these alcohols will be, make the run at a lower temperature (for 10° less, about one-half the rate). You will be assigned or will choose one of the alcohols to study.

b. *Run a reactive alcohol* at about 10° *and* 20° below the first temperature used. If possible do duplicates. Plot log k versus 1/K, and determine the energy of activation (see the treatment of data below).

c. *Explore the effect of changes in the concentration* of hydrobromic acid. Use the same conditions as a standard run but with 10% or 15% less HBr. Note that one of such modified reactions must be taken to completion for a new Pc. Compare this run with the standard run.

6. What effect does this change have on the rate? What are the mechanistic implications?

Question

Treatment of Data

Make a data table for all of the runs that includes the run number, the volume of each reagent, the time, the "length" (P_t) and ($P_c - P_t$) for each reading, the infinity "length" (P_c), and the k determined graphically. To determine k, plot log ($P_c - P_t$) versus t, and set the observed slope equal to $-k/2.303$ (equation 2 in the initial discussion for this chapter).

Question

7. How is it that plotting log $(P_c - P_t)$ is equivalent to plotting the log of the concentration of the alcohol at time t (log [A])?

If many of your classmates did the same alcohol and HBr concentration, pool the data and determine the mean value, the average value, and the average deviation from the average value.

Question

8. Discuss whether differences in k for runs with variations are significant in view of these class calculations.

Summary

1. Set up an apparatus for the measurement of volume change.

2. Dissolve alcohol in HBr.

3. Set up water bath for 96–98°C.

4. Add H_2SO_4 cautiously.

5. When this addition is complete, assemble and immerse in hot water bath immediately.

6. Make readings at appropriate intervals; let one go to completion.

7. Perform variations as directed.

8. Prepare table, graphs, and analysis of data.

Questions

ANSWER QUESTIONS 1–8 AS DIRECTED. MAKE PLOTS OF $\log(P_c - P_t)$ VERSUS t AND $\log k$ VERSUS $1/T$ AS PART OF YOUR REPORT.

Disposal of Materials

Alcohols (I, OS, OW, D)

conc. HBr (D, N)

conc. H_2SO_4 (D, N)

Mix acid reaction mixture (D, N)

Organic reaction layer (I, HO)

Additions and Alternatives

You could try a commercial HBr solution that is about 40% HBr and compare your results with runs using 48% HBr. You can distill the 40% acid to get a 48% acid or make the HBr from sodium bromide and phosphoric acid (Experiment 20). Use a computer graphics program to treat your data.

EXPERIMENT 69

The Hydrolysis of Sucrose

The sugar sucrose is a naturally occurring disaccharide. When hydrolyzed with appropriate catalysts, acids or enzymes, sucrose produces a mixture of glucose and fructose. All three of these sugars are optically active; sucrose and glucose are dextrorotatory (or +), and fructose is levrorotatory (or −). The latter has a specific rotation, $[\alpha]$, that is large enough to make the product mixture rotate a plane of polarized light to the left. (The initial mixture would rotate the plane of polarized light to the right.)

This experiment provides three interesting features: the reaction of a naturally occurring substance, the use of a polarimeter to follow the reaction, and the possibility of inferring something about the reaction mechanism.

This reaction has been studied with different enzymes, different acids, and different concentrations of acids. As expected, the reaction goes faster when more acid is present.

Two mechanisms have been suggested to account for the acid-catalyzed hydrolysis. Both involve a fast protonation in the first step and a slow second step. The *unimolecular* mechanism postulates that the protonated sucrose breaks down to products, whereas the *bimolecular* process requires collision with a water molecule. In the experiment presented here, either mechanism would predict first-order (or pseudo-first-order) kinetics. However, the plot of k, the specific rate constant, versus the hydrogen ion concentration should be a straight line if the kinetics represent a *bimolecular* process. If the process is unimolecular, further manipulation of the data would be needed to get a linear relationship. See *J. Chem. Educ.*, 1966, 43, 34.

Although different polarimeters may be used, the treatment of the data remains the same. It is not necessary to determine the specific rotation, $[\alpha]$, because differences in the observed rotation, α, are all that are needed. The symbol α_0 will be used for the rotation at time t = 0, α_t for some time t, and α_f for the rotation when the reaction is complete (t = infinity). The integrated form of a first-order rate equation is $kt = 2.303 \log A_0/A_t$ where A_0 is the initial concentration of substrate A (all of it) and A_t is the concentration of A remaining at time t. The total amount of A (A_0) can be represented by $(\alpha_0 - \alpha_f)$, the entire change of rotation, and A_t by $(\alpha_t - \alpha_f)$, the rotational change yet to occur. Thus,

$$kt = 2.303 \log(\alpha_0 - \alpha_f)/(\alpha_t - \alpha_f)$$

or

$$\log(\alpha_0 - \alpha_f)/(\alpha_t - \alpha_f) = k/2.303(t)$$

The value of k can be obtained from individual points or by plotting log $(\alpha_0 - \alpha_f)/(\alpha_t - \alpha_f)$ or log $(\alpha_t - \alpha_f)$ against t and getting k from the slope.

$$\log[(\alpha_0 - \alpha_f)/(\alpha_t - \alpha_f)] = k/2.303(t)$$

$$\text{slope} = k/2.303$$

or

$$\log(\alpha_0 - \alpha_f) - \log(\alpha_t - \alpha_f) = k/2.303(t)$$

$$\log(\alpha_t - \alpha_f) = -k/2.303(t) + \log(\alpha_0 - \alpha_f)$$

$$\text{slope} = -k/2.303$$

This experiment may be done in student teams, each team determining k for two or more different acid strengths. Acid assignments will depend upon the availability of polarimeters and polarimeter tubes. It is helpful if eight or more conditions are studied.

In general, two types of polarimeters are used: verticals, often with open tubes, and ones that are close to horizontal with closed tubes. Directions are given for both. *Carefully review the directions for the particular polarimeter you will use.*

Experimental Directions

Preparation of the sucrose solution: An aqueous solution with a concentration of 200 g/L may be made for your class. If your team makes its own, start this experiment a week ahead. Add 50 g of sucrose to a 250-mL volumetric flask, and add enough warm water to fill it to about 90% capacity. When the sucrose is in solution and the solution is at room temperature, add water to bring the solution to capacity.

Preparation of the acid solution: Your team will make up two or four acid solutions from one or two assignments in Table 15.1. Follow the directions for the *numbers* of the assignment.

Table 15.1 Acid Solution Assignments

| | "2" | | "4" |
Number	Volume of 12M HCl	Number	Volume of 12M HCl
2A	50 mL	4A	50 mL
2B	46	4B	46
2C	40	4C	40
2D	34	4D	34
2E	30	4E	27
2F	27	4F	20

For Assignments Starting with 2:   Transfer the listed volume of concentrated HCl (12M) from a buret into a 100-mL volumetric flask. Add enough distilled water to reach the neck of the flask with gentle mixing. Carefully, dropwise, add distilled water to the flask until the meniscus comes to the full mark. Label this flask with its molarity as the first solution for that assignment. With a transfer or automatic pipet, transfer exactly 50 mL of the first solution to a clean 100-mL volumetric flask. Complete the dilution as before, and label the flask properly.

For Assignments Starting with 4:   Follow the directions for "2" above to prepare the first solution. To make the second solution, transfer exactly 25 mL of the first solution to a clean, dry volumetric flask, and proceed with the dilution.

Kinetic Runs

A. Traditional Closed Tube Polarimeters   *Calibrate the polarimeter and introduce the reagents.* Determine the zero point of your polarimeter with an empty polarimeter tube. Record this reading in case corrections are needed. (*Note:* The table for recording data should already be in your notebook.) Place 20 mL of the acid solution to be used in one 20 × 200-mm test tube and 20 mL of the sucrose solution in another.

Initiate the reaction. Pour the sucrose solution into the acid, and then pour it all back into the other test tube. Record the time of first mixing as t = 0. As quickly as can be done with care, load this mixed solution into a polarimeter tube. Slide the cover glass on to avoid air bubbles! Read the optical rotation, and note the time. Record this value as the first reading.

Take readings. Take four to seven more readings at 2–5-minute intervals, depending upon how fast the optical rotations are changing. The polarimeter may be shared, but do not share the tubes. The first run should use the most dilute acid solution. After this run is well established, you may start the run with the other acid solution and "alternate" readings. You should be done or almost done before starting the second run. You may repeat these runs if needed or do these runs with another acid assignment. When you are finished with the readings, stopper the solution for storage (labeled) until the next week's laboratory period when the final optical rotation should be measured using the *same* tube. Wash the tubes thoroughly, and leave them to drain.

B. Open Tube Polarimeters (i.e., I^{2}R)   The directions are essentially the same as in (A) except that the volumes are 24 mL each and the sucrose solution starts in the polarimeter tube, so the mixed solution ends up in the polarimeter tube, which is then placed directly into the polarimeter. Do not store solutions in polarimeter tubes.

Recording and Treatment of Data

Your report should include: title, purpose, data tables, calculations, graphs, a discussion of results, and answers to the assigned questions.

Prepare a table in your notebook to record the readings for each run. The final version of this table(s) should be in your report. The table should include the molarity of the acid after mixing, the initial time, the time of each reading, delta t, the observed optical rotation (α_t), α_0, α_f, $\alpha_0 - \alpha_f$, $\alpha_t - \alpha_f$, the logs of these last two, the k for the reading, and the k for the run. Use the optical rotation of the first reading as α_0.

Plot (graph A) log $(\alpha_0 - \alpha_f)/(\alpha_t - \alpha_f)$ or log $(\alpha_t - \alpha_f)$ versus t for each run. From each plot, determine the slope and k for the run, and extrapolate the plot to t = 0 (the times used for the plot are delta t). Collect data from your class, and make a table including acid molarity *after* mixing and k (or average k) for each run. Plot (graph B) k versus the molarity of the acid. Does this plot pass through the origin?

Summary

1. Make or locate the sucrose solution.

2. Make up the acid solutions assigned.

3. Zero or calibrate your polarimeter.

4. Mix solutions and take the first reading.

5. Take several readings and prepare the next run.

6. When the first run nears completion, start the next and alternate readings.

7. Store the reaction mixtures to measure final values later.

8. Prepare table and graphs.

9. Analyze results.

Questions

1. Give the structures for sucrose, D-glucose and D-fructose.

2. What is the definition of specific rotation? Why is it not considered here?

3. Write the steps for the two proposed mechanisms. Why would the kinetics be first order or pseudo–first order?

4. Which of your two runs (which acid concentration) is expected to be faster? Why?

5. How good an approximation is the first optical rotation reading to α_0? Refer to graph A. Could you make a better approximation?

6. Is graph B a straight line? What does this suggest about the mechanism?

7. Why was a final reading taken a week later?

Disposal of Materials

conc. HCl (D, N)   dil. HCl (N)

Sucrose (S)   Sucrose solutions (N)

Additions and Alternatives

If a computer graphics program is available, use it to plot the data and analyze the graphs. Replot graph B with different functions, especially semi-log and log–log plots. See the reference above and/or textbooks of physical organic chemistry.

Use different concentrations of other moderately strong acids. Repeat this experiment with several acids of quite different pK_a values but at the same nominal concentration. This procedure should lead to a Brönsted relationship.

CHAPTER 16

Organic Quantitative Analysis

16.1   Introduction

Many students have the erroneous impression that the laboratory work in organic chemistry consists only of preparative syntheses. Although preparative organic chemistry, as illustrated in the previous chapters, is important, the significance of analytical methods in organic research should not be underestimated. Many of these methods are instrumental methods that have not been illustrated in this text. This chapter and the next, however, will introduce you to some of the techniques in both quantitative and qualitative organic chemistry.

Instrumental analysis and combustion analysis are the quantitative methods most widely used in organic chemistry. Reactions that take place quantitatively in solution are much less commonly employed. Limitations on this type of analysis are imposed by the relatively small number of organic reactions that are quantitative and by difficulties in detecting an end point or in isolating the product quantitatively. Some quantitative techniques have been required for Experiment 8. Although the ester in Experiment 47 should be formed quantitatively, it is not recovered quantitatively. Compare Experiment 70 with the saponification in Experiment 9B.

In Experiment 71, volumetric titration with an outside indicator will be used. This experiment is not included in Chapter 9, although it involves diazotization, because no attempt is made to form the diazonium salts or their derivatives as such. In this experiment, use is made of a reaction that is not only characteristic of primary aromatic amines but also quantitative for them; that is, one equivalent weight of primary aromatic amine will react completely in acidic solution with one mole of sodium nitrite or one liter of 1N sodium nitrite solution.

Question

1. How is the equivalent weight of a substance related to its molecular weight?

E X P E R I M E N T   7 0

Determination of Neutralization and Saponification Equivalents

Neutralization is the quantitative conversion of an acid to its salt by titrating it with a base. The **neutralization equivalent (N.E.)** is the weight of one equivalent of an acid (reacts with exactly one equivalent of base) and is a useful clue in the identification of an unknown acid. For a titration of a pure acid, the number of equivalents of acid equals the weight of acid used divided by its equivalent weight; the number of equivalents of base equals the number of milliliters of base used times the normality, divided by 1000. For quantitative neutralization, the number of equivalents of acid and base must be equal. Thus, the neutralization equivalent is equal to the weight of acid taken times 1000, divided by the mL of base required times its normality.

Question

2. Explain this calculation.

$$\frac{\text{Weight of acid}}{\text{Equivalent weight of acid (N.E.)}} = \text{no. of equivalents}$$

$$= \frac{\text{mL of base} \times \text{Normality (eq/L)}}{1000}$$

$$\therefore \text{N.E.} = \frac{\text{weight of acid} \times 1000}{\text{mL of base} \times \text{Normality}}$$

Saponification is the quantitative hydrolysis of an ester in the presence of excess base. The **saponification equivalent (S.E.)**, like the neutralization equivalent, is the weight of ester that will react with one equivalent of base and is used in the identification of unknown esters. Thus, S.E. = weight of ester × 1000/(mL of base × N of base). The **saponification number** is the number of milligrams of potassium hydroxide (56 g/equivalent) required to react with 1 g of fat or oil. In this experiment, the amount of base consumed is determined by back titration of excess base with standard hydrochloric acid.

Weigh the samples of acids and esters as follows: Place the material in a stoppered weighing bottle (with a dropper in the stopper for volatile liquids or in an open flask with a medicine dropper for nonvolatile liquids), and determine the total weight to the nearest 2 mg. Transfer the appropriate amount of material *quantitatively* to a clean, dry Erlenmeyer flask, and weigh the weighing bottle again with its contents and fittings. If the difference between these weighings is not in the right range, make suitable adjustments until the amount of material in the reaction flask is appropriate and accurately known.

Carefully clean all burets, and rinse them with the standard solutions before filling them with these solutions. Clean them again after use.

Part One: Saponification Equivalents
References: Sections 13.4 and Experiment 9B

$$RCOOCH_3 + KOH \longrightarrow CH_3OH + RCOOK$$

Make the stock solution. The determination of saponification equivalents requires a stock solution of approximately 0.5N potassium hydroxide in ethyl alcohol or in 80% diethylene glycol (*bis*-(β-hydroxyethyl)ether). Dissolve 1.4 g of potassium hydroxide in 50 mL of ethyl alcohol or in a mixture of 40 mL of diethylene glycol and 10 mL of water. If duplicate determinations are to be made, prepare a larger volume of base.

Introduce the reagents and initiate the saponification for the standard run. Weigh between 0.6 and 0.8 g of methyl or ethyl benzoate or ethyl butyrate accurately, and place it in a 25-mL round-bottom flask. Then carefully add 15 mL of the stock potassium hydroxide solution to the flask with a pipet. Reflux the mixture for 30–40 minutes with an air condenser (Figure 1.7, Unit I); if a solid salt forms, add a little water. If diethylene glycol is used in this determination, about half as much time is required for this reflux at a higher temperature.

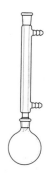

Figure 1.7, Unit I (page 24).

3. Explain why less time is required.

Question

Transfer the hydrolyzed solution to a clean 250-mL Erlenmeyer flask with a pipet. Avoid contact of the basic solution with the ground joint. Rinse the flask with 3–6 mL of water and transfer the wash water to the Erlenmeyer flask with the same pipet. Clean the flask immediately. Directions for titration are given below.

Introduce the reagents and initiate the saponification of unknowns. Weigh a sample of the unknown ester (0.6–0.8 g for simple esters or 1.0–1.4 g for fats and oils); add 15 mL of base and reflux as above. If duplicate trials are to be made, start them as soon as possible.

Standardize the stock solution. To standardize the stock potassium hydroxide solution, transfer a 5-mL portion of the stock solution to a clean 250-mL Erlenmeyer flask, and add 30 mL of water and 6–8 drops of phenolphthalein. Titrate the mixture with 0.1N standard acid solution with nearly constant swirling.

Determine the exact normality of the potassium hydroxide solution from the titration data. Calculate the effective normality from the standard run. Compare the two.

Titrate the refluxed sample. When the refluxing of the ester samples is completed, cool the mixture slightly, transfer as described above, and add 25–30 mL of water and 6–8 drops of phenolphthalein to each sample. Titrate the mixtures with 0.1N standard acid, with stirring or nearly continuous swirling. Acidification beyond the end point with concentrated acid solution should cause the separation of the free acid (odor or precipitate). If the acid component of the unknown ester separates as a solid, your instructor may direct you to collect it and determine its melting point as part of the identification.

Find the saponification equivalent. Calculate the number of equivalents of potassium hydroxide and/or the number of milliliters of potassium hydroxide solution consumed during saponification by subtracting the equivalents of acid from the equivalents of base. Use these calculations to find the saponification equivalent of each ester. Determine the accuracy of your determinations with the known esters.

Question

4. How could this calibration be used to improve the accuracy of the unknown determination?

Summary

1. Prepare the stock solution.

2. Weigh the samples of esters.

3. Add the stock solution and reflux for up to 45 minutes.

4. Carefully and completely transfer the solution to a titration flask.

5. Add water and indicator and back titrate with standardized acid.

6. Make appropriate calculations.

Part Two: Neutralization Equivalent

$$RCOOH + NaOH \longrightarrow RCOONa + H_2O$$

Determination of the neutralization equivalent should be done when time is available during the period (i.e., during reflux), because the saponification may require 45 minutes of refluxing and the saponified samples should be titrated as soon as possible.

Introduce the reagents and titrate with sodium hydroxide. Weigh and transfer samples (0.2–0.4 g each) of benzoic acid for a standard run and of the unknown acid to separate, clean 125-mL Erlenmeyer flasks. Add about 25 mL of water and a few drops of phenolphthalein to each flask, and titrate the solutions with 0.1N standard sodium hydroxide, with constant swirling, to a permanent pink end point. *If the acid does not dissolve in the 25 mL of water,* the addition of a bit more water, or some alcohol (up to 50% or so), or gentle warming (cool before titration) may facilitate the titration.

Question

5. Why would these changes help with solubility problems?

No end point should be accepted while there is still undissolved acid.

6. Why must all the acid be in solution?

Find the neutralization equivalent. Repeat each determination to get reproducible results and from these data calculate the neutralization equivalents of the acids. Compare the value obtained for benzoic acid with its known equivalent weight.

7. How could this calibration be used to improve the accuracy of the
unknown determination?

Summary

1. Weigh the samples of known and unknown acids.

2. Add water and indicator to the samples in their titration flasks.

3. Make sure all the acid dissolves.

4. Titrate the samples with standardized base.

5. Make the appropriate calculations.

QUESTIONS 1–7 ABOVE.

8. Why are alkaline solutions not allowed to stand in burets for extended periods?

9. Calculate the saponification number of the unknown ester even if it was not a fat or an oil.

10. What is the relationship between saponification and the manufacture of soap?

11. From the data for the known acid and ester, determine the accuracy of each method. Explain.

Disposal of Materials

Potassium hydroxide solutions (D, N, I)   Esters and acids (I, OW)

Standard acid (N)   Standard base (N)

Titrated solutions (I, D)

Additions and Alternatives

You may be directed to forego the determination of neutralization equivalents in favor of duplicate determinations of the saponification equivalents. When the equipment is available, you could do the extraction of trimyristin from ground nutmeg (Experiment 9B) and the determination of its saponification equivalent. You must work efficiently in this case, because two hours are required to extract 15–30 g of nutmeg with ether in a small Soxhlet apparatus and because the ester obtained by the evaporation of the ether should be recrystallized twice from alcohol before determining its melting point or saponification equivalent.

E X P E R I M E N T   7 1

Quantitative Diazotization

This experiment consists of three exercises. Part A is the preparation and standardization of 0.1N sodium nitrite stock solution. This part may be omitted because the stock solution, once prepared and standardized, is stable, and your time can be used to run duplicates. Part B is the determination of the amount of procaine hydrochloride (novocaine),

$$[p\text{-}NH_2\text{---}C_6H_4\text{---}COOCH_2CH_2NH(CH_3)_2]^+ Cl^-$$

in a sample of solution. This determination is analogous to determining the purity of an amine whose identity is known or to determining the concentration of novocaine in a saline anesthetic solution.

Question

1. Why might it be impractical to determine the concentration of this anesthetic in a saline solution by doing a chloride determination?

Part C is the determination of the equivalent weight of an amine whose formula is unknown.

Part One: Preparation and Standardization

$$H_2N\text{---}C_6H_4\text{---}SO_3H + H^+ + NO_2^- \longrightarrow HO\text{---}C_6H_4\text{---}SO_3H + N_2 + H_2O$$

Prepare the stock sodium nitrite solution by dissolving 3.5 g of sodium nitrite in distilled water, and dilute it to 500 mL.

Introduce the reagents. Accurately weigh about 0.7 g of dried sulfanilic acid (0.6927 g equals 0.004 mole); place this acid in a 250-mL Erlenmeyer flask, and dissolve it in 100 mL of water in which there is 0.2 g of sodium hydroxide. The sulfanilic acid (from the monohydrate) should have been dried for three hours at 120°C and allowed to cool before the laboratory period. Add to the sulfanilic acid–sodium hydroxide solution 20 mL of concentrated hydrochloric acid, dissolve 1 gram of sodium bromide in the mixture, and cool to 15°C.

Titrate the sulfanilic acid solution. Rinse a buret with stock solution, discard the rinsings, and fill the buret with the stock nitrite solution. Run about 35 mL of the stock sodium nitrite solution into the sulfanilic acid solution.

2. Why can so much be run in at once?

Question

Swirl the solution and transfer a small drop of it to a piece of starch–iodide paper, using a clean stirring rod. A blue spot will immediately appear if there is an excess of nitrite in the solution.

3. Explain the chemistry of this test, using the fact that the nitrite ion is reduced.

Question

If there is no blue spot, continue the addition of the nitrite solution, and test the sulfanilic acid solution after each addition (mix completely) until the immediate blue spot appears. As the end point is approached, each successive addition of sodium nitrite solution should be only a drop or two. All other end points should duplicate this immediate blue spot in color intensity.

Repeat the standardization. Use the information obtained the first time to speed up the procedure.

Make a blank determination. Place 105 mL of water, 15 mL of concentrated hydrochloric acid, and a trace of sodium bromide in a 250-mL Erlenmeyer flask. Add the sodium nitrite solution to this flask dropwise from the buret, and test the solution in the flask after each addition with external starch–iodide paper. Record the volume of stock solution required for a positive test as the volume of the blank, and subtract this volume from the volumes of stock solution used to titrate sulfanilic acid solutions before making other calculations.

Calculate the concentration of the nitrite solution for each determination. They should agree within about 1 percent. If possible, all the chemicals should be U.S.P. grade or better. If standardized sodium nitrite solution is available, this part may be studied and omitted.

Summary

1. Prepare the stock nitrite solution.

2. Weigh the dried sulfanilic acid.

3. Dissolve it in weak aqueous base.

4. Add conc. HCl, dissolve NaBr in the mixture, and cool to 10–15°C.

5. Titrate to an end point determined by an external indicator.

6. Repeat the standardization.

7. Make a blank determination.

8. Make appropriate calculations.

Part Two: Determination of Procaine Hydrochloride

$$R-C_6H_4-NH_2 + H^+ + NO_2^- \longrightarrow R-C_6H_4-OH + N_2 + H_2O$$

Introduce the reagents and titrate a blank. Place, in a 125-mL Erlenmeyer flask, 20–25 mL of water, 2 mL of concentrated hydrochloric acid, and a trace of sodium bromide. Titrate this blank according to the procedure used for the blank in Part A, and record the volume used.

Titrate the procaine hydrochloride solution. Place, in another 125-mL Erlenmeyer flask, 20–25 mL (accurately measured) of a procaine hydrochloride solution whose concentration is unknown. Use a pipet or buret to measure the transferred solution accurately. Add to this solution 2 mL of concentrated hydrochloric acid and a gram of sodium bromide, which acts as a catalyst in the diazotization. The solution should test acidic (ca. pH = 3) with external indicator paper. Titrate this mixture, at room temperature or below, to the usual immediate-blue-spot end point with the standardized nitrite solution. Make a duplicate determination with a fresh sample of amine solution.

It may be faster to do three determinations, using the first as a rough approximation that will allow the two accurate titrations to be done rapidly. Carry out the first titration with 3-mL, 4-mL, or 5-mL additions until the end point is passed. If one of these large additions is just short of the end point, it may be possible to use this as an accurate determination. In an accurate determination, the volume of the last addition, the one that produces a test for the end point, should be only a drop or two.

Question

4. Explain how a rough approximation will save time in the next titration.

Interpret the results. Include in your report the volumes of the blank and of the samples and nitrite solution used in each of the accurate titrations as well as in the calculations. In the calculations, use the net volumes of nitrite solution and its normality with the volumes of the samples to determine the normality or molarity of the amine hydrochloride solution and the weight of the anesthetic in a liter of solution (procaine hydrochloride, MW = 272.8).

Summary

1. Titrate a blank.

2. Place a carefully measured sample of unknown solution in the flask.

3. Add acid and catalyst.

4. Titrate to the usual end point.

5. Titrate one or two duplicates.

6. Make appropriate calculations.

Part Three: Determination of Equivalent Weight of an Amine

$$R—C_6H_4—NH_2 + H^+ + NO_2^- \longrightarrow R—C_6H_4—OH + N_2 + H_2O$$

Introduce the reagents and titrate. Place an accurately weighed sample (about 0.5 g) of an unknown amine in a 250-mL flask. Dissolve the sample in 10 mL of water and 4 mL of concentrated hydrochloric acid, with gentle warming. Add a little glacial acetic acid to complete the process if necessary. Dilute the solution with 75–100 mL of water; stir in 1 g of sodium bromide, and cool the mixture to 10–15°C. Check the acidity of the solution with external indicator paper. Titrate the solution with standardized sodium nitrite solution as outlined in Parts A and B. Note that the molecular weight of the amine will usually be between 100 and 200. Add about 20–25 mL of nitrite solution in one portion at the beginning of the titration.

5. Why so much added at once?

Question

Make a duplicate determination of the equivalent weight of the unknown amine.

Determine the blank correction factor for this part of the experiment by titrating a blank solution containing those components of the solutions titrated above, except the amine.

Interpret the results. Report the weights of the samples, the volumes used to titrate them, and the volume of the blank. Use the net volumes from the titrations to calculate the equivalent weight of the unknown amine. Show the complete calculations.

Summary

1. Titrate a blank.

2. Place a carefully weighed sample of unknown in the flask.

3. Dissolve in water and conc. HCl.

4. Dilute with water, dissolve in NaBr, and cool to 10–15°C.

5. Check acidity and make it acidic if it is not.

6. Titrate this sample and duplicate.

7. Make appropriate calculations.

Questions

QUESTIONS 1–5 ABOVE.

6. Give the complete equations for the organic reactions in Parts One and Two.

7. How is the equivalent weight of a substance related to its molecular weight?

8. Contrast these titrations with the diazotizations in Experiments 42 and 43.

Disposal of Materials

NaNO$_2$ (I, S)

NaOH (I, D, N)

NaBr (S)

Unknown amine (caution, HW, I)

aq. Sodium nitrite (I, react or D)

conc. HCl (D, N)

Procane hydrochloride solution (I)

Glacial acetic acid (D, N)

Titration solutions (I, evaporate, HW)

Additions and Alternatives

The titrations in this experiment have been suggested by Dr. Enno Wolthuis. He has also given an interesting discussion of the extension of this method in determining aromatic nitro compounds quantitatively (*Anal. Chem.*, **1954**, *26*, 1238), which you might wish to try.

CHAPTER 17

Organic Qualitative Analysis

17.1   Introduction

For many experiments in this book, you were directed to verify the identity of the product of a reaction. In these cases, the reactions yielded known products, i.e., expected compounds that have previously been identified. You had to determine certain physical properties, for which directions were given, in order to verify the identity of the compound in question. Occasionally, you also prepared derivatives.

In this chapter, you will be asked to take a more comprehensive approach to the identification of chemical compounds: organic qualitative analysis. In a research situation, you may have a compound to identify for which no directions are given. In this case, you must decide which physical and chemical properties to determine systematically, and then use this information to identify the substance. It is possible that the compound has already been identified and that its physical and chemical properties are reported in the literature. Then it is only necessary for you to match the experimental and reported values to verify the structure of the compound. On the other hand, it may be that the compound has never before been reported and that it is necessary for you, as the chemist, to identify the compound using your own experimental work and deductions.

In this chapter, you will be introduced to the identification of unknowns that have previously been reported. You will examine the compound in a systematic fashion and decide which of many possible tests will give the most information about its structure. Then, using the information gathered, you will match the data with known literature values using such references as Rapport's *Handbook of Tables for Organic Compound Identification*.

17.2   General Approach to Identifying Unknowns

The complete identification of a chemical compound *may* include seven activities.

1. Determination of physical properties: physical state, color, odor, refractive index, density, melting or boiling point.
2.. Purification and the determination of purity: melting point, recrystallization, distillation, thin layer chromatography, gas liquid chromatography.
3. Elemental analysis.
4. Determination of solubility group.
5. Spectroscopy: infrared to suggest the presence or absence of various functional groups and, possibly, nuclear magnetic resonance to suggest aspects of the structure.
6. Chemical tests to confirm which functional group(s) is (are) present.
7. Preparation of one or two derivatives whose melting points may be compared with known literature values.

For use in the last step, reference materials should be available in the laboratory or in the library. A summary of this general approach is shown as a flow chart in Table 17.1.

The goal of this process, then, is to obtain the data that, when compared with data found in the literature, will identify an unknown compound. The identification process may best be seen in reverse. The references commonly used list the possible compounds and their melting or boiling points, arranged by functional group class. To know what table to look in or what derivatives to make, you first need to know to what class of organic compounds your unknown belongs. This knowledge is established by using chemical, functional group tests. Because there are many such tests, it would be a waste of time to try them all on your unknown compound. In order to pick tests that will give significant results, you will need to gain hints from data collected by elemental analysis, solubility analysis, and spectroscopic analysis. As with the functional group tests, you need not always do all three of these; one or two may suffice, depending upon the knowledge gained from them. The results of these analyses may not be trusted, however, if the compound is not pure. First purify the compound, then determine the melting or boiling point, and then perform these analyses. You must know the melting or boiling points in order to look up derivative values in the reference tables. When looking up values for melting points, consider compounds melting five to eight degrees above and below your observed range. With liquids, consider an even greater spread of values.

To ensure as much flexibility as possible, the following material is divided into three review units (I, II, and III) and five experiments (72–76), each of which contains one or more exercises. Each unit or experiment may be done separately or as part of a complete unknown identification process. Each of the experiments, with appropriate knowns and unknowns, should occupy about one laboratory period. Before doing an

Table 17.1   General Flow Chart for Qualitative Analysis

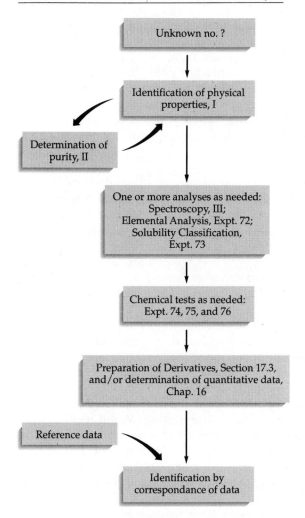

experiment, you should complete the equations for the chemistry of the tests involved, using known compounds. Test the knowns and the unknowns and compare their behavior under similar conditions. These comparisons should make it possible for you to obtain useful information about the unknown.

I.   Identification of physical properties
A.   *Physical State*
Note whether the compound being studied is a solid or a liquid. The physical state of the compound gives some very general information about

the compound because melting and boiling points generally increase with increasing molecular weight.

B. *Color*

Most pure organic compounds with which you will have contact are colorless liquids or white crystalline solids. However, the lack of color in a compound does not guarantee its purity nor does the presence of color necessarily imply a lack of purity. Color in an organic compound can be due to the presence of a functional group that imparts color and/or to the presence of extensive electron delocalization. Colored contaminants may form due to slow oxidation by air. A little colored impurity may make the sample appear quite colored.

C. *Odor*

There are no tables listing common organic compound odors because there are no precise descriptions of odors. However, many organic compounds have characteristic odors, and it is sometimes possible to match the odors of knowns and unknowns. Just remember, you should *be very cautious in smelling compounds* because they are frequently toxic.

D. *Ignition Test*

In a hood with the face pulled down or behind a safety shield, place 0.1 g of the compound in a porcelain crucible, and heat it, gently at first and then to a dull redness. Notice:

1. Is the compound flammable?
2. What is the color of the flame if the compound does burn? Is it sooty?
3. If the compound is a solid, does it melt?
4. Is the compound explosive?
5. Are gases evolved? What is the odor? (*Caution!*)
6. Does a residue remain after heating? Will it melt?

If a residue remains, add a drop of distilled water to it and test the solution with litmus paper. Add a drop of dilute hydrochloric acid. Is a gas evolved? Use a platinum wire to carry out a flame test on the acid solution to determine if a metal is present (Moeller, *Quantitative Analysis*, McGraw-Hill, New York, 1958). Always use known compounds in the test as comparison controls. Some suggested knowns include acetone, toluene, sucrose, barium benzoate, copper acetate, and sodium potassium tartrate.

E. *Refractive Index*

The index of refraction, a measurement of the velocity of a light wave in air compared with its velocity in a liquid, is a value characteristic of each individual compound. As such, the index of refraction of a liquid can be used as a means of identification by comparing an experimental value with a literature value. Specific directions for determining the index of refraction are given in Chapter 3, Section 3.8.

F. *Density*

Density equals the mass of a substance divided by its volume. The density of a liquid is a value characteristic of an individual compound and can be used to verify the identity of a compound by comparing its experimental value with the literature value. Specific directions for determining the density are given in Chapter 3, Section 3.7.

The experimental value obtained for the density of a liquid gives some information about the structure of the liquid. Alkanes, alkenes, and alkynes have densities less than that of water, 1 g/mL. Their densities increase very slightly as the carbon chain increases in length. Alkyl fluorides and chlorides have densities less than that of water, whereas alkyl bromides and iodides are expected to have densities greater than that of water. The densities of alkyl halides decrease as the length of the carbon chain increases. If a compound increases its number of halogen atoms, its density increases. For example, two chlorine atoms or a chlorine and an oxygen atom, or a chlorine atom and an aryl group may all cause the density of the compound to be greater than 1 g/1 mL.

The order of densities in compounds containing oxygen is generally:

$$ROR < RCH_2OH < R\overset{\displaystyle O}{\overset{\displaystyle \|}{C}}H < CH_3\overset{\displaystyle O}{\overset{\displaystyle \|}{C}}OR < R\overset{\displaystyle O}{\overset{\displaystyle \|}{C}}OH < RO\overset{\displaystyle O}{\overset{\displaystyle \|}{C}}{-}\overset{\displaystyle O}{\overset{\displaystyle \|}{C}}OR$$

The densities of the ethers, primary alcohols, aldehydes, and alkyl acetates are all less than 1 g/mL. The densities of the ethers, alcohols, and aldehydes generally increase with an increasing number of carbon atoms, whereas that of the alkyl acetates, carboxylic acids, and esters of oxalic acid all decrease with an increasing number of carbon atoms. The densities of the carboxylic acids, formic and acetic acid, are greater than 1 g/mL because of hydrogen bonding association, whereas the densities of the higher molecular weight carboxylic acids decrease to less than 1 g/mL. Esters containing a halogen, a ketone, the hydroxyl group, and possibly the aryl group generally have densities greater than 1 g/mL.

If compounds contain more than one functional group, generally they will have a density greater than 1 g/mL. If a compound doesn't have a halogen and its density is less than 1 g/mL, it probably contains one functional group. A compound that has a density greater than 1 g/mL is probably polyfunctional and/or contains halogens.

G.   *Melting Point*

The melting points of many organic solids occur over a narrow range, e.g., 1–2°C (the melting range is the number of degrees between the point at which melting begins and the point at which it ends). Generally, the narrower the range, the purer the compound. However, there are some organic compounds—e.g., amino acids, salts of acids or amines, and carbohydrates, among others—that generally melt over a wide range because of accompanying decomposition. Specific directions for determining the melting point of a solid are given in Chapter 2, Section 2.2. If the melting range is greater than 2–3°C, an attempt should be made to purify the compound by recrystallization.

H. *Boiling Point*

The boiling temperature of a compound, in addition to being characteristic of a compound, gives some information about the molecular weight and the amount of association that exists between molecules. The higher the molecular weight of the compound, the higher will be the boiling point. The greater the amount of intermolecular association that exists between the molecules, the greater will be the boiling point. For example, alcohols (ROH, hydrogen bonding) associate to a much grater degree than do thiols (RSH) and consequently boil at a higher temperature, even though the molecular weight of the thiol analog is greater. Neither oxygen ethers (ROR) nor thiol ethers (RSR) are capable of hydrogen bonding, and consequently the thiol ethers, because of their larger molecular weight, boil at a higher temperature than do the corresponding oxygen ethers.

Specific directions for determining the boiling point of a liquid are given in Chapter 3, Section 3.3. If a liquid boils over a range wider than 5°C, the material should be purified by distillation and its boiling point taken again. The distillation temperature is one approximation of the boiling point (Section 3.4).

II. **Determination of Purity (additional methods)**

A. *Thin Layer Chromatography (TLC)*

Using thin layer chromatography is one of the easiest and fastest ways to determine the purity of a compound. You merely place the sample on a thin layer plate and allow a solvent to carry the sample along much of the length of the plate in a closed system. Different compounds will be carried different distances along the plate by the solvent. If more than one spot is seen on the plate, the presence of an impurity is indicated. Refer to Chapter 4, Section 4.2C, for specific directions for using thin layer chromatography. This method is especially useful for solids, nonvolatile liquids, or heat-sensitive compounds, i.e., those compounds that can not be examined by gas chromatography.

B. *Gas Liquid Chromatography*

Gas liquid chromatography is a method used to separate volatile compounds. A sample is injected into a hot gas flow that carries the vapor along a column. Different compounds travel at different rates along the length of the column, depending upon their nature. If more than one component peak is observed, the compound may be said to be impure. Refer to Chapter 4, Section 4.2D, for more information about gas liquid chromatography.

III. **Spectroscopy**

A. *Infrared*

Infrared spectroscopy of an organic compound can provide a tremendous amount of information about the structure of the molecule. Absorption of infrared radiation is often related to the functional groups present in a molecule. For example, if a carbon–oxygen double bond, i.e., a carbonyl group, is present in a molecule, strong absorption of radiation occurs generally in the range of 1650 to 1850 cm^{-1}; the specific absorption values gives some information about the type of functional group containing the carbonyl. For example, a compound containing a carbonyl group could be a carboxylic acid, an ester, an acyl halide, an amide, an aldehyde, or a ketone. If strong absorption in the 1650 to 1850 cm^{-1} region is missing, all of the above types of compounds could be eliminated from consideration.

Combining infrared data with solubility test data and/or elemental analysis should suggest the functional group(s) present or absent in your unknown. These suggestions will lead you to possible confirmatory tests. Many of these tests are found in Experiments 74–76.

Refer to Chapter 5, Section 5.1, for a discussion of analysis using infrared spectroscopy.

B. *Nuclear magnetic resonance*

Nuclear magnetic resonance spectroscopy studies of an organic compound can provide vast amounts of information about the structure of the compound. Nuclear magnetic resonance (nmr) spectroscopy does not generally distinguish between various types of functional groups, as does infrared spectroscopy, but once the functional groups are known, nmr provides information on how various groups or atoms within the molecule are arranged in relation to each other. To use a very simple example, given a molecular formula of C_3H_8O, it would be a simple task to decide, using nmr, whether the component is 1-propanol ($CH_3CH_2CH_2OH$), 2-propanol ($CH_3CHOHCH_3$), or methyl ethyl ether ($CH_3OCH_2CH_3$).

Generally an nmr spectrum would be taken after one (or some) of the Exercises in Experiments 74–76 have been done to indicate or confirm the presence of a particular functional group(s). It may be used in conjunction with the preparation of derivatives, Section 17.2, for the positive identification of a particular member of a functional group family.

Directions for utilizing nmr spectroscopy are given in Chapter 5, Section 5.2.

E X P E R I M E N T   7 2

Elemental Analysis: Analysis for Nitrogen, Sulfur, and Halogens

A chemist gains much information about the identity of a compound by learning whether the common elements nitrogen, sulfur, and the halogens (fluorine, chlorine, bromine, and iodine) are present. It may be assumed that carbon and hydrogen are present in any organic compound. The performance of a sodium fusion test on the compound (to convert sulfur to S^{-2}, nitrogen to CN^{-1}, and halogen to X^{-1}), followed by tests to detect the presence or absence of these ions, provides the desired information.

WHEN CARRYING OUT THE FUSION, YOU SHOULD WEAR A FACE SHIELD OR GOGGLES AS WELL AS SAFETY GLASSES.

Assemble the apparatus and introduce the reagents. Support a clean, dry Pyrex test tube (6-inch, reserved or designated for this purpose) through a hole in a protective shield on an iron ring and/or by using a clamp lined with insulating material. Place a dry cube of freshly cut sodium, about 4 mm on the edge, in the test tube. Heat the tube with a hot flame. This should cause a mushroomlike cloud of sodium vapor to form in the lower end of the test tube. Drop 2–3 drops of the liquid or 0.2 g of the solid to be

tested into this cloud of vapor. THE INSTANTANEOUS REACTION MAY BE QUITE VIGOROUS (HEAT, LIGHT, SLIGHT EXPLOSION). Heating the tube to a bright redness for a minute should complete the reaction. When the tube has cooled to about room temperature, add enough methanol to destroy the excess sodium (about 2 mL). Add distilled water to bring the volume up to a little over half that of the test tube. Boil the tube for a few minutes to bring the ions into solution. Filtration should give 12–15 mL (10 mL minimum) of clear, colorless filtrate for your use in the anion tests.

$$\text{``C,H,O,N,S,X''} \xrightarrow{\text{Na}} NaX + Na_2S + NaCN + H_2O \qquad (X = \text{halogen})$$

Test for sulfur. Acidify one-tenth of the filtrate (or 1 mL) with acetic acid; add a few drops of 5% lead acetate to this solution, which should produce a black precipitate of lead sulfide if the sulfide ion is present.

$$Na_2S + Pb(O\overset{\overset{O}{\|}}{C}CH_3)_2 \longrightarrow PbS\downarrow + 2CH_3\overset{\overset{O}{\|}}{C}ONa$$
black

Test for nitrogen. Add 5 drops (or 7 if sulfur is found to be present) of freshly prepared 5% ferrous sulfate (or 0.05–0.1 g of powdered $FeSO_4$) and 5 drops of 10% potassium fluoride to a new 3 mL (or three-tenths) portion of the filtrate, and bring this solution to a boil. Filter the solution while hot if sulfur was indicated in the previous test. Acidify the cooled solution or filtrate with dilute sulfuric acid after 2 drops of 5% ferric chloride solution have been added. The formation of a deep blue precipitate or solution is a positive indication of nitrogen. The test is quite sensitive but occasionally does not give good results due to a lack of cyanide ion formation during fusion.

$$18CN^- + 3Fe^{+2} + 4Fe^{+3} \longrightarrow Fe_4[Fe(CN)_6]_3$$
blue

Test again for nitrogen. A more sensitive second test may be used to detect the presence of cyanide ion and ultimately of nitrogen in the organic compound. Use it if the previous test was negative. In a small test tube, place 1 mL of each of the following solutions: a 15% solution of *p*-nitrobenzaldehyde in 2-methoxyethanol and a 1.7% solution of *o*-dinitrobenzene in 2-methoxyethanol. Add 2 drops of a 2% sodium hydroxide solution in distilled water, with mixing, to the previous solutions. Finally, add 2 drops of sodium fusion filtrate to the test tube. The formation of a deep blue-purple color is evidence for the presence of nitrogen in the original compound. The appearance of a tan or yellow color is a negative test. If a yellow color is formed, check to make sure the solution is basic. If not, make it basic and observe the color.

blue-violet dianion

The test (Shriner et al., *The Systematic Identification of Organic Compounds*, 6th Ed., Wiley, New York, 1980, p. 80) is valid in the presence of sodium halide or sodium sulfide, which will be present if the original compound contained halogen or sulfur.

Test for the halogens. Since the general test for halogens (Cl, Br, I) involves precipitation with silver ion, any sulfur and cyanide ions present must be removed before the halogen test is carried out because silver sulfide and silver cyanide may also precipitate. **HOOD!** Acidify two-tenths of the filtrate (or 2 mL) with dilute nitric acid. Then, boil the solution for a minute or two to rid it of sulfide and cyanide ions, if the above tests were positive for those ions. A few drops of 0.1 M silver nitrate solution should cause a white or pale yellow precipitate that darkens rapidly with exposure to light, if the halogens are present.

CAUTION

$$AgNO_3 + X^- \longrightarrow AgX\downarrow + NO_3^-$$

AgCl is white
AgBr is pale yellow
AgI is yellow

If a halogen was indicated by this test, acidify four-tenths (4 mL) of the filtrate with sulfuric acid and boil it if H_2S and HCN are to be removed. Add about one-half of a milliliter of 1-octanol or methylene chloride and then fresh chlorine water dropwise (acidified commercial hypochlorite bleach will do nicely, provided the solution gives a positive result when tested with starch–iodide paper). The production of a purple color in the organic layer indicates the presence of iodine.

$$2\,NaI + Cl_2(H_2O) \longrightarrow 2\,NaCl + I_2 \text{ (organic solvent)}$$
$$\text{purple}$$

Continue adding the chlorine water dropwise with thorough shaking after each drop. The organic layer will become red brown if bromine is present. For other approaches, see the reference above.

$$2NaBr + Cl_2(H_2O) \longrightarrow 2\,NaCl + Br_2 \text{ (organic solvent)}$$
$$\text{red-brown}$$

You should test enough knowns so that you will have seen each positive test, including at least one compound that has two or more of these elements.

Summary

1. Support an insulated, shielded tube.

2. Introduce a dry cube of sodium with fresh-cut surfaces.

3. Heat the tube to begin the vaporization of the sodium.

4. *Caution*: Drop the sample into the vapor.

5. Heat the tube to redness for one minute.

6. Cool to room temperature or below and add about 2 mL of methanol.

7. Stir cautiously and then add 10–12 mL of water.

8. Filter the aqueous solution.

9. Test the acidity of 1 mL or so of the filtrate (8) (add drops of acetic acid until it is acidic).

10. Add lead acetate solution dropwise; a black solid suggests PbS.

11. Add, to 3 mL or so of the filtrate (8), 5 drops KF solution and $FeSO_4$.

12. Bring to boil, filter hot, and cool.

13. Add 2–3 drops of $FeCl_3$ solution and then acidity with dil. H_2SO_4.

14. Observe if a blue-black complex of cyanide is formed.
    (Alt. 11–14) Mix the organic reactants in their solvents.
    Add dil. NaOH.
    Add a few drops of filtrate (8).

15. *In the hood* add diluted HNO_3 to acidify 2 mL of filtrate (8).

16. If S or N have been suggested, boil the acid solution for two minutes.

17. Add $AgNO_3$ solution dropwise and look for a solid to form.

18. *Hood.* If (17) is positive, acidify 4 mL of filtrate with dil. H_2SO_4.

19. Boil if you did in (16).

20. Add an organic solvent and then Cl_2/H_2O dropwise.

21. Shake and allow to settle after every 2–3 drops.

22. Watch for successive color changes in the organic layer.

Questions

1. Why must enough methanol be added to react with all of the excess sodium?

2. Suggest a reason for using Pb^{++} ion instead of Cu^{++} ion to test for sulfide.

3. Why are the sulfide and cyanide ions removed before the general halogen test is made?

4. Describe the appearance and changes therein of AgCl, AgBr, AgI, Ag_2S, and AgCN.

5. If the general test for halogens is positive and those for bromine and iodine negative, what does this imply about chlorine?

6. If the general test for halogens is positive and that for bromine is positive, what does this imply about chlorine?

Disposal of Materials

Sodium (I, return to storage solvent (**CAUTION**), or
    react with 2-propanol, ethanol and N)

Methanol (OS, OW, I, D)

Lead acetate (HW, container, base, Na_2S, I)

$FeSO_4$ (S)

Lead acetate reaction mixture (I, evaporate, S)

aq. KF (I, container, D)

aq. $FeSO_4$ (I, container, D)

p-Nitrobenzaldehyde solution (I, OW, evaporate)

aq. $FeCl_3$ (I, container, D)

o-Dinitrobenzene solution (I, OW, evaporate)

aq. $AgNO_3$ (container for recovery)

Chlorine water (I, evaporate or bisulfite, N)

aq. NaOH (N)

Organic solvents (used and/or with halogen) (I, HO, OW)

dil. HNO_3 (N)

1-Octanol (OS, OW, I)

Methylene chloride (OS, HO, I)

EXPERIMENT 73

Solubility Classification

Determining which solvents dissolve unknown compounds provides valuable information about the structure of those compounds. Certain solvents are used in a systematic analysis of the solubility behavior of unknowns, which allows organic compounds to be divided into seven groups according to their solubilities, as shown in Table 17.2. An outline of the solubility classification procedure is given in Table 17.3.

Procedure for Determination of the Solubilities of Organic Compounds

Compounds are generally considered soluble if 0.2 mL of a liquid or 0.1 g of a solid dissolves completely in 3 mL of the appropriate solvent. This criterion is applied quantitatively with water and ether. With the remaining reagents, a clear indication of reaction constitutes a positive test.

Solubility in water *Place* 0.1 g of powdered solid in a small test tube, and add three successive 1-mL portions of water, shaking well after each addition. If the solid does not appear to dissolve, heat the test tube. If it

Table 17.2 Determination of Solubility Group

Group	Solubility	Types Often Found in Group
A	Soluble in water. Soluble in ether.	Lower members of a homologous series of oxygen- and nitrogen-containing compounds; acids and derivatives; phenols, aldehydes, ketones, alcohols, amines, and nitriles.
B	Soluble in water. Insoluble in ether.	Polyoxy, ionic, or easily ionized compounds; hydroxyacids, polyacids, polyols, sugars, amides, polyamines, amines, alcohols, sulfonic acids, and salts.
C_1	Insoluble in water. Soluble in 5% NaOH. Soluble in 5% $NaHCO_3$.	Strong organic acids, carboxylic acids with more than six carbons; phenols with electron-withdrawing groups in the ortho and para positions, and β-diketones.
C_2	Insoluble in water. Soluble in 5% NaOH. Insoluble in 5% $NaHCO_3$.	Weak organic acids, phenols, enols, oximes, imides, sulfonamides, five carbons; β-diketones; and nitro compounds with α-hydrogens.
D	Insoluble in water or 5% NaOH. Soluble in 10% HCl.	Basic compounds, amines, hydrazines.
E	Insoluble in water, 5% NaOH, or 10% HCl. Contain no N or S. "Soluble" in cold conc. H_2SO_4.	Unsaturated hydrocarbons, highly alkylated aromatic hydrocarbons, aldehydes, ketones, acid derivatives, alcohols, ethers, acetals, and halogen derivatives of these oxy compounds. (See G*).
F	No N or S. Insoluble in all of the above, including H_2SO_4.	Hydrocarbons, cyclic, aliphatic, and aromatic, and their halogen derivatives; diaryl ethers.
G*	Contain N or S. Not in Groups A–D.	The rest of the sulfur and nitrogen compounds.

*These compounds usually will appear in (E) if tested.

Table 17.3 Outline of Solubility Classification Procedure

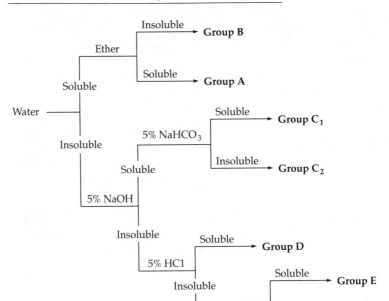

then dissolves, cool the solution to room temperature; shake it to prevent super saturation, and seed it to see if the compound is really soluble at room temperature. To test a liquid, add 0.2 mL (with a graduated pipet) to 3.0 mL of water, and shake well. Sometimes with two colorless insoluble liquids, it is possible to overlook the boundary of separation between the phases. To be certain that this is not the case, shake the test tube well after the liquid seems to have dissolved. If two phases are present, the solution will become cloudy.

If it does dissolve, determine the pH by testing with indicator paper.

Solubility in ether *Proceed as described above for solubility in water*, but do *not* heat.

Solubility in 5% NaOH Solution *Proceed as described above for solubility in water*, but do *not* heat.

Note whether there is an increase in temperature or other sign of reaction when the reagents are mixed. If the unknown does not appear to dissolve, it is probably a negative test, but it may be that the salt formed is not very soluble. If you are in doubt, remove some of the aqueous layer with a medicine dropper or pipet; place it in a small test tube, and add dilute HCl dropwise with stirring until the solution is acidic to litmus paper. If any precipitation or second layer forms (cloudiness?), assign the compound to Group C_1 or C_2.

Solubility in 5% NaHCO₃ solution   *Proceed as described above for solubility in water,* but do *not* heat.

Observe whether the unknown dissolves and, if so, whether carbon dioxide is generated immediately (evidence for carboxylic acids, sulfonic acids, negatively substituted phenols) or after a short while (as occurs with some amino acids).

Solubility in 5% HCl solution   *Proceed as described above for solubility in water,* but do *not* heat.

Note whether there is an increase in temperature or other sign of reaction when the reagents are mixed. If the unknown does not appear to dissolve, it is probably a negative test, but it may be that the salt formed is not very soluble. If you are in doubt, remove some of the aqueous layer with a medicine dropper or pipet, place it in a small test tube, and add 5–20% NaOH dropwise with stirring until the solution is basic to litmus paper. If any precipitation or second layer forms (cloudiness?), assign the compound to Group D.

Solubility in Concentrated H₂SO₄ Solution   *Place 3.0 mL of concentrated H₂SO₄ in a dry test tube, and add* 0.1 g of a solid or 0.2 mL of a liquid. If the unknown doesn't immediately dissolve, shake the tube. Note any changes in color or temperature, formation of a precipitate or of a gas. Such indications of a positive reaction classify the unknown substance as "soluble." Assign it to Group E even though the substance is not "soluble" in the usual sense.

All results should be recorded in a table similar to Table 17.4. Be sure to note the tests *not* done!

Table 17.4   Solubility Table

Compound	H_2O	Ether	5% NaOH	5% NaHCO₃	5% HCl	conc. H_2SO_4	Solubility Group
Benzoic acid	–	*	+	+	*	*	C_1
n-Hexane	–	*	–	–	–	–	F

*Test not done

Before beginning the solubility unknowns, you should test the following knowns: (1) ethyl benzoate or benzyl alcohol, (2) acetone, (3) bromobenzene, (4) phthalimide (only slightly soluble in cold water), (5) glucose or sucrose, (6) benzoic acid, and (7) dimethylaniline. This group contains an example of each class.

Summary

1. Test your compound with water.

2. If soluble, determine pH and go to (3) if it is *not* soluble, go to (5).

3. Test a new sample with ether.

4. If soluble: *Group A*; if not: *Group B*.

5. Test a new sample with 5–10% NaOH.

6. If it is soluble or reacts, go to (7); if *not* go to (9).

7. Test a new sample with 5–10% $NaHCO_3$.

8. If soluble or reactive: *Group C_1*; if not: *Group C_2*.

9. Test a new sample with 5–10% HCl.

10. If soluble or reactive; *Group D*; if not go to (11).

11. Test a new sample with conc. H_2SO_4.

12. If reaction occurs: *Group E or G*; if not: *Group F*.

NB: For all exercises, test each sample separately.

Questions

1. Why would a higher molecular weight alcohol be soluble in concentrated H_2SO_4?

2. How would *n*-amyl alcohol and neopentyl alcohol compare in water solubility?

3. What might you learn if you test a salt from Group B with NaOH and/or HCl?

4. What would be expected of a compound in Group D when tested with H_2SO_4? Would this be a good use of time?

5. Why is phenol, C_6H_5OH, more soluble in water than is naphthol, $C_{10}H_7OH$?

6. Explain why you did not do certain tests in Table 17.4.

Disposal of Materials

Knowns and unknowns (I, OW or HO)   Aqueous test solutions (I)

aq. NaOH and HCl (N)   aq. $NaHCO_3$ (I, D)

conc. H_2SO_4 (D, N)

E X P E R I M E N T   7 4

Functional group tests 1

Exercise 1: Test for Reaction with Br_2; Unsaturation

Reagent: 5% Br_2 in 1-octanol or CH_2Cl_2 (or 1% Br_2 in water)

Knowns: 0.2 g (or 0.2 mL = 4–5 drops) cyclohexene (or hexene or amylene), toluene (or xylene), phenol (or *p*-bromophenol), cinnamic acid, and benzaldehyde.

$$\underset{/}{\overset{\backslash}{C}}=\underset{\backslash}{\overset{/}{C}} + Br_2 \longrightarrow Br\overset{|}{\underset{|}{C}}-\overset{|}{\underset{|}{C}}Br$$

   Dissolve the sample to be tested in 2 mL or less of 1-octanol or CH_2Cl_2. Add the reagent dropwise until the red-purple color of bromine remains after shaking, or until 25 drops have been added. Be sure to distinguish between decolorization (a positive test) and dilution. *Is a gas evolved (substitution, not addition)?* Blow your breath across the mouth of the test tube or use moist blue litmus paper.

Exercise 2: Test for Oxidation ($KMnO_4$); Unsaturation

Reagent: 2% $KMnO_4$ in water

Knowns: Same as for Exercise 1.

$$\text{Unknown} + MnO_4^- \longrightarrow MnO_2 + \text{oxidized product}$$

   Dissolve the sample to be tested in a minimal amount of water or acetone (up to 2 mL). Add the reagent as in Exercise 1. The formation of a brown solid indicates a positive test. When it appears that the purple color of permanganate may obscure the brown precipitate (*What is this precipitate?*), a drop of solution on filter paper will reveal it if it is present along with the purple permanganate.

Exercise 3: Test for Labile Halogen

Reagent: 2% $AgNO_3$ in 95% ethanol

Knowns: 0.1 g or 2–3 drops chlorobenzene, *n*-butyl chloride, allyl bromide, benzyl chloride, benzoyl chloride, and *t*-butylbromide.

$$RX + Ag^+ + H_2O \longrightarrow ROH + H^+ + AgX$$

Add the samples to be tested to 2-mL portions of the reagent (nearly simultaneously for the knowns). In those cases where mixing at room temperature for a few minutes is not sufficient to cause appreciable precipitation, the reaction mixture may be heated in a hot water bath for up to five minutes. The differences between many of the knowns (and later unknowns) depend upon differences in rate; hence the differences are more often relative than absolute. Thus, the knowns should be arranged, at best, in the order of their ease of production of AgX (time and amount) from poorest to best, or, at least, in three groups: very productive, poor, and no reaction.

Exercise 4: Tests for Alcoholic Hydrogen (Active Hydrogen)

Reagent: Freshly cut and dried metallic Na

CAUTION

ACIDS, ACID CHLORIDES, ETC., *MUST BE* KNOWN TO BE *ABSENT* (SEE TEST FOR ORGANIC ACIDS AND SOLUBILITY GROUP C). *PROTECT YOUR EYES!*

Knowns: 10–12 drops methyl alcohol, *n*-butyl alcohol (*iso*-butyl), *s*-butyl alcohol, *t*-butyl alcohol, and *dry* toluene.

$$2ROH + 2Na \longrightarrow 2RONa + H_2$$

Add to the samples to be tested one or two small thin chips of sodium (with a maximum surface area) that have been cut under an inert solvent such as petroleum ether and dried just before use. If possible, start the first four knowns at the same time to compare rates.

Exercise 5: Test for Alcoholic Hydrogen

Reagent: 5–8 drops acetyl chloride

Knowns: 10–12 drops (0.5 mL or 0.2 g) of the knowns from Exercise 4.

$$ROH + CH_3COCl \longrightarrow CH_3COOR + HCl$$

Add the reagent dropwise with care (Protect your eyes!) to the samples to be tested. *Is there evidence of reaction?* If so the reaction is probably positive. After standing for a minute or so with occasional shaking, pour the reaction mixtures into 10 mL of 10% $NaHCO_3$. All samples should react because this reagent is intended to destroy the reagent. *When the acetyl chloride has been destroyed*, draw the vapors with your hand to your nose to search for esters. *What evidence is there for the formation of a new compound?*

Note: Your instructor may wish to introduce the Lucas and/or the iodoform test(s) (Exercise 10) at this point if, for the unknowns, you are expected to distinguish between 1°, 2°, and 3° alcohols or methyl carbinols from other alcohols.

Summary

Ex. 1   1. Dissolve the sample (knowns and unknowns).

       2. Add the reagent dropwise with swirling; continue until the dark color remains.

       3. Continue until the dark color remains or 20 drops are added.

Ex. 2   1. Choose the solvent based on solubility.

       2. Dissolve the sample (knowns and unknowns).

       3. Add the reagent dropwise (15–25 drops).

       4. Check for a brown-black solid.

Ex. 3   1. Add the sample (knowns and unknowns) to 2 mL of reagent.

       2. If no solid forms after a few minutes, warm the mixture.

       3. Compare ease of formation and amount formed.

Ex. 4   1. Place about a 0.5 mL of sample (knowns and unknowns) in a test tube.

       2. Add clean chip(s) of sodium.

       3. Compare the rate of gas evolution.

Ex. 5   1. Place about a 0.5 mL of sample (knowns and unknowns) in a test tube.

       2. Carefully add drops of the test reagent.

       3. Mix, let stand, and look for signs of reactions (heat, etc.).

       4. Destroy excess reagent with $NaHCO_3$.

       5. When it is all gone, cautiously smell for an ester.

Questions

1. Why would it be wise to use both Exercise 1 and Exercise 2 to demonstrate the presence of a double bond?

2. Why would it be wise to use both Exercise 4 and Exercise 5 to demonstrate the presence of an —OH group?

3. Why must compounds with acidic hydrogens be absent before you apply the test with metallic sodium?

4. Why was it specified that the toluene in Exercise 4 be dry? What happens when sodium is added to a beaker of water?

5. Write the equations for the positive reactions in Exercises 1–5 for which the chemistry has been discussed.

6. Compare Exercise 5 with Exercise 11. Explain.

Disposal of Materials

Bromine solution (HO, HW, I)	Ex. 1, test solutions (I, OW, HO)
1-Octanol (OS, OW, I)	Ex. 2, test solutions (I, or container, bisulfite)
aq. KMnO$_4$ (I, container)	Ex. 3, test solutions (I, container)
alc. AgNO$_3$ (I, container)	Ex. 4, test solutions (I, N)
Sodium (I, return or destroy)	Ex. 5, test solutions (I, N)
Acetyl chloride (I, HO, container)	aq. NaHCO$_3$ (I, D, or N)
Methylene chloride (OS, HO, I)	

EXPERIMENT 75

Functional group tests II

Exercise 6: Test for Organic Acids

Reagent: 2 mL 5% (or 10%) NaHCO$_3$

Knowns: 5 drops or 0.1 g benzoic acid, acetic acid, phthalimide, ethyl alcohol, phenol, and *p*-toluenesulfonic acid or aniline hydrochloride.

$$RCOOH + NaHCO_3 \longrightarrow RCOONa + CO_2 + H_2O$$

Add the sample to be tested to the reagent. Solubility and evolution of CO$_2$ indicate a positive test. A distinct hissing sound may often be detected when visual evidence is in doubt.

Exercise 7: Test for Aldehydes and Ketones

Reagents: 2 mL, 2,4-dinitrophenylhydrazine-diethylene glycol (or DMF), 4–6 drops of concentrated hydrochloric acid

Knowns: 2–4 drops (or 0.1 g) ethanol, acetone, acetophenone, benzaldehyde, and *n*-butyraldehyde (or isobutyraldehyde).

$$\underset{/}{\overset{\backslash}{C}}=O + O_2N-\underset{}{\overset{NO_2}{\bigcirc}}-NHNH_2 \longrightarrow \underset{/}{\overset{\backslash}{C}}=NNH-\underset{}{\overset{NO_2}{\bigcirc}}-NO_2 + H_2O$$

Add the sample to be tested to the 2,4-dinitrophenylhydrazine reagent, and allow them to dissolve (swirl). Then add the concentrated hydrochloric acid, swirl, and allow the mixture to stand. Heat, if needed, briefly and gently in a warm water bath. The precipitate and the solution may be quite similar in color. Note any differences in the colors of the products.

Exercise 8: Fehling's Test

Reagent: 1 mL Fehling's solution A* and 1 mL Fehling's solution B* *mixed*

Knowns: Same as for Exercise 7.

$$\text{Cu(complex)}^{+2} + \overset{O}{\underset{}{\overset{\|}{-C}}}-H + NaOH \longrightarrow Cu_2O + \overset{O}{\underset{}{\overset{\|}{-C}}}-ONa$$

Add the sample to the reagent as for Exercise 7, but heat the mixtures in a water or steam bath for five minutes or until a change (precipitate) occurs. This test and the next are really tests for ease of oxidation.

*Directions for the preparation of these reagents may be found in Wilcox, *Experimental Organic Chemistry: A Small Scale Approach*, Macmillan, New York, 1988, p. 455. Make solution A by adding water to 34.64 g of $CuSO_4 \cdot 5\ H_2O$ and dilute it to 500 mL. For solution B, dissolve 65 g of sodium hydroxide and 173 g of sodium potassium tartrate (Rochelle salt) in water and dilute it to 500 mL.

For an interesting discussion of cases in which Fehling's test fails, see *J. Chem. Educ.*, 1960, 37, 205.

Exercise 9: Tollens' Test

Reagent: Tollens' solution (prepare as described below)

TOLLENS' SOLUTION, USED OR UNUSED, SHOULD *NOT* BE WASHED DOWN THE DRAIN. Such solutions should be collected for silver recovery. Vessels that will contain or have contained this solution, especially test tubes, should be cleaned with hot HNO_3 (STRONG, OXIDIZING ACID) before use to encourage mirror formation and with cold dilute HNO_3 after use to destroy any fulminating silver, which can be explosive. The cleanings should be followed by a thorough rinsing with distilled water. All washes should be added to the silver recovery container.

CAUTION

CAUTION

Knowns: Same as for Exercise 7.

$$2Ag(NH_3)_2{}^+ + 2—CHO + 4OH^- \longrightarrow Ag + 4NH_3 + 2H_2O + 2—COO^-$$

Place in each specially cleaned test tube 2 mL of 5% aqueous $AgNO_3$ and 3 drops of 5% NaOH. To avoid an excess of ammonia, add dropwise, with stirring, a very dilute ammonia solution (5 drops of concentrated NH_3 solution in 5 mL of water) until one drop just dissolves the dark precipitate of silver oxide. Then proceed as in Exercise 7 except that gentle heating will usually be needed. Repeat this test if a black precipitate forms instead of a mirror.

Exercise 10: Test for Methyl Ketones and Carbinols (Iodoform Reaction)

Reagents: 10% NaOH; I_2-KI solution ($I_2 : KI : water = 1 : 2 : 10$)

Knowns: 5–6 drops (0.2 g) acetone, acetophenone, 3-pentanone, ethyl alcohol or isopropyl alcohol, *t*-butyl alcohol (or *n*- or *iso*-).

$$2—\overset{\displaystyle O}{\overset{\displaystyle \|}{C}}—CH_3 + 3I_2 + 2OH^- \longrightarrow 2—COO^- + 2CHI_3 \downarrow + 3H_2O$$

Dissolve the sample to be tested in 5 mL of water or dioxane (or a mixture), depending on its solubility, in a large test tube. Add a little less than a milliliter of the iodine solution and then the sodium hydroxide, dropwise with shaking. The iodine color should be destroyed, leaving a light yellow solution. If the solution is colorless, add more iodine, and repeat the process. Letting the solution stand may cause precipitation, or brief heating at 60–75°C may be needed (the solution should remain yellow). Continue adding the reagents as long as iodine is consumed. Cool the solution, and dilute it with 5–10 mL of water if dioxane was used. Observe the color and odor of the precipitate. See Experiment 23.

Summary

NB: For all exercises, test samples separately.

Ex. 6   1.   Place 2 mL of reagent in each test tube.

2.   Add the samples (knowns and unknowns) to them.

3.   Watch for signs of reaction.

Ex. 7   1.   Place 2 mL of hydrazine solution in each test tube.

2.   Add the samples (knowns and unknowns) to them.

3.   Dissolve by swirling, then add the conc. HCl.

    4. Allow to stand and warm if needed.

    5. Watch for a colored precipitate (note color).

Ex. 8 1. Place 2 mL of mixed reagent in each test tube.

    2. Add the samples (knowns and unknowns)

    3. Swirl and heat in a hot water bath.

    4. Watch for a brick red precipitate.

Ex. 9 1. Place 2 mL of $AgNO_3$ solution in each test tube.

    2. Add 3 drops of aq. NaOH.

    3. Add very dilute ammonia dropwise with stirring.

    4. Add just enough to dissolve the precipitate.

    5. Add the samples (knowns and unknowns).

    6. Heat if needed to form a mirror.

Ex. 10 1. Choose water or dioxane as a solvent, depending on solubility.

    2. Dissolve the sample in the chosen solvent in a large test tube.

    3. Add 0.6–0.9 mL of iodine solution and mix thoroughly.

    4. Add NaOH solution dropwise with shaking to discharge the I_2 color.

    5. If the solution is not yellow, repeat steps 3–5.

    6. If after 2 or 3 more additions, the solution is still colorless, heat briefly in a water bath at about 70°C.

    7. If the solution is colorless and no CHI_3 has formed, let cool and repeat steps 3–6.

    8. Cool the solution and add 5–10 mL of water if dioxane was used.

NB: 9. Test one positive and one negative known first to see how much is needed to produce CHI_3 (solid) or a permanent yellow-colored solution. Keep adding iodine and NaOH solutions until you get results.

Questions

1. Write equations for the positive reactions in Exercises 6–10, where the chemistry of the compounds has been previously studied.

2. How would a precipitate be confirmed in a colored solution?

3. Why, with an unknown, would Exercise 8 and Exercise 9 not be used until Exercise 7 had been tried? What is the basis of these tests?

4. Why do methyl carbinols, as well as methyl ketones, give positive iodoform tests? What are the other organic products in these tests?

5. Why should benzoic acid be soluble in bicarbonate solution?

Disposal of Materials

Knowns and unknowns (I, OW, HO)

2,4-dinitrophenylhyazine solution (HW)

conc. HCl (D, N)

Fehling's A (I, D)

Fehling's B (I, N)

aq. NaOH (D)

I_2-KI solution (I, HW, container)

aq. $NaHCO_3$ (I, D, N)

Ex. 6, test solutions (I, some N, D)

Ex. 7, tests solutions (I, HW)

Ex. 8, test solutions (I, some N, D)

Ex. 9, test solutions (I, container)

Ex. 10, test solutions (I, HW, HO)

E X P E R I M E N T   7 6

Functional Group Tests III

Exercise 11: Test for I° and II° Amines

Reagent: 5–8 drops of acetyl chloride

Knowns: 10–12 drops aniline, *N*-methylaniline, or *N,N*-dimethylaniline.

$$-N-H + CH_3COCl \longrightarrow CH_3CON- + HCl$$

Repeat the procedure for Exercise 5 with the amine knowns and unknowns. The sodium bicarbonate solution is not needed here, though it may be used in disposal. Compare the observations made in these two exercises. *What is the evidence of a positive test reaction for this exercise?*

Exercise 12: Test for Distinguishing Among Classes of Amines (Hinsberg)

Reagents: 6–8 drops benzenesulfonyl chloride; 5 mL of 10% NaOH

Knowns: 4–5 drops (0.15 g) aniline, *N*-methylaniline, or *N,N*-dimethylaniline.

$$-NH_2 + C_6H_5SO_2Cl \xrightarrow{\text{base}} [-NHSO_2C_6H_5] \xrightarrow{\text{base}} -N^-SO_2C_6H_5$$

$$\diagdown NH + C_6H_5SO_2Cl \xrightarrow{\text{base}} \diagdown NSO_2C_6H_5$$

Add the sample to be tested to the mixed reagents, and shake vigorously with cooling as needed. The odor of the benzenesulfonyl chloride should disappear. When this happens, if there still appears to be some oily amine left, add a few drops of the sulfonyl chloride, and shake the mixture as before. An excess of the chloride should be avoided because it may cause a primary amine to form a solid disulfonyl derivative. The solution should be distinctly alkaline throughout this test. However, additional information can sometimes be obtained by acidifying these solutions when the test has been completed. Explain the solubility changes observed. See also *J. Chem. Educ.*, 1952, *29*, 506 and *J. Chem. Educ.*, 1964, *41*, 280.

Exercise 13: Test for the Hydrolysis of Ammonium Salts

Reagents: 20% NaOH

Knowns: Two 0.25-g (two 0.5-mL) portions of each of the following: an ammonium salt of an organic acid, acetamide (or sulfanilamide), acetonitrile (or phenyl cyanide or mandelonitrile), and acetanilide.

$$NH_4^+ + OH^- \longrightarrow H_2O + NH_3$$

Proceed through the following steps or until ammonia is detected by a moistened strip of red litmus paper suspended where only gas can reach it. The odor may also be noticeable.

Place one 0.25-g sample on a watch glass or in a beaker, and add 5 drops of 20% NaOH.

If this does not produce ammonia, boil the other 0.25-g sample in a test tube with 1 mL of 20% NaOH.

Should this also fail to produce ammonia, add 1 mL of alcohol, transfer the mixture to a small flask or vial, and reflux the solution for 30 to 45 minutes with an air condenser (Figure 1.7 or 1.8A, Unit I). Moist, red litmus placed where the ammonia could escape to the atmosphere will usually detect it. If not, isolate the organic acid and demonstrate its presence. Observe the relative order of ease of hydrolysis.

Figure 1.7, Unit I (page 24).

Summary

Ex. 11 1. Place about 0.5 mL (0.25 g) of sample in a test tube.

       2. Carefully add 5–8 drops of reagent.

       3. Mix, let stand, and look for signs of reaction (heat, etc.).

Ex. 12 1. Place 5 mL of NaOH solution in a test tube.

       2. Add 6–8 drops of benzenesulfonyl chloride.

       3. Immediately add 4–5 drops (0.15 g) of sample.

4. Stir/mix vigorously.

5. If oily amine is seen, add 2–3 drops more of the chloride.

6. Test a drop of solution with pH paper.

7. If it is not basic, add a drop or two of NaOH solution to make it basic.

8. Observe the result.

9. If instructed to do so, make the solution acidic with 6N HCl, stir, and observe the results.

Ex. 13   1. Add 5 drops of NaOH solution to a small sample on a watch glass.

2. If ammonia is generated, stop.

3. If not, place a new sample in a test tube, vial, or flask with 1 mL of NaOH.

4. Heat to boiling; if ammonia is produced, stop.

5. Equip with an air condenser and reflux.

6. Moist red litmus may be used to detect the basic gas.

7. Rank the known in order of ease of hydrolysis.

Questions

1. Write equations for the positive reactions in Exercises 11–13, where the chemistry of the compounds has been studied.

2. Explain the differences between observations made in Exercise 5 and in Exercise 11.

3. What additional information would be gained by the acidification of the solutions in the Hinsberg tests? What should happen?

4. How may the order of ease of hydrolysis in Exercise 13 be explained?

Disposal of Materials

Knowns and unknowns (I, OW, HO)	Ex. 11, test solutions (I, HW, HO)
Benzenesulfonyl chloride (I, HO, container)	Ex. 12, test solutions (I, HW, HO)
	Ex. 13, test solutions (I, N)
Acetyl chloride (I, HO , container)	Ex. 13, test solids (I, N, OW, S)

Additions and Alternatives

You are directed to other sources for a discussion of these and other tests and for further information on qualitative organic analysis and derivative preparation, especially:

a. McElvain, *The Characterization of Organic Compounds*, Rev. Ed., Macmillan, New York, 1953.

b. Shriner, Fuson, Curtin, and Morrill, *The Systematic Identification of Organic Compounds*, Wiley, 6th Ed., New York, 1980.

c. Siggia, *Qualitative Organic Analysis via Functional Groups*, Wiley, New York, 1954.

d. Cheronis and Entrikin, *Semimicro Qualitative Organic Analysis*, Interscience, New York, 1957.

e. Fitton and Hill, *Selected Derivatives of Organic Compounds*, Chapman and Hall, London, 1970 (Barnes and Noble).

Some of the books in the preceding list may be available to you near the laboratory, in addition to some of the following:

f. *Tables for Identification of Organic Compounds*, 2nd Ed., Chemical Rubber Co., Cleveland, Ohio, 1964.

g. Extra copies of the derivative tables from (b).

h. Eastman Kodak Company, list of products by functional groups.

i. Heilbron, *Dictionary of Organic Compounds*.

j. Beilstein, *Handbuch der Organischen Chemie*.

You could include a separation experiment in this introduction to organic qualitative analysis. (See Laughton, *J. Chem. Educ.*, 1960, 37, 133.)

Although much information on the preparation of derivatives is available in the above sources, you will need some general guidance from your instructor. Directions for making some derivatives are given in Section 17.3 below. Experience has shown that you may need several sources of directions for the same derivative preparation.

17.3 Derivatives

You can prepare derivatives using techniques discussed in references such as (b) and (e) above, or you can try to form derivatives using the procedures given below.

17.3A   Alcohol Derivatives

1-Naphthylurethanes are formed according to the equation:

1-naphthylurethane

To prepare this derivative, place a mixture of 1 g of anhydrous alcohol or phenol and 0.5 mL of α-naphthyl isocyanate in a oven-dried test tube, small flask, or vial fitted with a drying tube containing calcium chloride pellets. To ensure that the alcohol used is anhydrous, dry it over an appropriate drying agent such as anhydrous magnesium sulfate. If the unknown is a phenol, add 2 or 3 drops of anhydrous pyridine (*Caution!*) to the mixture in order to catalyze the reaction. Shake or stir the reaction mixture. If you do not see an immediate reaction, warm the reaction mixture in a steam or sand bath for five minutes. If, after cooling the reaction mixture in an ice bath, crystal formation is not evident, scratch the walls of the test tube to induce crystallization. After collecting the product, dissolve it in 5–10 mL of hot petroleum ether (b.p. = 100–120°C). Filter the resulting solution hot to remove any di-α-naphthylurea, a by-product formed if any water was present in the reaction mixture (m.p. = 297°C). Cool the filtrate in an ice bath, and collect the crystals in a Hirsch funnel. If crystals do not form, evaporate half of the solution, and repeat the cooling process.

3,5-Dinitrobenzoates are formed according to the following reaction equation:

3,5-dinitrobenzoate

To prepare these derivatives, add 0.5 g of 3,5-dinitrobenzoyl chloride (check its melting point or make it fresh) to a test tube containing a mixture of 1 mL of alcohol or phenol and 3 mL of anhydrous pyridine. After allowing the initial reaction to subside, gently heat the mixture for about one minute using a low flame. Pour the mixture, still hot, into 10 mL of water. Stir the mixture vigorously. Collect the precipitate and return it

Table 17.5 Alcohols

Substance	b.p	m.p.	3.5-Dinitrobenzoate	1-Naphthylurethane
Methanol	65	—	108(107)	124
Ethanol	78	—	93	79
2-Propanol	82(83)	—	123(122)	106
2-Methyl-2-propanol	83	26	142	101
2-Propen-1-ol	97	—	49	108
1-Propanol	97	—	74	80(105)
2-Butanol	99(98)	—	75(76)	97
2-Methyl-2-butanol	102	—	116	72
2-Methyl-1-propanol	108	—	86(87)	104
3-Pentanol	116(114–115)	—	101(99)	95
1-Butanol	118(116)	—	64	71
2-Pentanol	119	—	62(61)	74
3-Methyl-3-pentanol	123	—	96(94)(62)	83(104)
1-Chloro-2-propanol	127	—	77	—
3-Methyl-1-butanol	131(132)	—	61	68
2-Chloroethanol	129(131)	—	95	101
4-Methyl-2-pentanol	132	—	65	88
2-Ethoxyethanol	135	—	75	67
1-Pentanol	138(136–138)	—	46	68
Cyclopentanol	140(141, 139–140)	—	115	118
2-Ethyl-1-butanol	146	—	51	—
2-Methyl-1-pentanol	148	—	51	75
2,2,2-Trichloro ethanol	151	19	142	120
1-Hexanol	157	—	58	60(59)
2-Heptanol	159	—	49	54
Cyclohexanol	161(160)	25	113(112)	129
Furfuryl alcohol	170(171)(172)	—	80	130
1-Heptanol	176(177)	—	47	62
2-Octanol	179(178)	—	32	63
1-Octanol	195(194)	—	61	67
1,2-Ethanediol	197	—	169(di)	176(di)
1-Phenylethanol	202(204)	—	92(93)(95)	106
Benzyl alcohol	203(204)(205)	—	113(112)	134
1,3-Propanediol	214	—	178(di)	164(di)
1-Nonanol	215	—	52	65
2-Phenylethanol	219(219–221)	—	108	119
1,4-Butanediol	230	—	—	199(di)
1-Decanol	231	—	57	73
3-Phenylpropanol	236	—	45(92)	—
Cinnamyl alcohol	257(250)	33(33–35)	121	114
1-Dodecanol	259	24	60	80
Glycerol	209(dec)	—	76(tri)	191(tri)
1-Tetradecanol	—	39(38–40)	67	82
(-)-Methanol	—	43(44)(41)	153(158)	119

(Continued)

Table 17.5 Alcohols *(continued)*

Substance	b.p	m.p.	3.5-Dinitrobenzoate	1-Naphthylurethane
1-Hexadecanol	—	49(45–50)	66	82
2,2-Dimethyl-1-propanol	113	69	—	100
1-Octadecanol	—	59	77(66)	89
Diphenylmethanol	288	69(68)	141	135(136)
Benzoin	—	133	—	140
Cholesterol	—	147(148)	195	176
Borneol	—	208	154	132(127)

See Figures 2.8A, B, C
(pages 56–58).

to the test tube. Wash the solid with 5 mL of 5% aqueous Na_2CO_3, making sure that you break up the solid as much as possible with a stirring rod. In this way, maximum contact occurs between the wash solution and all parts of the solid. Collect the washed solid using a Hirsch funnel, and recrystallize it using an ethanol–water mixture.

Directions are also available to use the acid directly.

17.3B Phenol Derivatives

1-Naphthylurethanes and **3,5-dinitrobenzoates** may also be derived from phenols according to the reactions and procedures given for alcohol in Section 17.3A.

Bromo derivatives may also be formed, according to the equation:

bromophenol

To prepare these derivatives, dissolve 0.1 g of the phenol in 1 mL of methanol or dioxane. To this mixture, add 1 mL of water and then 1 mL of a brominating solution. (If one has not been made, dissolve 0.75 g of potassium bromide in 5 mL of water. Then add 0.5 g of bromine to the mixture.) Swirl the resulting solution vigorously. Continue adding the brominating solution dropwise and swirling the resulting solution until the bromine reagent's color persists. Add 3–5 mL of water to the mixture and shake it well. Collect the solid product using a Hirsch funnel, washing it as thoroughly as possible with water. Recrystallize it using a methanol–water mixture as the solvent.

See Figures 2.8A, B, C
(pages 56–58).

Table 17.6 Phenols

Substance	b.p.	m.p.	1-Naphthy- lurethane	3,5-Dinitro- Benzoate
2-Chlorophenol	176(175)	—	120	48(49)(mono), 76(di)
3-Methylphenol	202(203)	—	128	84
2-Methylphenol	192(191)	31(32–33)	142	56
4-Methylphenol	202(203)	36(32–34)	146	49(di), 108(198)(tet)
Phenol	181(182)	42	133	95
4-Chlorophenol	217	43(43–45)	166	33(mono), 90(di)
2,4-Dichloro- phenol	210	45(42–43)	—	68
2-Nitrophenol	216	45(44–46)	113	117
4-Ethylphenol	219	47(42–45)	128	—
5-Methyl-2- isopropylphenol	234	51(49–51)	160	55
3,4-Dimethyl phenol	225(228)	64(63)	142(141)	171
4-Bromophenol	238	64(64–68)	169	95
3,5-Dichloro phenol	—	68	—	189
2,5-Dimethyl phenol	212	75(74)	173	178
1-Naphthol	278	96(94–96)	152	105
3-Nitrophenol	—	97	167	91
4-*t*-Butylphenol	—	100(99)	110	50(mono), 67(di)
1,2-Di-hydroxy- benzene	245	104–105	175	192(193)
1,3-Di-hydroxy- benzene	281	109(110)	275(206)	112
4-Nitrophenol	—	112–114	150	142(di)
2-Naphthol	286	123(121–124)	157	84
1,2,3-Tri-hydroxy- benzene	309	133–134	173	158
1,4-Di-hydroxy- benzene	—	172(171)	247	186

17.3C Aldehyde and Ketone Derivatives

2,4-Dinitrophenylhydrazones may be formed from both aldehydes and ketones according to the following equations:

2,4-dinitrophenylhydrazone

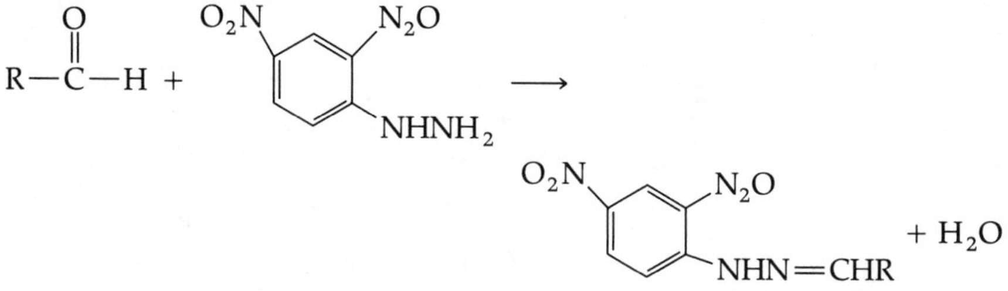

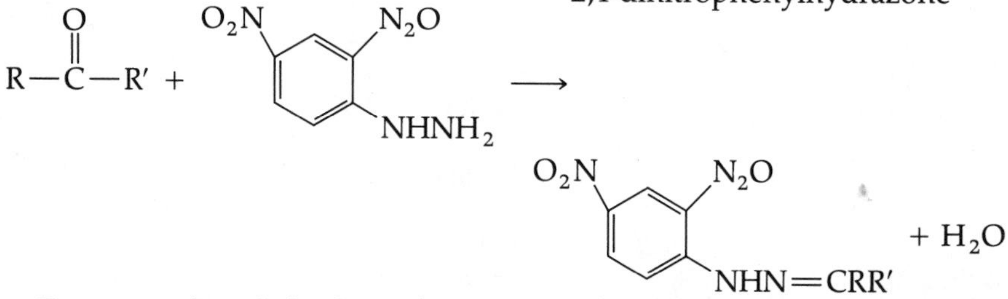

To prepare these derivatives, place 10 mL of a 2,4-dinitrophenylhydrazine solution (dissolve 2.0 g of 2,4-dinitrophenylhydrazine in 50 mL of 85% phosphoric acid with heating, cool this solution, and then add 50 mL of 95% ethanol to it, recooling it, and finally clarifying it using suction filtration; this solution can also be made in DMF with a few drops of concentrated hydrochloric acid) in a test tube with 1 millimole (about 0.1 g) of your unknown compound. If your unknown is a solid, dissolve it in a minimum amount of 95% ethanol or dioxane before you add it to the 2,4-dinitrophenylhydrazine solution. Crystallization should occur immediately. If not, gently warm the mixture on a steam bath for a few minutes, and set the test tube aside to allow crystallization to occur. Collect the crystals using a Hirsch funnel, wash them with cold 10% HCl, and recrystallize them using an alcohol or an alcohol–water mixture as a solvent.

See Figures 2.6–2.8 (pages 52–58).

For an alternative 2,4-dinitrophenylhydrazone preparation procedure, see Experiment 25A, 55, or 76, Exercise 7.

Semicarbazones, formed from both aldehydes and ketones, may be produced according to the following equations:

$$R{-}\overset{\overset{\displaystyle O}{\|}}{C}{-}H + H_2NNHCONH_2 \longrightarrow RCH = NNHCONH_2 + H_2O$$

semicarbazone

$$R{-}\overset{\overset{\displaystyle O}{\|}}{C}{-}R' + H_2NNHCONH_2 \longrightarrow RR'CH{=}NNHCONH_2 + H_2O$$

Table 17.7 Aldehydes and Ketones

Substance	b.p.	m.p.	Semicarbazone	2,4-Dinitrophenyl-hydrazone	Oxime
Aldehydes					
Ethanal	21	—	162(163)	168	47
Propanal	48(46–50)	—	89(154)	148(149)(155)(156)	40
Propenal	52	—	171	165	—
2-Methylpropanal	64(63)	—	125(119)	187(183)(182)	—
Butanal	75	—	95(106)(105)	123(122)	—
Trimethyl-acetaldehyde	75	—	190	209	41
3-Methylbutanal	95(90–92)	—	107	123	48
2-Methylbutanal	93	—	103	131	—
Chloral	98	—	90	131	56
Pentanal	103(102)	—	—	106(107)(98)	52
2-Butenal	104	—	199	190	119
2-Ethylbutanal	117	—	99	95(130)	—
Hexanal	130(131)	—	106	104	51
Heptanal	156(153)	—	109	106(108)	57
Cyclohexane-carboxaldehyde	161(162)	—	173(174)	172	90
2-Furaldehyde	162	—	202	212(230)	91(75)(90[syn, anti])
2-Ethylhexanal	163(173)	—	254(152)	114(120)	—
Octanal	171	14	101	106	60
Benzaldehyde	179(180)	—	222	237(239)	45(35)
Phenylethanal	135	33(−10)	153(155)(156)	121(110)	102(99)
2-Hydroxy-benzaldehyde	197	1	231	248(250)	57
2-Methyl-benzaldehyde	200	—	212	195	49
4-Methyl-benzaldehyde	204–205	—	234(215)	232(234)(239)	80
3,7-Dimethyl-6-octenal	207	—	82(84)(91)(92)	77(80)	—
2-Chloro-benzaldehyde	213(209–215)	17	146(225)(228)(229)	209(213)(214)(252)	70(76)(101)
4-Methoxy-benzaldehyde	247(248)	—	210	253	64 a133β
t-Cinnamaldehyde	250(dec)	—	215	255(dec)	64(138)
3,4-Methylene-dioxybenzaldehyde	263(264)	37	230	265(266)(dec)	110A(146S)
2-Methyoxy-benzaldehyde	238(245)	37(38)	215	253(254)	92(99)
4-Chloro-benzaldehyde	214	48(44–47)	230	254(270)(265)	110A(146S)
4-Bromo-benzaldehyde	—	56(57)	228	128(257)	111
3-Nitro-benzaldehyde	—	58	246	290(293)	120

(*Continued*)

Table 17.7 Aldehydes and Ketones *(continued)*

Substance	b.p.	m.p.	Semicarbazone	2,4-Dinitrophenyl-hydrazone	Oxime
4-Dimethyl-aminobenzal	—	74	222	325(236)	185
Vanillin	285(dec)	82(81–83)	230(229)	271(dec)	117(122)
3-Hydroxy-benzaldehyde	—	101(104)	198	257(dec)	88
4-Nitro-benzaldehyde	—	105(106)	221(211)	320(dec)(> 300)	113A(182S)
4-Hydroxy-benzaldehyde	—	116	224	270(280(dec))	72
d,l-Glyceraldehyde	—	142	160(dec)	167	—
Ketones					
Acetone	56	−95	187(190)	126	59
2-Butanone	80	−86	146(136)(186)	117(118)	—
Biacetyl	88	−2	235(mono), 278(di)	314(315)	75(mono), 245(di)
3-Methyl-2-butanone	94	—	112(113)	120(124)	—
2-Pentanone	100(101)(102)	—	111(112)(106)	143(144)	58
3-Pentanone	102	−40	138	156	69
Pinaclone	106	−50	157	125(127)	75(76)(79)
4-Methyl-2-pentanone	117(114–116)	−80	132(133)	95(81)	58
2,4-Dimethyl-3-pentanone	124	−80	160	88(95)	34
2-Hexanone	128(127)	—	125(123)	106(110)	40(49)
Cyclopentanone	130–131	—	210(203)(205)	146(144)	56
2,4-Pentanedione	139(133–135)	—	−209	122(mono), 209(di)	149(di)
4-Heptanone	144(145)	—	132	75	—
5-Methyl-2-hexanone	145	—	147	95	—
2-Heptanone	151(145–147)(149)	—	123(125)	89	—
3-Heptanone	146–149	—	101	81	—
Cyclohexanone	156(155)	—	166	162	90(91)
2-Methylcyclo-hexanone	162–163	—	195(196)	137	43
2,6-Dimethyl-4-heptanone	168(169)	—	122	92(66)	210
2-Octanone	173(172)	—	122(123)(124)	58	—
Cycloheptanone	181(180)	—	163	148	23
Acetophenone	202	19	198(203)	238(250)(338)	60
Phenyl-2-propanone	216	27	198	156	70
Propiophenone	218	18(21)	182(173)(174)	191	53(54)
4-Methyl-acetophenone	226	27(28)	205	258(260)	86(88)

Table 17.7 (*continued*)

Substance	b.p.	m.p.	Semicarbazone	2,4-Dinitrophenyl-hydrazone	Oxime
4-Chloro-acetophenone	232	12(20)	204(202)	236(233)	95
4-Phenyl-2-butanone	235	—	142	127	86
4-Methoxy-acetophenone	258	38(36–38)	198	228(220)	87
Benzophenone	305	48	167(164)	238(229)	142(144)
4-Bromo-acetophenone	225(255)	51	208	230(237)	128
Desoxybenzoin	320	56	148	204	98
3-Nitro-acetophenone	202	76(80)	257	228	132
9-Fluorenone	345(341)	83(79)(82–85)	234	283	195
Benzoin	344	134–136	206(205)(dec)	245	151
Camphor	205	179(177)	237	177	118

To form them, place 0.5 mL of a 2 M semicarbazide hydrochloride solution (prepared by dissolving 1.11 g of semicarbazide hydrochloride (MW = 111.5) in 5 mL of water) in a test tube with 1 millimole (about 0.1 g) of your unknown compound. If your unknown does not dissolve in the solution or if the solution becomes cloudy upon addition, add enough methanol to just dissolve the solid, producing a clear solution. Then add 0.5–0.6 g of sodium acetate or 10 drops of pyridine to the solution, using a Pasteur pipet. Gently heat the solution on a steam bath for about five minutes. The solid product should form at this time or with cooling. Collect the product using a Hirsch funnel; recrystallize it using ethanol if necessary. See also Experiments 24A and B.

See Figures 2.8A, B, C (pages 56–58).

Oximes may also be produced using aldehydes and ketones, according to these equations:

$$R-\overset{\overset{\displaystyle O}{\|}}{C}-H + NH_2OH \longrightarrow RCH{=}NOH + H_2O$$

$$\text{oxime}$$

$$R-\overset{\overset{\displaystyle O}{\|}}{C}-R' + NH_2OH \longrightarrow RR'C{=}NOH + H_2O$$

To form them, add 2 mL of a 10% aqueous NaOH solution to a solution comprised of 0.5 g of hydroxylamine hydrochloride and 3 mL of water. (Sodium hydroxide frees the hydroxylamine base.) Add 0.5 g of your unknown to this solution immediately. If it does not completely dissolve, add just enough ethanol to give a clear solution. Heat this solution, with stirring, on a hot water bath for ten minutes. Cool the solution on an ice bath. To further induce crystallization, you may wish to scratch the inner walls of the test tube with a glass stirring rod. If crystals still haven't formed, you may also add a few milliliters of water to the cold solution.

See Figures 2.6–2.8
(pages 52–58).

Collect the oxime using a Hirsch funnel. If you need to crystallize it, do so using water or an ethanol–water mixture.

17.3D Carboxylic Acid Derivatives

Carboxylic acids can be used to form **amides** as follows:

$$RCOOH \xrightarrow{SOCl_2} \underset{\text{acid chloride}}{RCOCl} \xrightarrow{NH_3} \underset{\text{amide}}{RCONH_2} + NH_4Cl$$

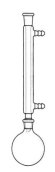

Figure 1.7, Unit I
(page 24).

See Figures 2.6–2.8
(pages 52–58).

To begin the formation procedure, place 0.5 g of the acid and 2 mL of thionyl chloride in a small, round-bottomed flask or a reaction vial. Attach this flask to a reflux condenser, and reflux the mixture on a steam bath for about 30 minutes. Allow the mixture to cool, and transfer it dropwise with a Pasteur pipet, swirling it into a beaker or conical flask containing 10 mL of ice-cold, concentrated ammonium hydroxide. Stir this mixture well. When the reaction is complete, collect the product using a Hirsch funnel, wash with a little ice water, and recrystallize it, if necessary, using a water or a water–ethanol mixture.

Anilides may also be derived from carboxylic acids.

$$R\text{—}COOH \rightleftharpoons H^+ + R\text{—}COO^- \xrightarrow{PhN=C=S}$$

$$\text{PhN}=C\overset{\frown}{-}\overset{..}{S} \longrightarrow \text{PhN}\overset{\frown}{-}C=S \xrightarrow{-COS}$$

$$O=C-O \qquad \qquad {}^-O-C-O$$
$$| \qquad \qquad \qquad |$$
$$R \qquad \qquad \qquad R$$

$$RCO\overset{-}{N}\text{—}Ph \xrightarrow{H^+} R\text{—}CONHPh$$

To form them, place 0.15 g (or about 1 millimole) of phenyl isothiocyanate and 1 millimole (about 0.15 g) of your unknown carboxylic acid in a test tube, and heat it at 160–170°C for 10–15 minutes, mixing it thoroughly with a glass stirring rod. Triturate (grind a solid in a liquid to disperse it) the viscous substance collected with aqueous ethanol. Filter this substance using a Hirsch funnel, and recrystallize it using aqueous ethanol as the solvent. In order to improve the yield, you may wish to add a few drops of anhydrous pyridine (*Caution!*) as a catalyst. Please note that you will need to add pyridine to some unknowns such as acetic, butyric, and *o*-nitrobenzoic acid reaction mixtures in order to collect a satisfactory amount of product. See *J. Chem. Educ.*, 1983, *60*, 508.

See Figures 2.8A, B, C
(pages 56–58).

Table 17.8   Carboxylic Acids

Substance	b.p.	m.p	*p*-Toluidide	Anilide	Amide
Formic acid	101	8	53	47(50)	43
Acetic acid	118	17	148	114	82
Propenoic acid	139	13	141	104	85
Propanoic acid	141	—	124	103	81
2-Methylpropanoic acid	154	—	104	105	128
Butanoic acid	162	—	72	95	115
2-Methylpropenoic acid	163	16	—	87	102
Timethylacetic acid	164	35	120	127	178
Pyruvic acid	165(dec)	14	109	104	124
3-Methylbutanoic acid	176	—	109(107)	109	135
Pentanoic acid	186	—	70(74)	63	106(105)
2-Methylpentanoic acid	186(196)	—	80	95	79
2-Chloropropanoic acid	186	—	124	92	80
Dichloroacetic acid	194	6	153	118	98
Hexanoic acid	205	—	75	95	101
2-Bromopropanoic acid	205	24	125	99	123
Heptanoic acid	224	—	80	71	96
Octanoic acid	237(239)	16	70	57	107
Nonanoic acid	254	12	84	57	99
Cyclohxyl carboxylic acid	—	30	—	146	186
Decanoic acid	268	32	78	70	108
4-Oxopentanoic acid	246	33	108	102	108(dec)
Dodecanoic acid	299	43(44)	87	78	100(110)(99)
3-Phenylpropanoic acid	279	48	135	98(92)	105(82)
Bromoacetic acid	208	50	—	131	91
Tetradecanoic acid	—	54	93	84	103
Trichloroacetic acid	198	57	113	97	141
Hexadecanoic acid	—	62	98	90	106
Chloroacetic acid	189	63	162	137	121(120)
Octadecanoic acid	—	69(70)	102	95	109
t-2-Butenoic acid	—	72	132	118	158(160)
Phenylacetic acid	—	77	136	118	156
Glycolic acid	—	80	143	97	120
Pentanedioic acid	—	97	—	224(di)	175(di)
Phenoxyacetic acid	—	99	—	99	101
2-Methoxybenzoic acid	200	101	—	131	129
Oxalic acid dihydrate	—	101(102)(186d)	169(mono), 268(di)	148(mono), 257(246)(254)(di)	219(mono), 400(dec) (419(dec))(di)

(Continued)

Table 17.8 Carboxylic Acids *(continued)*

Substance	b.p.	m.p	*p*-Toluidide	Anilide	Amide
2-Methylbenzoic acid	—	104	144	125	142
Nonanedioic acid	—	106(104)	210(di)	107(mono), 186(di)	93(mono), 175(di)
3-Methylbenzoic acid	263	110(112)	118	126	94
d,l-Phenyl hydroxyacetic acid	—	118(120)	172	151	133
Benzoic acid	249	122	158	163	130
2-Benzoylbenzoic acid	—	127	—	195	165
2-Furoic acid	—	129	107	123	143
Maleic acid	—	130(133)	142(di)	198(mono), 187(di)	172(181)(mono), 260(266)(di)
Decanedioic acid	—	133	201(di)	122(mono), 200(di)	170(mono), 210(di)
t-Cinnamic acid	300	133	168	153	147
d,l-Malic acid	—	133	178(mono), 207(di)	155(mono), 198(di)	163(di)
Acetylsalicylic acid	—	136	136	136	138
Malonic acid	—	137	86(mono), 253(di)	132(mono), 230(di)	170(di)
m-Tartaric acid	—	140	—	193(mono)	190(di)
2-Chlorobenzoic acid	—	140	131	118	139(140)
3-Nitrobenzoic acid	—	140	162	155	143(174)
2-Aminobenzoic acid	—	146	151	131	109
Diphenylacetic acid	—	148	172	180	167
2-Bromobenzoic acid	—	150	—	141	155
Benzilic acid	—	150	190	175	154
Hexanedioic acid	—	152	239(238)	151(mono), 241(di)	125(mono), 220(224)(di)
Citric acid	—	153	189(tri)	199(tri)	210(tri)
4-Chloro phenoxyacetic acid	—	158	—	125	133
3-Chlorobenzoic acid	—	158	—	122	134
2-Hydroxybenzoic acid	—	159	156	136	142
2-Iodobenzoic acid	—	162	—	141	110
1-Naphthoic acid	—	162	—	161	205
Methylenesuccinic acid	—	166(dec)	—	152(mono)	191(di)
d-Tartaric acid	—	173(169)	—	180(mono), 264(di)	171(mono), 196(di)
4-Methylbenzoic acid	—	180	160(165)	145	160

Table 17.8 (*continued*)

Substance	b.p.	m.p	*p*-Toluidide	Anilide	Amide
Butanedioic acid	235(dec)	188	180(mono), 255(di)	143(mono), 230(di)	157(mono), 260(di)
N-Benzoylglycine	—	190	208	—	183
3-Hydroxybenzoic acid	—	201	163	157	170
3,5-Dinitrobenzoic acid	—	202(205)	—	234	183
Phthalic acid	—	210(dec)	150(mono), 201(di)	169(mono), 253(di)	149(144)(mono), 220(di)
4-Hydroxybenzoic acid	—	214	204	197	162
Pyridine-3-carboxylic acid	—	236(238)	150	132	128
4-Nitrobenzoic acid	—	240	204	211	201
4-Chlorobenzoic acid	—	242	—	194	179(170)

p-**Toluidides** can be formed using carboxylic acids in addition to those derivatives mentioned previously.

$$R-\overset{\overset{\displaystyle O}{\|}}{C}-Cl + CH_3-\!\!\left\langle\!\!\!\bigcirc\!\!\!\right\rangle\!\!-NH_2 \longrightarrow R-\overset{\overset{\displaystyle O}{\|}}{C}-NH-\!\!\left\langle\!\!\!\bigcirc\!\!\!\right\rangle\!\!-CH_3$$

toluidide

Place 0.5 g of your unknown carboxylic acid in a round-bottomed-flask, and add 1 drop of dimethylformamide (DMF). **Do *not* add DMF to a hot solution.** Set up an apparatus for reflux (Figure 1.7, Unit I) in the hood, and attach it to the round-bottomed flask. Pour 4 mL of thionyl chloride (**very reactive and corrosive; its reactions generate a corrosive gas**) through the condenser, and warm the resulting mixture gently on a steam bath for 15–20 minutes. Then allow the solution to cool to room temperature. During this time, dissolve 1 g of *p*-toluidine in 30 mL of dichloromethane in a 125-mL Erlenmeyer flask. To this flask, add 20 mL of a 25% NaOH solution. Briefly cool this mixture in an ice bath.

After allowing the acid chloride to form completely and when the reaction mixture has cooled down to room temperature, dilute it with 10 mL of dichloromethane. Add this diluted acid chloride solution, vigorously swirling and cooling after each addition, to the 125-mL Erlenmeyer flask in several 1–2-mL portions with a Pasteur pipet. Make sure that the aqueous phase has a pH of about 10 by checking this upper layer with pH paper after each addition of diluted acid chloride solution. You should observe, upon the addition of each portion of the acid chloride solution, a vigorous reaction. In some cases, the heat produced by this reaction will be sufficient to cause the dichloromethane solution to briefly boil. If you

CAUTION

CAUTION

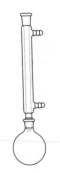

Figure 1.7, Unit I
(page 24).

Figure 4.1   (page 129).

See Figures 2.6–2.8
(pages 52–58).

lose a significant amount of dichloromethane in this manner, add more solvent to the reaction mixture.

After completing this addition, swirl the flask vigorously for ten minutes at room temperature. You may observe a yellow color in the dichloromethane layer at this state. After transferring the mixture to a separatory funnel, remove the lower, dichloromethane layer, and discard the basic, aqueous phase. Wash the organic layer with 15 mL of a 5% HCl solution; then wash it with 15 mL of water. Under the hood, using a steam bath, evaporate the dichloromethane solution to dryness (Section 1.3I). Recrystallize the solid residue collected using aqueous alcohol as the solvent. You may also use a mixture of acetone and petroleum ether as the solvent if necessary.

17.3E   Amine Derivatives

Acetamides are formed using amines as shown in this equation:

$$RNH_2 + (CH_3CO)_2O \longrightarrow CH_3CONHR + CH_3COOH$$
$$\text{acetamide}$$

To prepare these derivatives, heat a small Erlenmeyer flask containing 1 millimole (ca. 120 mg) of your unknown amine and 0.5 mL of acetic anhydride for about five minutes. Add 5 mL of water to this solution, stirring it well to facilitate product precipitation and acetic anhydride hydrolysis. If a solid product does not form, you may wish to scratch the inner walls of the flask with a glass stirring rod. Collect the solid by filtering it through a Hirsch funnel and by washing it with several

See Figures 2.8A, B, C
(pages 56–58).

portions of cold, 5% hydrochloric acid. To recrystallize this derivative, use a mixture of methanol and water as the solvent.

If your unknown amine is aromatic, the above procedure might require 2 mL of pyridine (*Caution!*) acting as a catalyst. With pyridine, increase the heating time to about one hour and add a reflux condenser to your reaction setup. To remove pyridine, wash the solid product with 5–10 mL of 5% sulfuric acid or hydrochloric acid as above. For an alternative version of this procedure, see Experiment 3C.

Benzamides may also be prepared using amines.

$$RNH_2 + C_6H_5COCl \longrightarrow C_6H_5CONHR + HCl$$
$$\text{benzamide}$$

To prepare this derivative, add 0.5 g (or 0.4 mL) of benzoyl chloride to a test tube containing a suspension of 1 millimole (about 120 mg) of your amine in 1 mL of 10% sodium hydroxide solution. Place a cork stopper on the test tube, and rapidly shake it for about ten minutes. Then, reduce the pH of the solution to about 6 or 7 by adding dilute hydrochloric acid. Collect your product by suction filtration using a Hirsch funnel; wash it

Table 17.9 Amines

Substance	b.p.	m.p.	Benzamide	Acetamide
Primary Amines				
Isopropylamine	33–34	—	71(100)	—
t-Butylamine	46(44)	—	134	101
Popylamine	48	—	84	—
sec-Butylamine	63	—	76	—
Isobutylamine	69(64–71)	—	57	—
Butylamine	78(77)	—	42	—
Ethylenediamine	118	—	244(di)	172(di)
1-Aminohexane	129	—	40	—
Cyclohexylamine	135(134)	—	149	104
Benzylamine	184(182–185)	—	105	60
Aniline	184	—	163	114
1-Phenylethylamine	187	—	120	57
2-Phenylethylamine	198	—	116	114
2-Methylaniline	200(199–200)	—	144	110
3-Methylanilide	203–204	—	125	65
2-Chloroanilide	208–210	—	99	87
2-Ethylaniline	210	—	147	111
4-Ethylaniline	216	—	151	94
2,6-Diethylaniline	216(215)	11	168	177
2,4-Diethylaniline	218	—	192	133
2,5-Diethylaniline	218	—	140	139
2-Methoxyaniline	225	6	60(84)	85
3-Chloroaniline	230	—	120	74(72)(78)
2-Ethoxyaniline	231–233	—	104	79
4-Chloro-2-methylaniline	241	29	142	140
4-Methylaniline	200	43(41–44)	158	147
2-Aminobiphenyl	—	49	102(86)	119
2,5-Dichloroaniline	251	50(49–51)	120	132
1-Naphthylamine	—	50	160	159
4-Aminobiphenyl	—	53	230	171
4-Methoxyaniline	240	58	154	130
2-Aminopyridine	—	57–60	165(di)	71
4-Bromoaniline	245(dec)	64(66)	204	168
2,4,5-Trimethyl aniline	—	64	167	162
3-Phenylenediamine	—	64–66	125(mono), 240(di)	86(mono), 191(di)
4-Iodoaniline	—	67	222	184
4-Chloroaniline	232	68–71(72)	192	179(172)
2-Nitroaniline	—	72(71–73)	110(98)	92
Ethyl *p*-amino benzoate	—	89	148	110
2-Phenylenediamine	258	102	301(di)	185(di)
2-Methyl-5-nitroaniline	—	106(104–107)	186	151

(Continued)

Table 17.9 Amines *(continued)*

Substance	b.p.	m.p.	Benzamide	Acetamide
2-Chloro-4-nitroaniline	—	108(107–109)	161	139
3-Nitroaniline	—	112–114	157(155)(150)	155(76)
4-Methyl-2-nitroaniline	—	115–116	148	99
2,4,6-Tribromo-aniline	300	120–122	200	232(mono)
3-Aminophenol	—	122	174(198)(153)	148(mono), 101(di)
2-Methoxy-4-nitroaniline	—	138–140	149	153
4-Phenylenediamine	267	140(138–142)	128(mono), 300(di)	162(mono), 304(di)
2-Aminobenzoic acid	—	147	182	—
4-Nitroaniline	—	148–149	199	215
2-Aminophenol	—	174	182	201
2,4-Dinitroaniline	—	180(176–178)	202(220)	120
4-Aminophenol	—	184	234(di)	168(mono), 150(di)

Secondary Amines

Substance	b.p.	m.p.	Benzamide	Acetamide
Diethylamine	56	—	42	—
Piperidine	106	—	48	—
Morpholine	129	—	75	—
Diisobutylamine	137–139	—	—	86
N-Methylcyclo-hexylamine	148(149)	—	85	—
N-Methylaniline	196	—	63	102
N-Ethylaniline	205	—	60	54
N-Benzylaniline	298	37	107	58
Indole	254	52	68	157
Diphenylamine	302	52	180	101
N-Phenyl-1-naphthylamine	335	60–62	152	115
N-Phenyl-2-naphthylamine	—	108	148(136)	93
4-Nitro-N-Methylaniline	—	152	112	153

well with cold water. Use an ethanol–water mixture as the solvent for recrystallization.

See Figures 2.6–2.8 (pages 52–58).

Bromo derivatives may be formed from aromatic amines in addition to those derivatives mentioned here. For a reaction equation and an experimental procedure, see Experiment 30.

Other *quantitative methods* may be used in place of the preparation of one derivative. These might include specific rotation, neutralization or saponification equivalents, and other methods discussed in Chapter 16.

Some More Spectra

Spectrum A.1

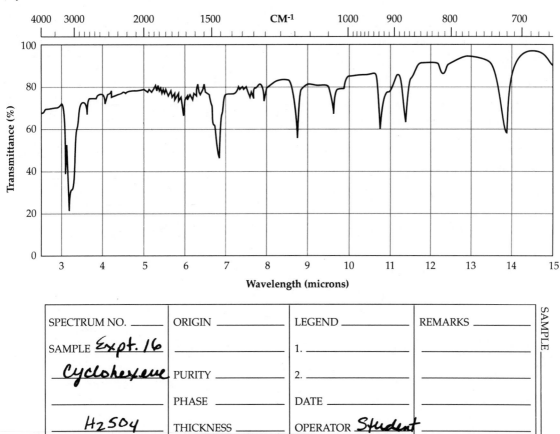

SPECTRUM NO. _____	ORIGIN _____	LEGEND _____	REMARKS _____
SAMPLE *Expt. 16*	_____	1. _____	_____
Cyclohexene	PURITY _____	2. _____	_____
_____	PHASE _____	DATE _____	_____
H₂SO₄	THICKNESS _____	OPERATOR *Student*	_____

SAMPLE

Spectrum A.2

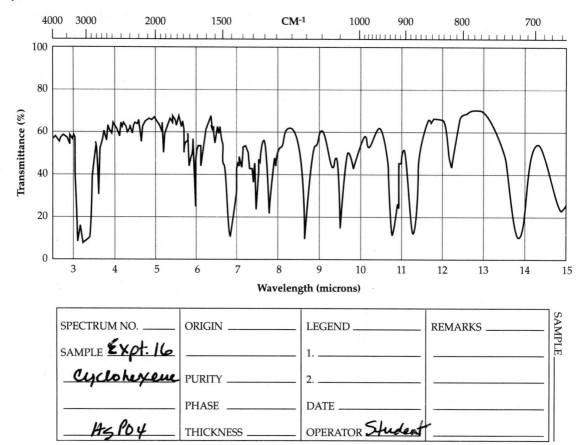

SPECTRUM NO. _____	ORIGIN _____	LEGEND _____	REMARKS _____
SAMPLE **Expt. 16**		1. _____	_____
Cyclohexene	PURITY _____	2. _____	_____
_____	PHASE _____	DATE _____	_____
As PO4	THICKNESS _____	OPERATOR **Student**	_____

SAMPLE _____

Spectrum A.3

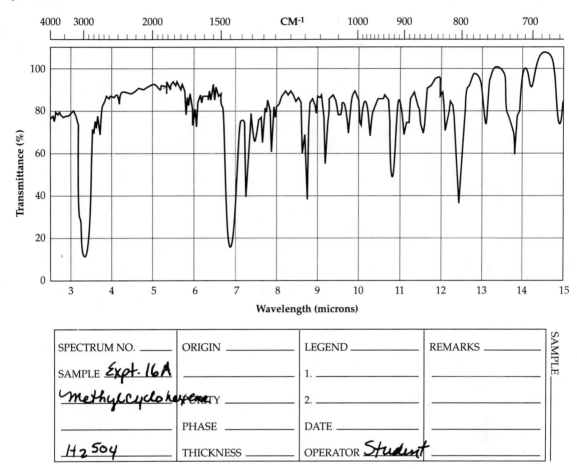

SPECTRUM NO. _____ ORIGIN _____ LEGEND _____ REMARKS _____

SAMPLE *Expt. 16A* _____ 1. _____ _____

methylcyclohexene ~~TY~~ _____ 2. _____

_____ PHASE _____ DATE _____

H₂SO₄ _____ THICKNESS _____ OPERATOR *Student* _____

SAMPLE

Spectrum A.4

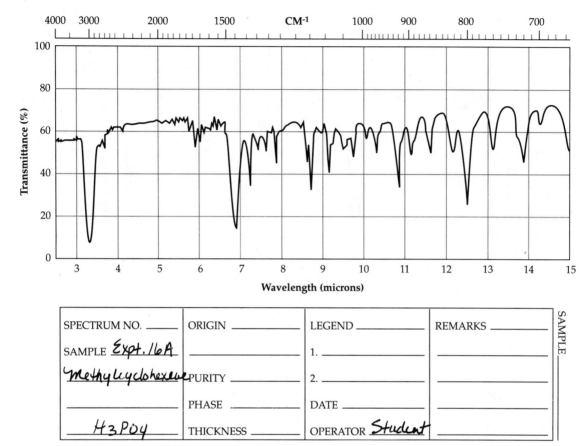

SPECTRUM NO. _____   ORIGIN _____   LEGEND _____   REMARKS _____

SAMPLE *Expt. 16A*   _____   1. _____   _____

methylcyclohexene PURITY _____   2. _____   _____

_____   PHASE _____   DATE _____   _____

_____ *H3PO4* _____   THICKNESS _____   OPERATOR *Student*   _____

SAMPLE

Spectrum A.5

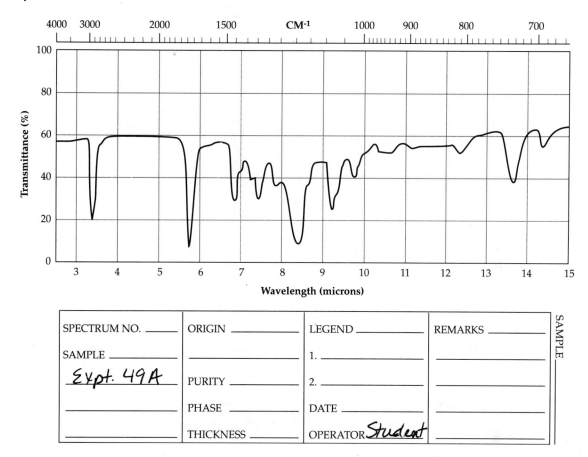

SPECTRUM NO. _____ ORIGIN _____ LEGEND _____ REMARKS _____

SAMPLE _____

Expt. 49A PURITY _____ 1. _____

_____ PHASE _____ 2. _____

_____ THICKNESS _____ DATE _____

OPERATOR *Student*

SAMPLE

Spectrum A.6

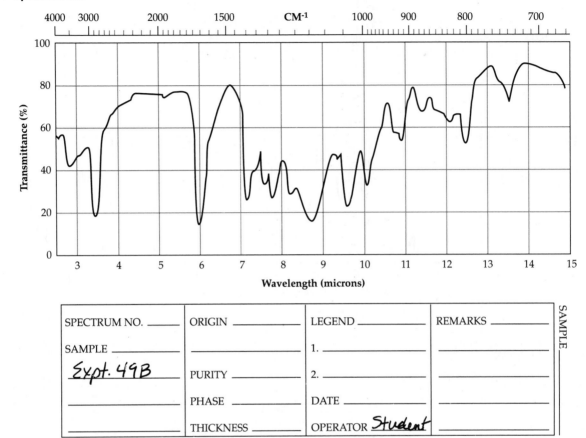

SAMPLE

Spectrum A.7

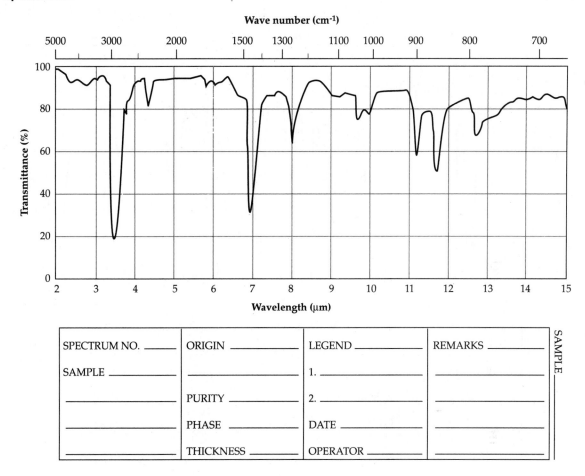

Wave number (cm⁻¹)

Wavelength (μm)

Transmittance (%)

SPECTRUM NO. _____	ORIGIN _____	LEGEND _____	REMARKS _____
SAMPLE _____		1. _____	
_____	PURITY _____	2. _____	_____
_____	PHASE _____	DATE _____	_____
_____	THICKNESS _____	OPERATOR _____	_____

SAMPLE

Spectrum A.8

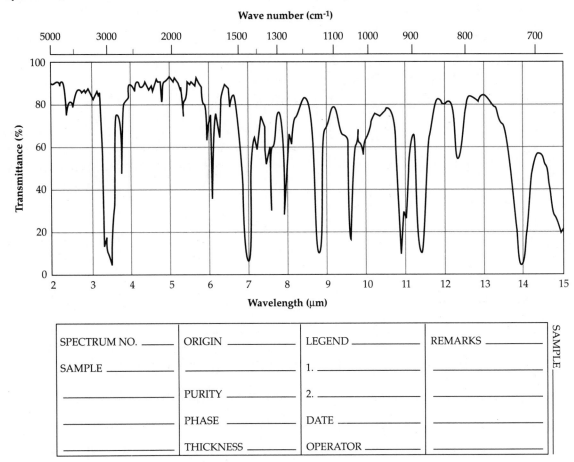

Spectrum A.9

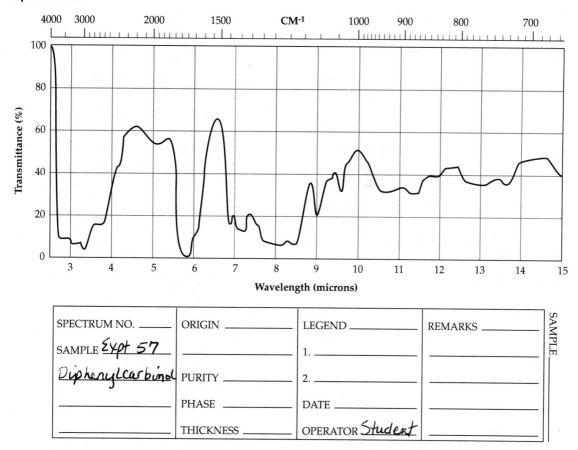

SPECTRUM NO. _____ ORIGIN _____ LEGEND _____ REMARKS _____

SAMPLE *Expt 57* _____ 1. _____ _____

Diphenylcarbinol PURITY _____ 2. _____

_____ PHASE _____ DATE _____

_____ THICKNESS _____ OPERATOR *Student* _____

SAMPLE

Preparation and Reporting
Student Report by G. Whiz

Expt. XX
The Esterification of Malonic Acid

PURPOSE Preparation of dibutyl malonate by the esterification of malonic acid with *n*-butyl alcohol (1-butanol).

EQUATION

$$\underset{\substack{|\\ CH_2 \\ |\\ COOH}}{COOH} + 2CH_3CH_2CH_2CHOH \longrightarrow \underset{\substack{|\\ CH_2 \\ |\\ COOCH_2CH_2CHCH_3}}{COOCH_2CH_2CH_2CH_3} + 2H_2O$$

Compound	MW	Density g/mL	m.p., °C	b.p., °C	Solubility (g/100 mL)	
Malonic acid	104.1	1.631	130–135(d)	—	H_2O $138^{16°}$	ether $8^{15°}$
1-Butanol	74.1	0.810	—	118	$9^{15°}$	v. sol.
Benzenesulfonic acid	158.2	—	65–66	—	v. sol.	i
Dibutyl malonate	216.3	0.981	—	251	i	s
Dibutyl ether	130.2	0.773	—	143	< 0.05	v. sol.
1-Butene	56.1	0.60	—	−6.3	i	v. sol.
2-Butene	56.1	0.60–0.62	—	0.88–3.7	i	v. sol.
Benzene	78.11	0.879	—	80.1	0.07	v. sol.

PROCEDURE

In a 100-mL, round-bottomed flask place 10.4 g of malonic acid, 30 mL of 1-butanol, 1 g of benzenesulfonic acid, 25 mL of benzene, and a boiling stone. Attach a Barrett trap (or Dean–Stark separator) and condenser, and reflux the mixture for an hour at which time about 2 mL of water collects in the trap. (The Barrett trap collects the water that was azeotropically distilled and separated after condensation.)

After the reflux, wash the mixture with two 15-mL portions of water. Extract the water-combined layers twice with 10-mL portions of benzene.

Combine the organic layers and wash with 15 mL of saturated sodium chloride solution. Fractionally distill the organic layer.

RESULTS

Forerun: b.p. 68–245° (plateaus at 78–80°, 117–119°, and 140–144°), mostly benzene, about 45 mL. Product: b.p. 245–250°, 17.0 g

Calculations for This Experiment

Malonic acid:

$$10.4 \text{ g} \times \frac{1 \text{ mole}}{104 \text{ g}} = 0.10 \text{ mole}$$

1-Butanol:

$$30 \text{ mL} \times .81 \text{ g/mL} \times \frac{1 \text{ mole}}{74 \text{ g}} = 0.328 \text{ mole}$$

Benzene is the solvent; benzenesulfonic acid is the catalyst. Consider 0.1 mole of malonic acid × 2 moles alc/1 mole acid = 0.2 mole of 1-butanol needed, but there is 0.328 mole available = plenty plus some left over.

∴ malonic acid is the limiting reagent and must be used in all calculations.

$$0.1 \text{ mole acid} \times \frac{1 \text{ mole ester}}{1 \text{ mole acid}} \times 216.3 = 21.63 \text{ g/mole (theoretical yield)}$$

$$\frac{\text{actual yield}}{\text{theo yield}} \times 100 = \frac{17.0 \text{ g}}{21.63 \text{ g}} \times 100 = 78.6\% \text{ (percent yield)}$$

Possible identity of fractions in the forerun:

68–80°	benzene and benzene–water azeotrope
117–119°	unreacted 1-butanol
140–144°	by-product, dibutyl ether

Product Label for This Experiment

G. Whiz   12/3/93
Dibutyl malonate
B.p. 245–250°C
Yield = 17.0 g or 78.6%

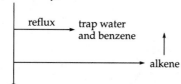

Dibutyl Malonate

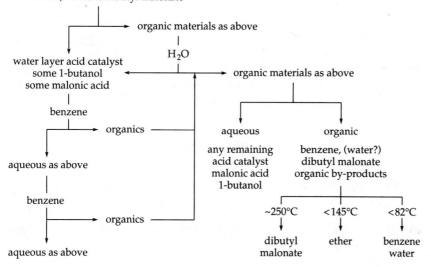

Malonic acid, 1-butanol,
benzene, acid catalyst

reflux → trap water
and benzene

→ alkene

water, malonic acid, acid catalyst
1-butanol, benzene dibutyl malonate

→ organic materials as above

H_2O

water layer acid catalyst
some 1-butanol
some malonic acid

→ organic materials as above

benzene

organics

aqueous as above

benzene

organics

aqueous as above

aqueous
any remaining
acid catalyst
malonic acid
1-butanol

organic
benzene, (water?)
dibutyl malonate
organic by-products

~250°C → dibutyl
malonate

<145°C → ether

<82°C → benzene
water

Report for Experiment 19A, Using the Standard Report Form

Experiment No. _19A_

Preparation of _1- Bromobutane_

Name _G. Whiz_

Date _10/17/93_

Section _1_

Reaction equation:

$$CH_3CH_2CH_2CH_2OH + Br^-/H_2SO_4 \rightarrow CH_3CH_2CH_2CH_2Br + H_2O$$

Properties of compounds involved (if pertinent):

Compound	1- butanol	NaBr	H₂SO₄	nC4H9Br	dibutyl ether	no extraction solvent used	butenes
Mol. wt.	74.1	102.9	18M	137.03	130.2		56.1
Vol. (liq.)	5ml	—	6mL	-	—		-
Density	0.810	—	1.834	1.275	0.773		0.6-0.62
Weight	4.05	8.4	-	—	—		-
Moles	0.055	0.082	0.108	—	-		-
MP/BP (lit.)	(118)	755	(340d)	(101.6)	(143)		(<RT)
Solubility	sl sol/H₂O	sol/H₂O	sol/H₂O	insol/H₂O	V.sol/H₂O		insol/H₂O

Results:

Limiting reagent _1- butanol_

Theoretical yield _7.53_ g

Actual yield _3.1_ g

Percent yield _41_ %

MP/BP (observed range) _100-102_ °C

$4mL \times 0.773 = 3.1g$

$(3.1/7.53)(100)$

Calculation of theoretical and percent yield:

$18M \times 0.006L = 0.108 \text{ moles}$

$5mL \times 0.81g/mL \times \frac{1}{74} = 0.055 \text{ moles}$

$0.055 \times \frac{1 BuBr}{1 BuOH} \times 137.03 = 7.53 g$

This space and the reverse side of this form may be used for the answers to questions,
a flow chart or process chart, and for comments on difficulties, deviations, explanations, and tests.

Sample flow chart for Experiment 19A, preparation of 1-bromobutane

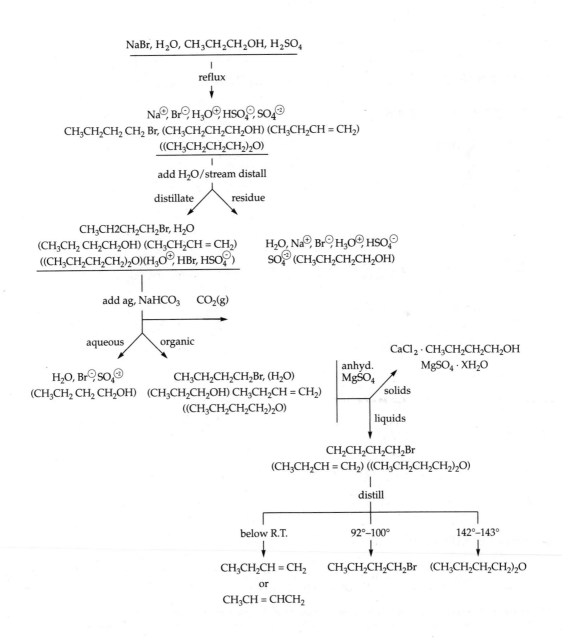

Other Reports by a Student

Name: Dylan Stewart   **Date: 03-16-91**

Report of Unknowns

UNKNOWN NUMBER 3   **COMPOUND (ASSUMED):**   adipic acid

1.   PHYSICAL EXAMINATION: (*solid:* fine, white crystals)

2.   PHYSICAL CONSTANTS: *melting point* determination (3 trials)

Trial 1 $\Rightarrow$ 148–154°C

Trial 2 $\Rightarrow$ 152.5–154°C

Trial 3 $\Rightarrow$ 151.5–153°C

3.   NEUTRALIZATION EQUIVALENT: helps determine the *formula weight*

Concentration of standardized NaOH = 0.984 M

Mass of UNKNOWN No. 3	Volume of NaOH [mL]	Milliequiv. of UNKNOWN No. 3.
0.2171 grams	30.8 mL	3.031 milliequivalents
0.2285 grams	32.4 mL	3.188 milliequivalents

GRAM EQUIVALENT MASS(ES):

Trial 1 $\Rightarrow$ 71.63 grams/equivalent

Trial 2 $\Rightarrow$ 71.68 grams/equivalent

4.   POSSIBLE COMPOUNDS   (Carboxylic Acids listed by *melting point*)

Carboxylic Acid	m.p.	Mol. Weight	#of — COOH's	m.p. Anilide	m.p. Amide
diphenylacetic	148	212.2	1	180	167
o-bromobenzoic	150	201.0	1	141	155
p-nitrophenylacetic	152	181.1	1		198
2,5-dichlorobenzoic	153	191.1	1		155
adipic acid	**154**	**146.1**	**2**	**235**	**220**
m-bromobenzoic	155	201.0	1	146	155
salicylic	157	138.1	1	134	139

Since the gram/equivalent weight is so low (below anything on the list), I took a guess and said it might be **adipic acid**. Since it has two —COOH groups, the Molecular Weight would be double the gram/equiv weight and $71.65 \times 2 = 143.3$. This disagrees with the accepted value of 146.1 by only 1.9%.

In addition, the experimental melting points lie <u>very</u> close to the accepted melting point of 154°C—supporting my hypothesis.

5. PREPARATION OF DERIVATIVES *(Amide)*

The amide was prepared from the unknown carboxylic acid (as detailed in the lab manual) and the yield was extremely low. I believe it to be due to the fact that if the unknown *is* adipic acid, then since it has two —COOH groups, then it would take twice as much $SOCl_2$ as listed in the manual, to get a good yield.

Enough crude product was obtained to make a melting point determination. Although it was impossible to purify the amide, the melting point was still very high...about **215–219°C**. This agrees with the expected melting point for the amide derivative of adipic acid.

Name: Dylan Stewart   Date: May 3, 1991

Report of Unknowns

UNKNOWN NUMBER 20

COMPOUND (ASSUMED):
o-Toluidine (2-Methylaniline)

1. PHYSICAL EXAMINATION: (*liquid:* orange-brown: crude; pale yellow: pure)

2. PHYSICAL CONSTANTS: *distillation point* 195–197°C

3. ELEMENTARY ANALYSIS: took a liquid I.R. after purification (enclosed)

$$3500\text{–}3000 \text{ cm}^{-1} = \text{N—H stretch (hydrogen bonding)}$$

$$1600 \,\&\, 1500 \text{ cm}^{-1} = \text{C=C (aromatic rings)}$$

$$1350\text{–}1150 \text{ cm}^{-1} = \text{C—N stretch}$$

$$1375 \text{ cm}^{-1} = \text{CH}_3 \text{ stretch}$$

$$770\text{–}735 \text{ cm}^{-1} = \text{disubstituted benzene ring (ortho)}$$

4. SOLUBILITY TESTS: *Group D* = basic compounds, amines, hydrazines

water (H$_2$O)	ether	5% NaOH	5% NaHCO$_3$	5% HCl	conc. H2SO$_4$	GROUP
–	X	–	X	+	X	D

5. POSSIBLE COMPOUNDS: (1° and 2° *amines* listed by *boiling point*)

Source: CRC Handbook of Tables for Organic Compounds (3rd Ed.) Zvi Rappoport, CRC Press Inc. (1967)

Amines (liquid)	b.p.	m.p. Acetamide	m.p. Benzamide	m.p. Phenyl Thiourea
3-Aminopropyl alcohol	188	—	—	—
1,2,3-Triaminopropane	190	200–2	217–8	—
4-Amino-2,6-dimethyl- piperidine	195	—	—	—
N-Methylaniline	196	102	63	87
1-Phenylisopropylamine	196–7	—	159	—
β-Phenylethylamine	198	51	116	135
Benzyl ethyl amine	199	—	—	—
o-Toluidine	**200**	**110–1**	**146**	**136**
n-Nonylamine	201	34–5	49	—
3-Methylaniline	203	65	125	95
Di-*n*-amylamine	205	—	—	—

From the solubility test done above, the compound proved to dissolve with vigorous shaking in 5% HCl, which let me believe that it was *Group D* (maybe an amine?) By checking the measured distillation point (boiling) with the boiling points for amines, the above compounds became possible candidates. Checking the liquid I.R. spectrum strengthened my opinion that it could be an amine. I followed up on this theory and checked various I.R. spectra of the above amines. Two spectra from *The Aldrich Library of I.R. Spectra* (2nd Ed.) [Charles J. Pouchert Aldrich [Chemical Co. Inc. (1975)] fit my observed I.R. pretty closely, **N-Methylaniline** and **o-Toluidine**. Checking in the *Merck Index* (9th Ed.) [M. Windholz et al., Merck Co. Inc. (1976)] shows that *N*-Methylaniline and *o*-Toluidine are both light yellow but darken on exposure to air, light, and heat.

6. PREPARATION OF DERIVATIVES:
Procedures for formation of the derivatives was obtained from *The Systematic Identification of Organic Compounds.* (6th Ed.), Shriner, Curtin, Fuson, and Morill, John Wiley and Sons (1979).

I. Formation of Phenylthioureas from Amines (pg 234)

Phenylthiourea

This reaction produced a yellowish white solid when heated for 3 minutes over a low flame (*Note:* lack of an instantaneous reaction suggested that the amine was *aromatic* rather than aliphatic). After purification, the white solid gave an observed **melting point of 135–138°**.

II. The Schotten–Baumann Reaction (pg 219)

Benzamide

The reaction produced gas and a smelly solid that turned white after washing and filtration. When **melting point** was taken the observed range was 139–43.

The melting points observed in the preparation of the derivatives closely resembles that of **o-Toluidine**. Therefore I predict the identity of *unknown* 20 to be that of *o*-Toluidine.

Experiment No. ―――――

Preparation of ―――――――――――

Name ―――――――――――――――――

Date ―――――――――――――――――

Section ―――――――――――――――――

Reaction equation:

Properties of compounds involved (if pertinent):

Compound ―――― ―――― ―――― ―――― ―――― ―――― ――――

Mol. wt. ―――― ―――― ―――― ―――― ―――― ―――― ――――

Vol. (liq.) ―――― ―――― ―――― ―――― ―――― ―――― ――――

Density ―――― ―――― ―――― ―――― ―――― ―――― ――――

Weight ―――― ―――― ―――― ―――― ―――― ―――― ――――

Moles ―――― ―――― ―――― ―――― ―――― ―――― ――――

MP/BP (lit.) ―――― ―――― ―――― ―――― ―――― ―――― ――――

Solubility ―――― ―――― ―――― ―――― ―――― ―――― ――――

Results:

Limiting reagent ―――――――――――――――――

Theoretical yield ――――――――――――――――― g

Actual yield ――――――――――――――――― g

Percent yield ――――――――――――――――― %

MP/BP (observed range) ――――――――――――― °C

Calculation of theoretical and percent yield:

This space and the reverse side of this form may be used for the answers to questions,
a flow chart or process chart, and for comments on difficulties, deviations, explanations, and tests.

Index

Experiment No. _____

Preparation of _____

Name _____

Date _____

Section _____

Reaction equation:

Properties of compounds involved (if pertinent):

Compound	_____	_____	_____	_____	_____	_____	_____
Mol. wt.	_____	_____	_____	_____	_____	_____	_____
Vol. (liq.)	_____	_____	_____	_____	_____	_____	_____
Density	_____	_____	_____	_____	_____	_____	_____
Weight	_____	_____	_____	_____	_____	_____	_____
Moles	_____	_____	_____	_____	_____	_____	_____
MP/BP (lit.)	_____	_____	_____	_____	_____	_____	_____
Solubility	_____	_____	_____	_____	_____	_____	_____

Results:

Limiting reagent _____

Theoretical yield _____ g

Actual yield _____ g

Percent yield _____ %

MP/BP (observed range) _____ °C

Calculation of theoretical and percent yield:

This space and the reverse side of this form may be used for the answers to questions, a flow chart or process chart, and for comments on difficulties, deviations, explanations, and tests.

Experiment No. _____

Preparation of _____

Name _____

Date _____

Section _____

Reaction equation:

Properties of compounds involved (if pertinent):

Compound	_____	_____	_____	_____	_____	_____	_____
Mol. wt.	_____	_____	_____	_____	_____	_____	_____
Vol. (liq.)	_____	_____	_____	_____	_____	_____	_____
Density	_____	_____	_____	_____	_____	_____	_____
Weight	_____	_____	_____	_____	_____	_____	_____
Moles	_____	_____	_____	_____	_____	_____	_____
MP/BP (lit.)	_____	_____	_____	_____	_____	_____	_____
Solubility	_____	_____	_____	_____	_____	_____	_____

Results:

Limiting reagent _____

Theoretical yield _____ g

Actual yield _____ g

Percent yield _____ %

MP/BP (observed range) _____ °C

Calculation of theoretical and percent yield:

This space and the reverse side of this form may be used for the answers to questions, a flow chart or process chart, and for comments on difficulties, deviations, explanations, and tests.

Experiment No. _____

Preparation of _____

Name _____

Date _____

Section _____

Reaction equation:

Properties of compounds involved (if pertinent):

Compound	_____	_____	_____	_____	_____	_____	_____
Mol. wt.	_____	_____	_____	_____	_____	_____	_____
Vol. (liq.)	_____	_____	_____	_____	_____	_____	_____
Density	_____	_____	_____	_____	_____	_____	_____
Weight	_____	_____	_____	_____	_____	_____	_____
Moles	_____	_____	_____	_____	_____	_____	_____
MP/BP (lit.)	_____	_____	_____	_____	_____	_____	_____
Solubility	_____	_____	_____	_____	_____	_____	_____

Results:

Limiting reagent _____

Theoretical yield _____ g

Actual yield _____ g

Percent yield _____ %

MP/BP (observed range) _____ °C

Calculation of theoretical and percent yield:

This space and the reverse side of this form may be used for the answers to questions, a flow chart or process chart, and for comments on difficulties, deviations, explanations, and tests.

Experiment No. _____

Preparation of _____

Name _____

Date _____

Section _____

Reaction equation:

Properties of compounds involved (if pertinent):

Compound							
Mol. wt.							
Vol. (liq.)							
Density							
Weight							
Moles							
MP/BP (lit.)							
Solubility							

Results:

Limiting reagent _____

Theoretical yield _____ g

Actual yield _____ g

Percent yield _____ %

MP/BP (observed range) _____ °C

Calculation of theoretical and percent yield:

This space and the reverse side of this form may be used for the answers to questions,
a flow chart or process chart, and for comments on difficulties, deviations, explanations, and tests.

Experiment No. _____

Preparation of _____

Name _____

Date _____

Section _____

Reaction equation:

Properties of compounds involved (if pertinent):

Compound	___	___	___	___	___	___	___
Mol. wt.	___	___	___	___	___	___	___
Vol. (liq.)	___	___	___	___	___	___	___
Density	___	___	___	___	___	___	___
Weight	___	___	___	___	___	___	___
Moles	___	___	___	___	___	___	___
MP/BP (lit.)	___	___	___	___	___	___	___
Solubility	___	___	___	___	___	___	___

Results:

Limiting reagent _____

Theoretical yield _____ g

Actual yield _____ g

Percent yield _____ %

MP/BP (observed range) _____ °C

Calculation of theoretical and percent yield:

This space and the reverse side of this form may be used for the answers to questions, a flow chart or process chart, and for comments on difficulties, deviations, explanations, and tests.

Experiment No. ———————

Preparation of ———————————

Name —————————————————————

Date —————————————————————

Section ——————————————————————

Reaction equation:

Properties of compounds involved (if pertinent):

Compound	———	———	———	———	———	———	———
Mol. wt.	———	———	———	———	———	———	———
Vol. (liq.)	———	———	———	———	———	———	———
Density	———	———	———	———	———	———	———
Weight	———	———	———	———	———	———	———
Moles	———	———	———	———	———	———	———
MP/BP (lit.)	———	———	———	———	———	———	———
Solubility	———	———	———	———	———	———	———

Results:

Limiting reagent ——————————————————————

Theoretical yield ———————————————————— g

Actual yield ———————————————————— g

Percent yield ———————————————————— %

MP/BP (observed range) ———————————————— °C

Calculation of theoretical and percent yield:

This space and the reverse side of this form may be used for the answers to questions, a flow chart or process chart, and for comments on difficulties, deviations, explanations, and tests.

Experiment No. _____

Preparation of _____

Name _____

Date _____

Section _____

Reaction equation:

Properties of compounds involved (if pertinent):

Compound	_____	_____	_____	_____	_____	_____	_____
Mol. wt.	_____	_____	_____	_____	_____	_____	_____
Vol. (liq.)	_____	_____	_____	_____	_____	_____	_____
Density	_____	_____	_____	_____	_____	_____	_____
Weight	_____	_____	_____	_____	_____	_____	_____
Moles	_____	_____	_____	_____	_____	_____	_____
MP/BP (lit.)	_____	_____	_____	_____	_____	_____	_____
Solubility	_____	_____	_____	_____	_____	_____	_____

Results:

Limiting reagent _____

Theoretical yield _____ g

Actual yield _____ g

Percent yield _____ %

MP/BP (observed range) _____ °C

Calculation of theoretical and percent yield:

This space and the reverse side of this form may be used for the answers to questions, a flow chart or process chart, and for comments on difficulties, deviations, explanations, and tests.

Table of Figures

(Continued)